Build Your Own Electric Vehicle

About the Authors

Seth Leitman (Briarcliff Manor, NY) is currently President and Managing Member of the ETS Energy Store, LLC, which sells organic, natural, and sustainable products for business and home use (from energy efficient bulbs to electric vehicle conversion referrals). Previously, he worked for the New York State Power Authority and the New York State Energy Research and Development Authority, where he helped develop, market, and manage electric and hybrid vehicle programs serving New York State and the New York metropolitan area. Seth is the consulting editor for a series of upcoming titles called the "Green Guru Guides," which focus on implementing environmentally friendly technologies and making them work for you.

Bob Brant was the author of the first book, and some might say ahead of his time in his passion to convert to electric. The first edition of this book was published in 1993. While there have obviously been updates and technological advances since that time, many of the concepts in that first book are still in use today.

Bob grew up in New York City, got a BSEE, and worked on NASA projects such as Apollo, Lunar Excursion Module, and the Earth Resources Technology Satellite. He then went on to get an MSEE and MBA, and worked for a company that worked on the Lunar Rover. Bob was always fascinated with every electric vehicle breakthrough, was convinced of its personal and environmental benefits, and was curious why stronger steps had not been taken to make electric vehicles a reality. Seth Leitman and everyone at McGraw-Hill would like to think of this updated edition as a tribute to Bob and other forward thinkers like him.

Build Your Own Electric Vehicle

Seth Leitman
Bob Brant

Second Edition

New York Chicago San Francisco
Lisbon London Madrid Mexico City
Milan New Delhi San Juan
Seoul Singapore Sydney Toronto

The McGraw-Hill Companies

Cataloging-in-Publication Data is on file with the Library of Congress

Leitman, Seth.
Build your own electric vehicle / Seth Leitman, Bob Brant.— 2nd ed.
p. cm.
Includes bibliographical references and index.
ISBN 978-0-07-154373-6 (alk. paper)
1. Rev. ed. of: Build your own electric vehicle / Bob Brant. c1994. 2. Electric vehicles—Design and construction—Amateurs' manuals. 3. Automobiles, Home-built. I. Brant, Bob. II. Brant, Bob. Build your own electric vehicle. III. Title.
TL220.B68 2009
629.22'93—dc22

2008023313

McGraw-Hill books are available at special quantity discounts to use as premiums and sales promotions, or for use in corporate training programs. To contact a representative please visit the Contact Us pages at www.mhprofessional.com.

Build Your Own Electric Vehicle, Second Edition

1 2 3 4 5 6 7 8 9 0 DOC/DOC 0 1 4 3 2 1 0 9 8

ISBN 978- 0-07-154373-6
MHID 0-07-154373-2

The pages within this book were printed on acid-free paper containing 15% post-consumer fiber.

Sponsoring Editor
Judy Bass

Acquisitions Coordinator
Rebecca Behrens

Editorial Supervisor
David E. Fogarty

Project Manager
Patricia Wallenburg

Copy Editor
Claire Splan

Proofreader
Paul Tyler

Indexer
Karin Arrigoni

Production Supervisor
Pamela A. Pelton

Composition
Patricia Wallenburg

Art Director, Cover
Jeff Weeks

Contents

Preface

The electric vehicle's time has come.

The electric vehicle (EV) movement has broadened to multiple levels of the public debate. Al Gore and Leonardo DiCaprio have recently made movies about the need to assist the environment and how oil and energy have created the global warming problems that our world currently faces. Al Gore just won a Nobel Peace Prize for his discussion and call to action about global warming/climate change. The price of oil has recently fluctuated around $140 per barrel and $4.50 per gallon.

Well, EVs solve a lot of problems quickly. Electric vehicles bypass high energy prices. Electric cars cost pennies to charge. Electric cars have zero tailpipe emissions. While they charge up on electricity from power plants, they can also charge on electricity from solar, wind, and any other renewable resource. Also, if you compare emissions from power plants for every car on the road versus gasoline emissions, electric cars are always *always* cleaner. In addition, as power plants get cleaner and our power plants reduce emissions, electric cars will only get cleaner.

Electric cars also help develop the economy. We all know that we need to increase the number of electric cars. Hybrid electrics, plug-in hybrids, and low-speed vehicles all expand electric transportation. We as a country—no, we as world—are increasing our involvement in this industry. From China to India, to Great Britain to France, and back here in the United States, electric transportation can only create a new industry that will increase our manufacturing sector's ability to build clean efficient cars. I recently spoke with an owner of an electric car company who said that the UAW was more than excited to build electric cars since the traditional car companies were leaving Detroit in single file. This can only increase domestic jobs in the United States and help our economy. Our world is depending on fossil fuels from countries that predominantly have not supported the best financial interests of the United States. It is a national security issue for all of us to be sending over billions of dollars to countries that are politically unstable and/or antagonistic with Western nations. Another way to ask the questions is: Should we be sending more money to Iran and Venezuela, or should we keep it in our own pockets? That is why I believe in a pollution-free, oil-free form of transportation. My first company's tag line use to be "Pollution-Free, Oil-Free, It's Good To Be Free." That is the mantra I would like to provide for this book! When you drive an electric vehicle, that is how you feel—*free*.

Who I Am

I'm Managing Member of this store, called the ETS Energy Store, LLC (www.etsenergy.com), which sells organic, natural, and sustainable products for businesses, homes, and families. It also provides electric vehicle consulting services for companies needing marketing, engineering and technical expertise on their respective product lines. In addition, I also run a blog called www.greenlivingguy.com which will discuss sustainable living, energy efficiency and electric cars—green living!

I used to own a company called Electric Transportation Solutions, LLC, which sold all forms of electric vehicles, consulted for other companies on improving their perception in the market, and helped them determine strategic partnerships. I shut down that company to focus on the ETS Energy Store, LLC.

I am also the Consulting Editor for McGraw-Hill on a new series called the Green Guru Guides. These books should be coming out in 2009.

Just like Bob Brant, I am a New Yorker who rode the electric-powered subway trains. In fact, when I worked for the New York Power Authority, which powers those subways, I gained a new appreciation for electric transportation every time I took the train into and around New York City.

My interest in renewable energy and energy efficiency, however, began in graduate school at the Rockefeller College of Public Affairs and Policy in Albany, New York, where I received a master of public administration degree. I concentrated on comparative international development, which focused on the World Bank and the International Monetary Fund. After I read about the World Bank funding inefficient and environmentally destructive energy projects, such as coal-burning power plants in China or dams in Brazil that had the potential to destroy the Amazon, I decided to take my understanding to another level. For my master's thesis I interviewed members of the World Bank, the International Monetary Fund, and the Bretton Woods Institutions. I was fortunate enough to be able to ask direct questions to project managers who oversaw billions that went to China to build coal-burning power plants. I asked them how the Bank could fund an environmentally destructive energy project when there were no traps or technologies to recapture the emissions and use that energy or recapture it back into the plant to use as energy. The answers were not good, but since I researched the Bank, attention to environmental issues has expanded by leaps and bounds and the Bank is starting to work toward economic and environmental efficiencies. While they still have a lot of work to do, it's clear that progress is possible.

This passion to understand how organizations could create positive environmental, energy-efficient, and economic development programs (all in one) led me to work for the New York State Energy Research and Development Authority (NYSERDA). While at NYSERDA, I was the lead project manager for the USDOE Clean Cities Program (a grass roots–based program that develops alternative-fueled vehicle projects across the country) for five of the seven Clean Cities in New York. I was also the lead project manager for the Clean Fueled Bus Program for the Clean Air/Clear Water Bond Act, which provided over $100 million in incremental cost funding for transit operators to purchase alternative-fueled buses. When I funded programs and realized the benefits of electric cars or hybrid electric buses versus their counterparts, I was transformed. I saw that electric transportation was the way to go.

I fell so in love with electric transportation that I went to work for the New York Power Authority electric transportation group. In total, I helped to bring over 3,500 alternative-fueled vehicles and buses into New York.

I was the Market Development and Policy Specialist for the New York Power Authority, the nation's largest publicly owned utility. I worked on the development, marketing, and management of electric and hybrid vehicle programs serving the New York metropolitan area. I developed programs that expanded the NYPA fleet from 150 to over 700 vehicles, while enhancing public awareness of these programs.

I was the lead manager of the NYPA/TH!NK Clean Commute Program™, which under my leadership expanded from 3 to 100 vehicles. This program was the largest public/private partnership of its time. I secured and managed a $6.5 million budget funded by the federal government, project partners, participants (EV drivers), and Ford Motor Company. I developed an incentive program that offered commuters up-front parking at the train stations with electric charging stations. We provided insurance rebates and reduced train fares. This program secured media coverage in *USA Today*, *Associated Press*, *Reuters*, *The New York Times*, *CNN*, *Good Morning America*, the *Today Show*, and other media sources. To date, this was the largest electric vehicle station car program in the world. Figures 1, 2, and 3 show station parking and chargers for cars leased in Chappaqua, New York. This was one of the most successful train stations (excluding Huntington and Hicksville in Long Island, New York) for the program. Figure 1 is a great overhead shot of the cars lined up in the areas we set up right next to the train station for prime parking incentives. Figure 2 shows one of the charging stations up close using an AVCON charger with an overhead light. The station connector cables were designed to be like a regular gas station (thanks to Bart Chezar, former manager of the Electric Transportation Group, and Sam Marcovicci, who was the electric charger specialist for NYPA at the time) and ETEC out in Arizona (listed in the Sources section of this book). Figure 3 shows how close it was to the train station since you can see the platform and station name on the platform behind the row of TH!NK City EVs. It was a great program. On a related note, TH!NK is reemerging into the international automotive marketplace and is starting to open offices in the United States. The TH!NK is coming back!

FIGURE 1 An electric car dream. TH!NK City electric cars at Chappaqua train station in the NYPA/TH!NK Clean Commute Program™. (Photo courtesy of Town of New Castle.)

Figure 2 TH!NK City cars charging with AVCON chargers at Chappaqua train station. (Photo courtesy of Town of New Castle.)

I also led and worked with multiple state and local agencies to place over 1,000 GEM and TH!NK Neighbor low-speed vehicles for their respective donation programs to meet zero emission vehicle credits for New York State.

I've published reports with the Electric Power Research Institute, NYSERDA, and the U.S. Department of Energy on the NYPA/TH!NK Clean Commute Program and Green Schools.

Figure 3 Look how close the electric cars are to the train station platform. Talk about incentive! (Photo courtesy of Town of New Castle.)

My intention in rewriting this book was to not just tell people how to convert their cars to electric. I also wanted to tell people why they should convert their car. The intent was to create a useful guide to get you started; to encourage you to contact additional sources in your own "try-before-buy" quest; to point you in the direction of the people who have already done it (i.e., electric vehicle associations, consultants, builders, suppliers, and integrators); to familiarize you with the electric vehicle components; and finally, to go through the process of actually building/converting your own electric vehicle. My intention was never to make a final statement, only to whet your appetite for electric vehicle possibilities. I hope you have as much fun reading about the issues, sources, parts, and building process as I have had.

Another timely point to add concerns the price of oil and its relation to the entire book. When Bob Brant first wrote this book, he determined that $100 for a barrel of oil was the worst-case scenario. While this book was being produced, oil was trading at $142 a barrel, and I saw at my local gas station $4.50 for gas and $5.03 for diesel. My friends in the financial markets are telling me that $170 to $200 for a barrel of oil can happen in the future. In addition, when I submitted the first manuscript, the price of gas was $3.50 per gallon. As we go to print, we are at $113.00. Bob Brant thought that $100 was going to be the worst case scenario. Therefore, since the price of gas is a moving target, I left all mathematical equations assuming $4.50 per gallon. You might not be at this price today, but there is nothing saying it can't get back to that level and above in the future.

—Seth Leitman

Acknowledgments

To paraphrase an actor who just won an Emmy, "There are so many people to thank." However, I want to first dedicate this book to my family. They have watched my involvement in electric cars expand over the years and, with global warming and green becoming the new black, they really appreciate what I am doing. More importantly, I am in this movement so that my children can have a better earth to live in, since climate change and global warming is a reality.

There are also so many people to thank. Don Francis from Southern Company has known me since my time with the State of New York and when I worked in the Electric Transportation Group of the New York Power Authority. He has seen my involvement expand over the years and has even helped out with this book with a simple phone call. He is a great friend and colleague.

Thanks also go to Ron Freund, Chairman of the Electric Auto Association. When I told him about this book he immediately said, "Look no further." He wasn't interested in anything but making sure the book was a success. Not for personal glory, but for the electric car movement. "They don't build them like that anymore," a friend once told me and they are right. This organization represents all the clubs across the country that promote electric cars and they need all of our support. There are some great, honorable people involved in these clubs with a love and desire to do the right thing: to make sure electric cars flourish. To me, that deserves our appreciation and thanks.

I'd also like to thank Chelsea Sexton, Executive Director of Plug-In America. We all know her from the movie, *Who Killed The Electric Car*. When I asked her for her assistance, she was more than willing. More than that, there was no ego or posturing. I say this because she came to help out on the book with a smile, decency, and greatness. Chelsea founded Plug-In America, which is working to make sure plug-in hybrid electric cars become a reality. We'll talk about plug-in hybrids later. What I like is that the organization does their work in a bipartisan manner and they do it for all the right reasons. Even Felix Kramer from Plug-In America once told me that he always tells people that the best type of car is an electric car. They deserve to receive kudos too.

I also want to thank Carl Vogel and Steve Clunn. Carl Vogel started Vogelbilt Corporation. He developed an electric motorcycle with the assistance of the original version of this book. He has added his comments on the chapters in this book and has always been a great friend to me. More importantly, he is writing a book called *Build Your Own Electric Motorcycle*—stay tuned!!

Steve Clunn assisted with the book and is an electric vehicle conversion specialist. He owns a company in Florida called Grass Roots EV and we have been talking about car conversions for the past three years. I have always supported him and believed he was doing great things. Now, he has so many cars to convert he has a waiting list. That is so great for him and a great sign for the conversion industry.

Credit must also be extended to the many other electric vehicle enthusiasts who made this book possible, such as Lee Hart, one of the experts of the electric vehicle industry. I also want to thank, Lynne Mason from electric-cars-are-for-girls.com for showing some great organization of an electric vehicle conversion on her site. Some of the pictures we approved by the original sources of the image, but I got the idea to get that image of add a company from her website. Also, thank s go to Tom Gage from AC Propulsion, Rick Woodbury from Commuter Cars Corporation, Ian from the Zero Emission Vehicles Australia, and Chris Isles. He was one of the clean commuters in Chappaqua and this past year, Chris and I connected again. He said he wished the program was back. I just hope this book helps create more programs and electric vehicles are used more regularly.

I also want to thank Bob Brant for writing this masterpiece, as well as his wife, Bonnie Brant.

I want to thank Judy Bass from McGraw-Hill. She is such a sweet loving person who believed in me. I have always believed that people come into your life for a reason. Her reason was to bless me with the opportunity to update this book and give this industry the jolt it needs. She believed in my willingness to change the book for mass appeal while keeping the technical approach that Bob Brant used and everyone appreciates.

I also need to thank the best copywriter and editor that I could have ever been blessed with, Patty Wallenburg. She has been so supportive of the book and also so great with her entire team of copy editors and proofreaders. Their job is to make the book great and she really did that!

At the time when Judy approached me to do this book, Art Buchwald, a great political syndicate columnist for the *Washington Post* and a relative of mine, had passed away. He was a great inspiration to me. In addition, he wrote a letter of reference for me that helped me to get an internship between my undergraduate and graduate school years for Senator Daniel Patrick Moynihan, former U.S. Senator for New York. Art also autographed his book, *I'll Always Have Paris*, for me. He wrote, "Dear Seth, You deserve this book because of any Buchwald, you are the only one who should go to Paris." I will never forget him and he was a special person in my life. When Judy Bass asked me to rewrite this book, I told her I was grateful because I could now work to continue the Buchwald tradition.

I dedicate this book to my beautiful wife, Jessica, and my beautiful sons, Tyler and Cameron. While sometimes things have been difficult for a person starting a company after the realities of living the Northeast version of the movie *Who Killed the Electric Car?*, Jessica always supported me, loved me, wished for my happiness and was a rock during the tough times. My sons Tyler and Cameron are two great boys who only inspire me with their happiness, love, and intellect. As I see them grow up in the world we live in today, I am glad that this book can help to do its part for their future.

Build Your Own Electric Vehicle

CHAPTER 1

Why Electric Vehicles Are Still Right for Today!

"What is desirable and right is never impossible."
—Henry Ford (inscribed on plaque in Ford Fairlane Mansion)

Why should anyone buy, convert, or build an electric vehicle today? Simply put, they are the cleanest, most efficient, and most cost-effective form of transportation around—*and* they are really fun to drive. When I worked for the State of New York, we always used to say that electric cars were almost maintenance-free: they never require oil changes, new spark plugs, or any other regular repairs. When a person would say, "Really?" I would then say, "Well, not quite—you need to change the washer fluid for the windshield."

Electric vehicles (EVs) are highly adaptable and part of everyday society: Electric cars are found on mountain tops (railway trams, cable cars), at the bottom of the sea (submarines, Titanic explorer), on the moon (Lunar Rover), in tall buildings (elevators), in cities (subways, light rail, buses, delivery vehicles), hauling heavy rail freight or moving rail passengers fast (Pennsylvania Railroad Washington to New York corridor). Are they all electric vehicles? Yes. Do they run on rails or in shafts or on tethers or with nonrechargeable batteries? Yes.

EVs were designed to do whatever was wanted in the past and can be designed and refined to do whatever is needed in the future. What do you need an EV to be: big, small, powerful, fast, ultra-efficient? Design to meet that need. General Motors' EV1 is an excellent example of what can be done when starting with a clean sheet of paper. Closer to home and the subject of this book, do you want an EV car, pickup, or van? You decide.

As car companies continue producing sport utility vehicles (SUVs) that cannot meet federal fuel standards or reduce emissions that are harmful to our environment, think about some of the statistics and facts from the U.S. Department of Energy (USDOE) and various notable sources: The USDOE states that more than half of the oil we use every day is imported. This level of dependence on imports (55 percent) is the highest in our history. The USDOE even goes on to say that this dependence on foreign oil will increase as we use up domestic resources. Also, as a national security issue, we should all be concerned that the vast majority of the world's oil reserves are concentrated in the Middle East (65 to 75 percent), and controlled by the members of the OPEC oil cartel (www.fueleconomy.gov/feg/oildep.shtml).

Further, USDOE goes on to state that 133 million Americans live in areas that failed at least one National Ambient Air Quality Standard. Transportation vehicles produce 25 to 75 percent of key chemicals that pollute the air, causing smog and health problems. All new cars must meet federal emissions standards. But as vehicles get older, the amount of pollution they produce increases.

In addition, only about 15 percent of the energy in the fuel you put in your gas tank gets used to move your car down the road or run useful accessories like air conditioning or power steering. The rest of the energy is lost. Clearly the potential to improve fuel economy with advanced technologies is enormous.

What can we do?

Drive electric cars.

Here are some reasons why.

1. Although they are only at a relatively embryonic stage in terms of market penetration, electric cars represent the most environmentally friendly vehicle fuel, as they have absolutely no emissions (www.greenconsumerguide.com/governmentll.php?CLASSIFICATION=114&PARENT=110). The energy generated to power the EV and the energy to move the vehicle is 97 percent cleaner in terms of noxious pollutants.
2. Another advantage of electric motors is their ability to provide power at almost any engine speed. Whereas only about 20 percent of the chemical energy in gasoline gets converted into useful work at the wheels of an internal combustion vehicle, 75 percent or more of the energy from a battery reaches its wheels.
3. One of the big arguments made by car companies against electric cars is that EVs are powered by power plants, which are powered primarily by coal. Less than 2 percent of U.S. electricity is generated from oil, so using electricity as a transportation fuel would greatly reduce dependence on imported petroleum (www.alt-e.blogspot.com/2005/01/alternative-fuel-cars-plug-in-hybrids.html).
4. Even assuming that the electricity to power the EV is not produced from rooftop solar or natural gas (let's assume it comes 100 percent from coal), it is *still* much cleaner than gasoline produced from petroleum! (www.drivingthefuture.com/97pct.html)
5. The power plants are stationary sources that can be modified over time to become cleaner.

The major concerns facing the electric vehicle industry are range, top speed, and cost. Ultimately, it's the batteries that will determine the cost and performance of EVs. The only way electric vehicles are going to make a big difference in people's lives is if they can do everything a gas car can do and more. They have to look great (almost be an extension of the person buying the car), and they have to be safe.

Conversions use currently approved frames and have been going on for years with gas cars using performance-based engines and motors. With controllers (basically, the engine) reaching 1,000 to 2,000 amps, high-end car batteries, and the light weight of a

Porsche 911 chassis, electric cars can provide respectable performance. They are fun to drive, virtually silent, and they coast very easily when you let off the accelerator pedal (www.worldclassexotics.com/Electriccaronv.htm). In other words, you can convert an old Porsche 911 to go over 100 mph with a 50-mile range using lead-acid batteries alone! With lithium ion technology you can get the car to go 180–200 miles and the cost is still less than some brand new SUVs on the market.

In an effort to move the market toward the electric car, some people are trying other alternatives, including:

- Driving hydrogen/fuel cell cars
- Converting hybrid cars to either grid-connected or plug-in hybrids
- Buying hybrid-electric cars in droves
- Purchasing low-speed electric vehicles, such as the GEM car
- Driving the last remaining car company–built electric cars, such as the EV1 or the Toyota RAV4 EV or the TH!NK City (manufactured with Ford Motor Company)

Sooner or later, we will get to an electric car by the car companies. Whether it happens in my lifetime is not the question. My point is that you can get an electric vehicle today. You can also take any vehicle you want and convert it to an electric vehicle. We can also encourage the fix-it guy down the street to help us with our conversion so that more mechanics across the country are building electric cars.

Convert That Car!

Electric vehicles are not difficult to build and they are easy to convert. There are so many reasons to convince you of the need to go electric:

- The cost of a gallon of gas
- Higher asthma rates
- Our need to reduce our reliance on imported oils
- The prospect of owning a car that is cost-effective, fun, and longer-lasting than most cars on the road today.

Once again: How about the fact that you can convert an electric vehicle today? Right now! And it would cost you less than some new cars on the market. There are so many television shows showing people tricking out their cars, adding better engines, or doing just about anything to make it go fast and be safe. Conversions of an electric vehicle do all that and more. They allow the next generation to have a safer world without relying on foreign sources of oil while giving our kids really clean and cool cars to drive.

In very practical terms, the 2001 Porsche and the 1993 Ford Ranger pickup electric vehicle conversions shown in Figures 1-1 and 1-2 (which you'll learn how to convert in Chapter 11) goes 75 mph, gets 60 miles (or better) on a charge, uses conventional lead-acid batteries and off-the-shelf components, and can be put together by almost anyone. Its batteries cost about $2,200 and last about three years, its conversion parts cost about $5,500 to $7,000, and it costs $1.25 to recharge, or "fuel up." In New York terms, that's

Figure 1-1 A converted electric Porsche.

Figure 1-2 A converted electric Ford Ranger.

less than a token on a subway. Also, the maintenance costs are negligible compared to the oil changes, radiator repairs, and all the other additional maintenance costs of an internal combustion engine version.

In this chapter you'll learn what an electric vehicle is, and explore the change in consciousness responsible for the upsurge in interest surrounding it. You'll discover the truths and untruths behind electric vehicle myths. You'll also learn about the EV's advantages, and why its benefits—assisted by technological improvements—will continue to increase in the future.

To really appreciate an electric car, it's best to start with a look at the internal combustion engine vehicle. The difference between the two is a study in contrasts.

Mankind's continued fascination with the internal combustion engine vehicle is an enigma. The internal combustion engine is a device that inherently tries to destroy itself: numerous explosions drive its pistons up and down to turn a shaft. A shaft rotating at 6,000 revolutions/minute produces 100 explosions every second. These explosions in turn require a massive vessel to contain them—typically a cast-iron cylinder block. Additional systems are necessary:

- A cooling system to keep the temperatures within a safe operating range.
- An exhaust system to remove the heated exhaust products safely.
- An ignition system to initiate the combustion at the right moment.
- A fueling system to introduce the proper mixture of air and gas for combustion.
- A lubricating system to reduce wear on high-temperature, rapidly moving parts.
- A starting system to get the whole cycle going.

It's complicated to keep all these systems working together. This complexity means more things can go wrong (more frequent repairs and higher repair costs). Figure 1-3 summarizes the internal combustion engine vehicle systems.

Unfortunately, the internal combustion engine vehicle's legacy of destruction doesn't just stop with itself. The internal combustion engine is a variant of the generic combustion process. To light a match, you use oxygen (O_2) from the air to burn a carbon-based fuel (wood or cardboard matchstick), generate carbon dioxide (CO_2), emit toxic waste gases (you can see the smoke and perhaps smell the sulfur), and leave a solid waste (burnt matchstick). The volume of air around you is far greater than that consumed by the match; air currents soon dissipate the smoke and smell, and you toss the matchstick.

Today's internal combustion engine is more evolved than ever. However, we still have a carbon-based combustion process that creates heat and pollution. Everything about the internal combustion engine is toxic, and is still one of the least-efficient mechanical devices on the planet. Unlike lighting a single match, the use of hundreds of millions (soon to be billions) of internal combustion engine vehicles threatens to destroy all life on our earth. You'll read about environmental problems caused by internal combustion engine vehicles in Chapter 2.

While an internal combustion engine has hundreds of moving parts, an electric motor only has one. That's one of the main reasons why electric cars are so efficient. To make an electric vehicle out of a car, pickup, or van you are driving now, all you need

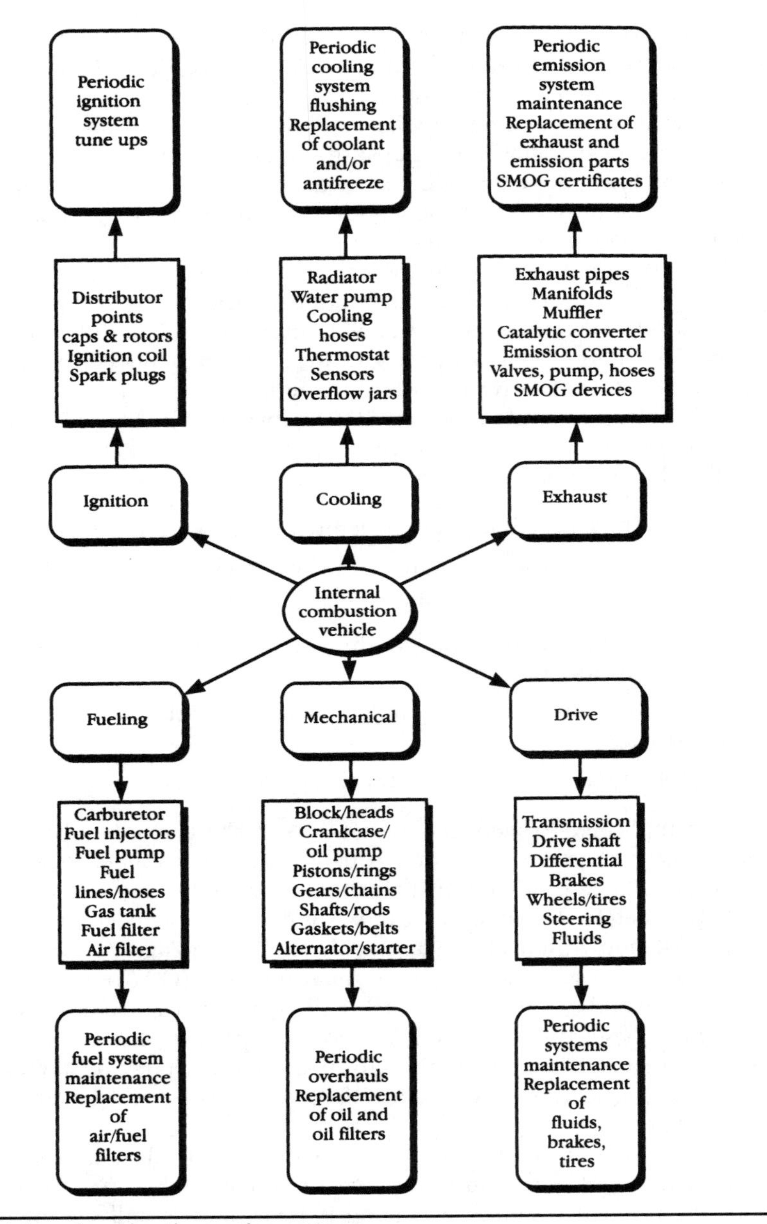

FIGURE 1-3 Internal combustion engine systems.

to do is take out the internal combustion engine along with all related ignition, cooling, fueling, and exhaust system parts, and add an electric motor, batteries, and a controller. Hey, it doesn't get any simpler than this!

Figure 1-4 shows all there is to it: Batteries and a charger are your "fueling" system, an electric motor and controller are your "electrical" system, and the "drive" system

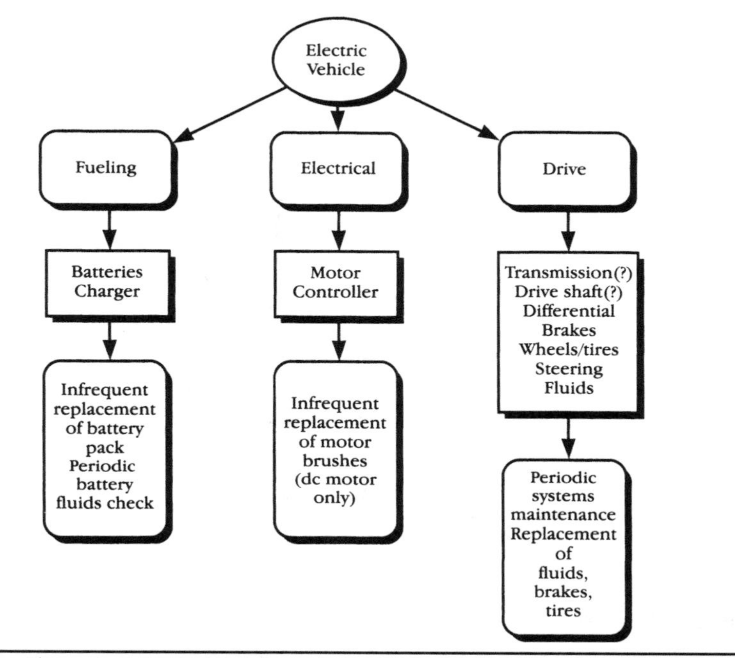

FIGURE 1-4 Electric vehicle systems.

was as before (although today's advanced electric vehicle designs don't even need the transmission and drive shaft).

A simple diagram of an electric vehicle looks like a simple diagram of a portable electric shaver: a battery, a motor, and a controller or switch that adjusts the flow of electricity to the motor to control its speed. That's it. Nothing comes out of your electric shaver and nothing comes out of your electric car. EVs are simple (therefore highly reliable), have lifetimes measured in millions of miles, need no periodic maintenance (filters, etc.), and cost significantly less per mile to operate. They are highly flexible as well, using electric energy readily available anywhere as input fuel.

In addition to all these benefits, if you buy, build, or convert your electric vehicle from an internal combustion engine vehicle chassis as suggested in this book, you perform a double service for the environment: You remove one polluting car from the road and add one nonpolluting electric vehicle to service.

You've had a quick tour and side-by-side comparison of electric vehicles and internal combustion engine vehicles. Now let's take a closer look at electric vehicles.

What Is an Electric Vehicle?

An electric vehicle consists of a battery that provides energy, an electric motor that drives the wheels, and a controller that regulates the energy flow to the motor. Figure 1-5 shows all there is to it—but don't be fooled by its simplicity. Scientists, engineers, and inventors down through the ages have always said, "In simplicity there is elegance." Let's find out why the electric vehicle concept is elegant.

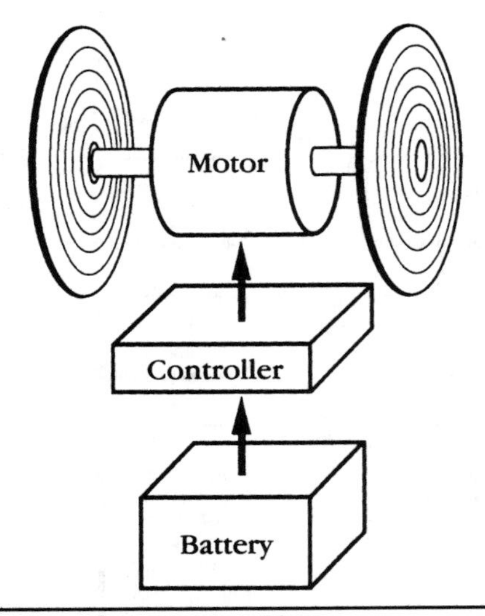

Figure 1-5 Simple block diagram of an electric vehicle.

Electric Motors

Electric motors can be found in so many sizes and places, and have so many varied uses, that we tend to take them for granted. Universal in application, they can be as big as a house or smaller than your fingernail, and can be powered by any source of electricity. In fact, they are so reliable, quiet, and inexpensive that we tend to overlook just how pervasive and influential they are in virtually every civilized person's life.

Each of us encounters dozens, if not hundreds, of electric motors daily without even thinking about them: The alarm clock that wakes you; the television you turn on for the news; you remove coffee beans from the refrigerator and put the coffee beans in a grinder; in the bathroom you use an electric shaver, electric toothbrush, or hair dryer; breakfast might be assisted by your electric juicer, blender, or food mixer; you might clean your home with your vacuum cleaner or clean your clothes with your washer and dryer; next you're into your automobile, subway, bus, or light rail transit to ride to work, where you might go through an automatic gate or door or take an elevator or escalator to your floor; at home or at work you sit down at your computer, use the Internet, e-mail, cell phone, or Blackberry, and use your fax or copier after you adjust the fan, heater, or air conditioner. Back at home in the evening, you might use an electric garage door opener, program your TiVO, or use an electric power tool on a project. On and on, you get the picture.

Why are electric motors ubiquitous? In one word—*convenience*. Electric motors do work so that you don't have to. Whether it's pulling, pushing, lifting, stirring, or oscillating, the electric motor converts electrical energy into motion, which is further adapted to do useful work.

What is the secret of the electric motor's widespread use? Reliability. This is because of its simplicity. Regardless of type, all electric motors have only two basic components: a rotor (the moving part) and a stator (the stationary part). That's right—it has only one moving part. If you design, manufacture, and use an electric motor correctly, it is virtually impervious to failure and indestructible in use.

In internal combustion automobiles, in addition to your all-important electric starter motor, you typically find electric motors in the passenger compartment heating/cooling system, radiator fan, windshield wipers, electric seats, windows, door locks, trunk latch, outside rear view mirrors, outside radio antenna, and more.

Batteries

No matter where you go, you cannot get away from batteries either. They're in your pocket tape recorder, portable radio, telephone, cell phone, laptop computer, portable power tool, appliance, game, flashlight, camera, and many more devices. Batteries come in two distinct flavors: rechargeable and nonrechargeable. Like motors, they come in all sorts of sizes, shapes, weights, and capacities. Unlike motors, they have no moving parts. The nonrechargeable batteries you simply dispose of when they are out of juice; rechargeable batteries you connect to a recharger or source of electric power to build them up to capacity. There are different types of batteries. There are rechargeable lead-acid, nickel metal hydride, and lithium-ion batteries (as some examples), which can be used in your car to manage the recharging process invisibly via an under-the-hood generator or alternator that recharges the battery while you're driving.

Why are batteries ubiquitous? In a word—convenience. The battery, in conjunction with the starter motor, serves the all-important function of starting the automobile powered by the conventional internal combustion engine. In fact, it was the battery and electric starter motor combination, first introduced in the early 1920s and changed very little since then, that put the internal combustion engine car on the map—it made cars easy to start and easy to use for anyone, anywhere.

Another great thing about the promise for electric cars is lithium-ion battery technology. It is moving rapidly into the marketplace and dropping in price. Over the next few years we can expect further drops in price, making EV conversions more affordable. Soon enough, the standard will be lithium-ion batteries in any conversion kit.

Rechargeable lead-acid automotive batteries perform their job very reliably over a wide range of temperature extremes and, if kept properly charged, will maintain their efficiency and deliver stable output characteristics over a relatively long period of time—several years. A lead-acid automotive battery is unlikely to fail unless you shock it, drop it, discharge it completely, or allow a cell to go dry. The only maintenance required in lead-acid batteries is checking each cell's electrolyte level and periodically refilling them with water. Newer, sealed batteries require no maintenance at all.

Controllers

Controllers have become much more intelligent. The same technology that reduced computers from room-sized to desk-sized allows you to exercise precise control over an electric motor. Regardless of the voltage source, current needs, or motor type, today's controllers—built with reliable solid-state electric components—can be designed to meet virtually any need and can easily be made compact to fit conveniently under the hood of your car.

Why are electric vehicles elegant? When you join an electric motor, battery, and controller together, you get an electric vehicle that is both reliable and convenient. Perhaps the best analogy is that when you "go EV" you can drive your entire car from an oversized electric starter motor, a more powerful set of rechargeable batteries, and a very sophisticated starter switch. But it's only going to get better.

Back in the early 1990s when the first edition of this book was published, electric vehicles resembled your battery-operated electric shaver, portable power tool, or kitchen appliance. Today and tomorrow's electric vehicles more closely resemble your portable laptop computer in terms of both sophistication and capabilities.

Have You Driven an Electric Vehicle Lately?

Besides all the discussion of electric cars and the California Air Resources Board (CARB) mandates to incentivize electric cars, hybrids, and fuel cells that are replicated in several of the United States, very few people have bought, built, or converted an electric vehicle because they wanted to save planet earth.

Here are some of the reasons why people *do* get into electric vehicles.

Electric Vehicles Offer the "Total Experience"

Word of mouth and personal experience make a difference. Another way (more recently) is from the documentary movie, *Who Killed The Electric Car?* The cumulative effect of numerous people attending EV symposiums, rallies, and Electric Automobile Association meetings and movies all over the world—and experiencing first-hand what it's like to ride or drive one—has gradually done the job. Almost universally, people enjoy their consciousness-raising electric vehicle experience, are impressed by it, and tell a friend. That's the real reason for the resurgence in interest in electric cars.

Electric Vehicles Are Fun to Drive

Imagine turning on a car and hearing nothing! The only way you can tell that the car is on is by looking at the battery/fuel gauge on the dashboard. This is only the first surprise of many when you get into an electric vehicle.

When I used to work for the New York Power Authority (NYPA) and do ride-and-drives for the public, I always used to say that once you get in, you are changed forever. It's true!! Every single time a person got out of the car, there would be a smile on their face, a sense of real excitement, and then the inevitable first question ("Where can I get one?!") would always pop up.

Electric vehicles are first and foremost practical—but also fun to own and drive. Owners say they become downright addictive. Tooling around in breezy electric vehicle silence gives you all the pleasure without the noise. As I liked to put it to my friends, *"You can really hear your stereo."*

Electric Vehicles Make a Difference by Standing Out

Electric vehicles are a great way to drive and make a real contribution to the country. By driving an oil-free, gasoline-free car, you reduce our country's reliance on imported oil; that will make you friends.

Whether you've owned or even driven an EV1 from GM (what a ride!), Toyota RAV4, TH!NK City, Solectria Force, or a converted Porsche 914 or Ford Ranger pickup truck, or a built-from-scratch chassis with custom kitcar body, your electric vehicle is a sexy, quiet, technologically spiffy show-stopper.

Believe me, if the words "electric car" or "electric vehicle" appear prominently on the outside of your car, pickup, or van, you will not want for instant friends at any

FIGURE 1-6 "Hey, where's the engine?"

stoplight, shopping center, gas station, or just playing stop and go on the highway. One time I was driving a Toyota RAV4 EV on the West Side Highway in Manhattan with the window down. All different types of people were intrigued. ("Is that an electric car? When can I get one? How far does it go? Is it as good as people say?" Or the best was, "That is one phat ride!") Or you can park it in a conspicuous spot, lift the hood, and wait for the first passerby to ask questions, as Figure 1-6 suggests.

There's a level of respect you receive, a pride in riding in the car, and a feeling of leading the pack in those experiences, at shows and demonstrations. One suggestion too: have plenty of literature always available on hand so you can keep your electric vehicle discussions to under five minutes in length. On the other hand, if you just want to meet people, make the letters on your sign real big and you will never want for company.

The first owners of anything new always have an aura of prestige and mystique about them. You will be instantly coronated in your own neighborhood. You are driving what others have only talked about. While hopefully everyone will own one in the future, you are driving an electric vehicle *today*. When TV sets were introduced in the 1950s, the whole neighborhood crowded into the first houses with the first tiny black and white screens. Expect the same with your electric vehicle project. This will show you can reduce your carbon footprint with an electric car and it won't cost hundreds of thousands of dollars.

Electric Vehicles Save Money

All this emotional stuff is nice, but let's talk out-of-pocket dollars. Ask any electric vehicle conversion owner, and they'll tell you it transports them where they want to go, is very reliable, and saves them money. Let's examine separately the operating, purchase, and lifetime ownership costs and summarize the potential savings.

Operating Costs

Electric vehicles only consume electricity. In between charge-ups, there are no other consumables to worry about except an occasional watering of the batteries. These figures are covered in more detail later, but the Ford Ranger electric vehicle pickup conversion of Chapter 10 averages about 0.44 kWh (kilowatt-hours, a measure of energy consumption) per mile. At $.165 per kWh for electricity in New York (check your electric utility monthly statement for the prevailing rate in your area) that translates to

$$0.44 \text{ kWh/mile} \times \$.165/\text{kWh} = .0726 \text{ (7.3 cents) per mile}$$

(Note: Does not include charging cost and 3.3 cents per mile for battery replacement.)

Let's compare these costs with the EV's gasoline-powered internal combustion engine counterpart in a pickup chassis. The latter consumes gasoline; its ignition, cooling, fueling, and exhaust systems require filters, fluids, and periodic maintenance. The gasoline-powered pickup chassis (equivalent to the previous example) averages 20 miles per gallon or 0.05 gallons per mile. At $4.50 per gallon for gas, that translates to

$$0.05 \text{ gallons/mile} \times \$4.50/\text{gallon} = .225 \text{ (22.5 cents) per mile}$$

Consumables and periodic maintenance must still be added. Assuming these cost $41.67 per month (oil change averaged over three months, fuel additives, aligning and balancing tires), and annual mileage is 12,000 miles per year, this translates to

$$\$500/\text{year} \div 12{,}000 \text{ miles/year} = .0416 \text{ (4.2 cents) per mile}$$

Adding the two figures together, you're looking at 27 cents per mile operating cost for a gasoline-powered vehicle versus 7.3 cents per mile for its electric vehicle equivalent—almost three times the cost of the electric vehicle. While your average EV conversion—made with off-the-shelf components—might consume about 0.4 kWh per mile, General Motor's Impact electric vehicle is rated at 0.1 kWh per mile (0.07 kWh/km). This drops your electric vehicle operating costs to 0.5 cents per mile! (Note: cost per mile varies with driver.)

Purchase Costs

Commercially manufactured electric vehicles are prohibitively expensive today—if you can find one at all. Tomorrow's electric vehicle costs will obviously drop to become equal to or less than internal combustion–powered vehicles as more units are made (and manufacturing economies-of-scale come into play) because they have far fewer (and much simpler) parts.

But this book advocates the conversion alternative—*you* convert an existing internal combustion engine vehicle to an electric vehicle. You remove the internal combustion engine and all systems that go with it, and add an electric motor, controller, and batteries. If you start with a used internal combustion engine vehicle chassis you can save even more (with the advantage of having the drive train components already broken in, as later chapters point out). (Even if you buy a Porsche or other high-end car, if you were planning to buy a Porsche in the first place, you would already be willing to spend more money than the average consumer.)

To this must be added the cost of your internal combustion engine vehicle chassis. If you start with a brand new vehicle, this could mean $10,000 or more (less any credit for removed internal combustion engine components). A good, used chassis might cost you just $2,000 to $3,000 (or less if you take advantage of special situations as mentioned in Chapter 5). So your total purchase costs are in the $8,000 to $25,000 ballpark. This is substantially less than already converted electric vehicles! Obviously, you can do better if you buy carefully and scrounge for parts. Equally obviously, you can also spend more if you elect to have someone else do the conversion labor, decide you must have a brand new Ferrari Testarossa chassis, or elect to build a Kevlar-bodied roadster with titanium frame from scratch.

This book promotes building it yourself. As a second vehicle choice, logic (and Parkinson's law—the demand upon a resource tends to expand to match the supply of the resource) dictates that the money spent for this decision will expand to fill the budget available—regardless of whether an internal combustion engine vehicle or EV is chosen. So second vehicle purchase costs for an internal combustion or electric vehicle are a wash—they are identical.

Electric Vehicles Are Customizable

EVs are modularly upgradable. See a better motor and controller? Bolt them on. Find some more efficient batteries? Strap them in. You don't have to get an entirely new vehicle; you can adapt new technology incrementally as it becomes available.

Electric cars are easily modified to meet special needs. Even when EVs are manufactured in volume, the exact model needed by everyone will not be made because it would be prohibitively expensive to do so. But specialist shops that add heaters for those living in the north, air conditioners for those living in the south, and both for those living in the heartland will spring up. Chapter 4 will introduce you to the conversion specialists that exist today.

Safety First

Electric vehicles are safer for you and everyone around you. EVs are a boon for safety-minded individuals. Electric vehicles are called *ZEVs* (*zero emission vehicles*) because they emit nothing, whether they are moving or stopped. In fact, when stopped, electric vehicle motors are not running and use no energy at all.

This is in direct contrast to internal combustion engine–powered vehicles that not only consume fuel but also do their best polluting when stopped and idling in traffic. EVs are obviously the ideal solution for minimizing pollution and energy waste on congested stop-and-go commuting highways all over the world, but this section is about saving yourself: as an electric vehicle owner, you are not going to be choking on your own exhaust fumes. Electric vehicles are easily and infinitely adaptable. Want more acceleration? Put in a bigger electric motor. Want greater range? Choose a better power-to-weight design. Want more speed? Pay attention to your design's aerodynamics, weight, and power.

When you buy, convert, or build an EV today, all these choices and more are yours to make because there are no standards and few restrictions. The primary restrictions regard safety (you want to be covered in this area anyway), and are taken care of by using an existing internal combustion engine automotive chassis that has already been safety qualified. Other safety standards to be used when buying, mounting, using, and servicing your EV conversion components are discussed later in this book.

It gets better: EVs carry no combustible fuels, 20,000-volt spark plug ignition circuits, hot exhaust manifolds, catalytic converters, or hot radiators on board.

Those who enjoy:

- Engine compartment fires (caused by ignition or hot manifolds—as seen by the side of the road),
- Hot radiator coolant explosions (caused by improper radiator cap removal; sadly many of us have experienced this firsthand), or
- Starting forest fires (caused by hot catalytic converters parked over dry grass)

simply have to go elsewhere for their entertainment, because it ain't happening here!

Even better, you save wear and tear on yourself by not having to do needless chores associated with vehicle ownership. Contrast the numerous periodic internal combustion engine vehicle activities shown in Figure 1-3 with the far simpler electric vehicle requirements shown in Figure 1-4.

In the "yuck" but not really dangerous category—electric vehicle owners don't have to mess with oil (no dark slippery spots on your garage floor), antifreeze (no lighter, slippery spots on your garage floor), or filters (the kind you hold in a rag far away from your body because they are filthy or gunky).

For critics who comment that lead-acid batteries emit potentially dangerous hydrogen gas when charging, and point out that electric vehicles have multiple batteries: When is the last time you heard of a death or injury resulting from charging a battery? It's possible but very unlikely.

How about the acid part in lead-acid batteries? You'll learn about battery details in Chapter 8, but the acid is diluted sulfuric acid. It definitely hurts if it spills on you or anything else, but it doesn't explode or catch fire and can readily be counteracted by flushing with water.

On another safety side, while electric vehicles do not emit noise pollution, there has been concern about hybrid vehicles being unsafe for seeing-impaired pedestrians because the engines don't make noise.

However, the Baltimore-based National Federation of the Blind presented written testimony to the United States Congress asking for a minimum sound standard for hybrids to be included in the state's emissions regulations. As the president of the group, Marc Maurer, mentioned, he's not interested in returning to gas-guzzling vehicles, they just want fuel-efficient hybrids to have some type of warning noise.

"I don't want to pick that way of going, but I don't want to get run over by a quiet car, either," Maurer said.

Manufacturers are aware of the problem but have made no pledges yet. Toyota is studying the issue internally, said Bill Kwong, a spokesman for Toyota Motor Sales USA.

"One of the many benefits of the Prius, besides excellent fuel economy and low emissions, is quiet performance. Not only does it not pollute the air, it doesn't create noise pollution," Kwong said. "We are studying the issue and trying to find that delicate balance."

The Association of International Auto Manufacturers Inc., a trade group, is also studying the problem, along with a committee established by the Society of Automotive Engineers. The groups are considering "the possibility of setting a minimum noise level standard for hybrid vehicles," said Mike Camissa, the safety director for the manufacturers' association." (Source: Michael d'Estries, Groovy Green.)

Electric Vehicles Save the Environment

EV ownership is visible proof of your commitment to help clean up the environment. Chapter 2 will cover in detail the environmental benefits of this choice. EVs produce no emissions of any kind to harm the air, and virtually everything in them is recyclable. Plus, every electric vehicle conversion represents one less polluting internal combustion vehicle on the road. Electric vehicles are not only the most modern and efficient forms of transportation, but they also help reduce our carbon footprint today!

Electric Vehicle Myths (Dispelling the Rumors)

There have been four widely-circulated myths/rumors about electric vehicles that are not true. Because the reality in each case is the 180-degree opposite of the myth, you should know about them.

Myth #1: Electric Vehicles Can't Go Fast Enough

Well, this is probably true if you are talking about a four-ton van carrying 36 batteries. The reality is that EVs can go as fast as you want—just choose the electric vehicle model (or design or build one) with the speed capability you want. One example of how fast they can accelerate was when I was driving a TH!NK City (really small City EV) in New York City, as shown in the introduction. I was at a traffic light next to a Ford cab *(how appropriate since Ford owned TH!NK at the time)*. The cabbie wanted to see how fast it could go so I said, "I know it can beat you." (Please note that all of this was done well within the legal speed limits on the road in Manhattan!) He said, "You're crazy!" So the light turned green and I hit the accelerator. The look on the cabbie's face was worth a million dollars. He was more than surprised at the torque and acceleration. People on the street were screaming, "Go, go, go." I blew him away. We met up at the next traffic light and he said, "Where can I get one?" Enough said.

Current technology EVs use nickel batteries, such as the Toyota RAV4 or even today's hybrid electric cars such as the Prius or the Honda Civic hybrid. Other conversion companies are starting to use lithium-ion batteries; however, most still use lead acid.

The speed of an electric vehicle is directly related to its weight, body/chassis characteristics such as air and rolling resistance, electric motor size (capacity), and battery voltage. The more voltage, the more batteries you have, the faster any given electric motor will be able to push the vehicle—but adding batteries adds also to the vehicle weight. All of these factors mean you can control how much speed you get out of your EV, and you're certainly not limited in any way. If speed is important, then optimize the electric vehicle you choose for it. It's as simple as that.

Myth #2: Electric Vehicles Have Limited Range

Nothing could be further from the truth but, unfortunately, this myth has been widely accepted. The reality is that electric vehicles can go as far as most people need. Remember, this book advocates an electric vehicle conversion only as your second vehicle. While lithium-ion batteries will expand your range dramatically and there are some people that are travelling cross country in EVs, it is not yet the best use for a massive road trip at this time.

But what is its range? The federal government reports that the average daily commuter trip distance for all modes of vehicle travel (auto, truck, bus) is 10 miles, and this figure hasn't changed appreciably in 20 years of data-gathering. An earlier study showed that 98 percent of all vehicle trips are under 50 miles per day; most people do all their driving locally, and only take a few long trips. Trips of 100 miles and longer account for only 17 percent of total miles. General Motors' own surveys in the early '90s (taken from a sampling of drivers in Boston, Los Angeles, and Houston) indicated:

- Most people don't drive very far.
- More than 40 percent of all trips were under 5 miles.
- Only 8 percent of all trips were more than 25 miles.
- Nearly 85 percent of the drivers drove less than 75 miles per day.

Virtually any of today's 120-volt electric vehicle conversions will go 75 miles—using readily available off-the-shelf components—if you keep the weight under 3,000 pounds. This means an EV can meet more than 85 percent of the average needs. If you're commuting to work—a place that presumably has an electrical outlet available—you can nearly double your range by recharging during your working hours. Plus, if range is really important, optimize your electric vehicle for it. It's that simple.

Myth #3: Electric Vehicles Are Not Convenient

The myth that electric cars are not effective as a real form of transportation or that they are not convenient is a really silly myth/rumor. Car companies and others have complained that there is not enough recharging infrastructure across the country or that you cannot charge the car anywhere you would like as with fueling up a car. A popular question is, "Suppose you're driving and you are not near your home to charge up or you run out of electricity; what do you do?" Well, my favorite answer is, "I would do the same thing I'd do if I ran out of gas—call AAA or a tow truck."

The reality is that electric vehicles are extremely convenient. Recharging is as convenient as your nearest electrical outlet, especially for conversion cars using 100-volt charging outlets. Here are some other reasons:

- You can get electricity anywhere you can get gas—there are no gas stations without electricity.
- You can get electricity from many other places—there are few homes and virtually no businesses in the United States without electricity. All these are potential sources for you to recharge your electric vehicle.
- Over time, as electric vehicles and plug-in hybrid electric cars become prevalent, fuel providers will increase their infrastructure for charging stations and we will all love the price of that versus a gallon of gas.
- As far as being stuck in the middle of nowhere goes, other than taking extended trips in western U.S. deserts (and even these are filling up rapidly), there are

only a few places you can drive 75 miles without seeing an electric outlet in the contiguous United States. Europe and Japan have no such places.

- Plug-in-anywhere recharging capability is an overwhelming electric vehicle advantage. No question it's an advantage when your electric vehicle is parked in your home's garage, carport, or driveway. If you live in an apartment and can work out a charging arrangement, it's an even better idea: a very simple device can be rigged to signal you if anyone ever tries to steal your car.
- How much more convenient could electric vehicles be? There are very few places you can drive in the civilized world where you can't recharge in a pinch, and your only other concern is to add water once in a while. Electricity exists virtually everywhere; you just have to figure out how to tap into it. If your electric vehicle has an onboard charger, extension cord, and plug(s) available, it's no more difficult than going to your neighbor's house to borrow a cup of sugar. Except, of course, you probably want to leave a cash tip in this case.

While there are no electrical outlets specifically designated for recharging electric vehicles conveniently located everywhere today, and though it's unquestionably easier and faster to recharge your electric vehicle from a 110-volt or 220-volt kiosk, the widely available 120-volt electric supply does the job quite nicely. When more infrastructure exists in the future, it will be even more convenient to charge your batteries. In the future, you will be able to recharge quicker from multiple voltage and current options, have "quick charge" capability by dumping one battery stack into another, and maybe even have uniform battery packs that you swap and strap on at a local "battery station" in no more time than it takes you to get a fill-up at a gas station today. Just as it's used in your home today, electricity is clean, quiet, safe, and stays at the outlet until you need it.

Myth #4: Electric Vehicles Are Expensive

While perhaps true of electric vehicles that are manufactured in low volume today—and partially true of professionally done conversion units—it's not true of the do-it-yourself electric vehicle conversions this book advocates. The reality, as we saw earlier in this chapter, is that electric vehicles cost the same to buy (you're not going to spend any more for it than you would have budgeted anyway for your second internal combustion engine vehicle), the same to maintain, and far less per mile to operate. In the long term, future volume production and technology improvements will only make the cost benefits favor electric vehicles even more.

Disadvantages

Well, there had to be a downside. If any one of the factors below is important to you, you might be better served by taking an alternate course of action.

Extended trips as already mentioned; the electric vehicle is not your best choice for transcontinental travel at this time, or long trips in general. Not because you can't do it. Alternate methods are just more convenient. As mentioned, this book advocates the use of the convert-it-yourself electric vehicle as a *second* vehicle. When you need to take longer trips, use your first vehicle, take an airplane, train, or bus, or rent a vehicle.

Time to Purchase/Build

Regardless of your decision to buy, build, or convert an electric vehicle, it is going to take you time to do it, but certainly less than it used to. There is a growing network of new and used electric vehicle dealers and conversion shops. However, the supply for the highest grade controllers and motors do take time to produce and will slow down your conversion process. Although, as demand for these products increases, the supply will increase too and the time that it takes for these products to be built will get reduced. (Check the Sources section at the end of the book.) But plan on taking a few weeks to a few months to arrive at the electric vehicle of your choice.

Electric vehicle resale (if you should decide to sell your EV) will take longer—for the same reason. While a reasonably ready market exists via the Electric Auto Association chapter and national newsletters, it is still going to take you longer and be less convenient than going down to a local automobile dealer.

Repairs

Handy electric vehicle repair shops don't exist yet either. Although the build-it-yourself experience will enable your rapid diagnosis of any problems, replacement parts could be days, weeks, or months away—even via expedited carriers. You could just stockpile spare parts yourself, but the time to carefully think through this or any other repair alternative is before you make your electric vehicle decision.

The Force Is with You

What you are seeing today in electric vehicles is just the tip of the iceberg. It is guaranteed that future improvements will make them faster, longer-ranged, and even more efficient. There are five prevailing reasons that guarantee electric vehicles will always be with us in the future. The only one not previously discussed is technological change. All the available technology has just about been squeezed out of internal combustion engine vehicles, and they are going to be even more environmentally squeezed in the future. This will hit each buyer right in the pocketbook. Incremental gains will not come inexpensively. Internal combustion engines are nearly at the end of their technological lifetime. Almost all improvements in meeting today's higher Corporate Average Fuel Economy (CAFE) requirements have been achieved via improving electronics technology. CAFE is the sales weighted average fuel economy, expressed in miles per gallon (mpg), of a manufacturer's fleet of passenger cars or light trucks with a gross vehicle weight rating (GVWR) of 8,500 lbs. or less, manufactured for sale in the United States, for any given model year. Fuel economy is defined as the average mileage traveled by an automobile per gallon of gasoline (or equivalent amount of other fuel) consumed as measured in accordance with the testing and evaluation protocol set forth by the Environmental Protection Agency (EPA) (Source: National Highway Traffic Safety Administration).

By 2020, CAFE mandates all new cars built will have the fuel economy of 35 miles per gallon as approved by the United State Congress. Since the fuel efficiency of an electric vehicle is more than that, an EV will always be the best approach.

In addition, once lithium battery technology becomes the standard, electric vehicles will be able to go 300 to 600 miles (depending on the technology). Unquestionably, the future looks bright for electric vehicles because the best is yet to come.

The four remaining reasons were covered in this chapter, and are summarized in Figure 1-7:

- Fun to drive and own
- Cost efficient
- Performance efficient
- Environmentally efficient

Any one of the four reasons above is compelling by itself; the benefits of all these reasons taken together are overwhelming.

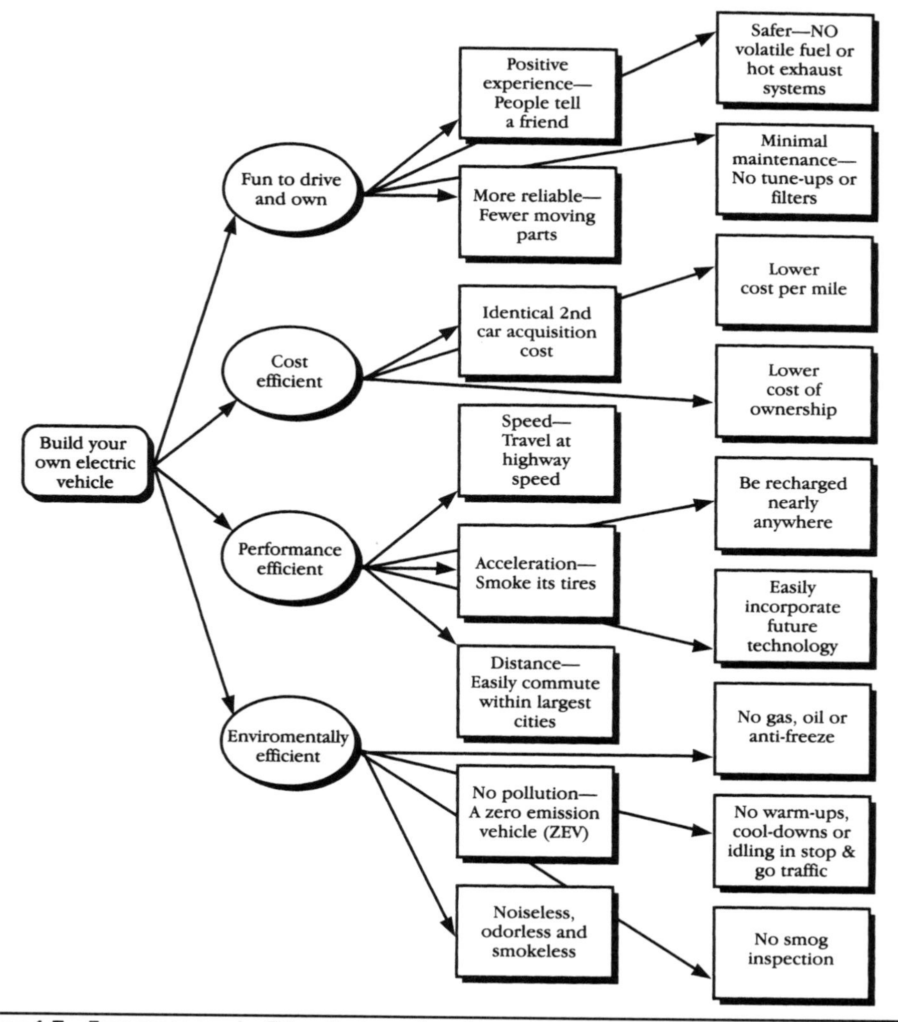

FIGURE 1-7 Four reasons why EVs will always be with us in the future.

CHAPTER 2

Electric Vehicles Save the Environment and Energy

"The needs of the many outweigh the needs of the few."
—Mr. Spock (from *Star Trek, The Motion Picture*)

Besides the fact that the consumer has been consistently interested in the electric car (despite popular reporting by car companies), there is a newly sparked (no pun intended) excitement in either plug-in hybrid electric cars (which are more electric car than hybrid) and electric cars (Tesla, TH!NK City, RAV4, EV1). This only means great things for the planet. Specifically, zero tailpipe emissions and greater air quality in our major metropolitan cities. And with this is a significant reduction in overall energy use.

Why Do Electric Vehicles Save the Environment?

Electric vehicles are zero emission vehicles (ZEVs). They do not emit toxic compounds into our atmosphere. Even the power plants that generate the power for EVs are held to a higher standard (meaning a lower level of toxic emissions) compared to the emissions related to gasoline-powered vehicles. Everything going into and coming out of an internal combustion vehicle, on the other hand, is toxic and it's still classified among the least efficient mechanical devices on the planet.

Far worse than its inefficient and self-destructive operating nature is the legacy of environmental problems (summarized in Figure 2-1) created by internal combustion engine vehicles when multiplied by hundreds of millions/billions of vehicles. The greatest of these problems include the following:

- Dependence on foreign oil (environmental and national security risk)
- Greenhouse effect (atmospheric heating)
- Toxic air pollution
- Wasted heat generated by its inefficiency

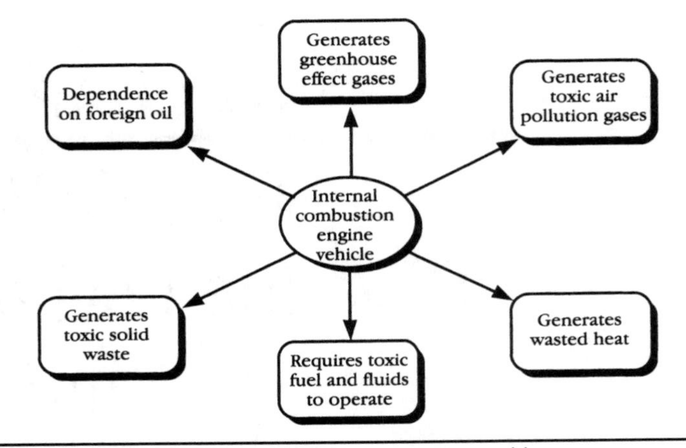

FIGURE 2-1 Internal combustion engine vehicles create many problems.

Save the Environment and Save Some Money Too!

Because electric vehicles use less energy then gasoline-powered vehicles, their effect on the environment is much less than vehicles powered by fossil fuels. Because electric vehicles are more efficient than gasoline-powered vehicles, they cost less to run.

Electric vehicles have existed for more than a hundred years (predating internal combustion engine vehicles). The aerospace-derived technology to improve them has existed for decades. Unquestionably, EVs will be the de facto transportation mode of choice for years to come, if life on our planet is to continue to exist in its modern form with the conveniences we count on today.

Figure 2-2 shows the reasons why. In direct contrast to the problems created by internal combustion engine vehicles, EVs:

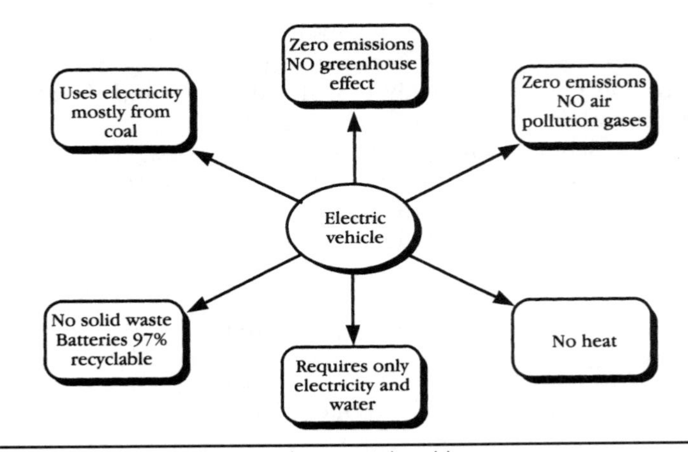

FIGURE 2-2 Electric vehicles create no environmental problems.

- Use electricity (today, electricity comes mostly from coal).
- Are zero emission vehicles; they emit no pollutants.
- Generate little toxic waste (lead-acid batteries are 98 percent recyclable).
- Require no toxic input fluids such as water and wash fluid (only occasional watering of batteries, if you choose cheaper lead-acid batteries)
- Are highly efficient (motors and controllers 90 percent, batteries 75 to 80 percent).
- Benefit electric utilities (a market for electricity sales).

Even the cost advantage goes to the electric vehicle since the electric vehicle converts about 70 percent of the charging energy into motor energy, whereas a typical gasoline-powered vehicle converts only about 20 percent of the energy in gasoline into engine energy. An automatic transmission represents another significant loss, as do auxiliaries such as power steering and air conditioning. (Around town, air conditioning can consume 40 percent or more of available power.) For the conversion, 100,000 BTU (1 therm) = 29.3 kWh = 39.3 hp/hr (horsepower per hour). Thus for the same price, you have 47 hp/hr for the electric and 21 hp/hr for the vehicle powered by California gasoline. In actual use you can have a RAV4-EV costing $0.04 per mile and a gasoline-powered RAV4 costing $0.16 per mile just for fuel.

Fuel-Efficient Vehicles

Car companies are now making an effort to develop or provide more fuel-efficient cars. The market wants them, the large organizations have responded to the market, and the car companies that are doing well (Toyota and Honda) are producing fuel-efficient and hybrid cars. Now, Ford, GM, and the other car companies are either producing or developing these type of cars as well. This is to be expected in business.

Much has been said about creating a new type of car that can get 35 mpg. In other countries where fuel is very expensive, such cars already exist. The cars are much smaller and have smaller engines. What has not been said is the great extent to which drivers control the range of these vehicles. In countries where fuel is expensive, drivers tend to drive at slower speeds. Driving twice as fast requires four times the energy to overcome aerodynamic losses. To go from 50 mph to 100 mph increases the rate fuel or electrical energy is used almost by a factor of eight. (Since you get there in half the time, total energy used is increased by a factor of four.) It is the driver's right foot more than anything else that controls mpg or miles/kWh for a particular vehicle. *Even if you do not have an electric car, plan on converting a car, or plan on buying a hybrid-electric car, one thing to take away from this book is that driving more efficiently will reduce your carbon footprint.*

So Who's to Blame?

No one and everyone. The line from *Dr. Zhivago* about "one Russian ripping off wood from a fence to provide heat for his family in winter is pathetic, one million Russians doing the same is disaster" applies equally well to all of us and our internal combustion engine vehicles. Applied collectively, the legacy of the internal combustion engine is greenhouse effect, foreign oil dependence, and pollution. Let's succinctly define the problem and its solution.

United States Transportation Depends on Oil

Although small amounts of natural gas and electricity are used, the United States transportation sector is almost entirely dependent on oil. A brief look at a few charts will demonstrate the facts (see Figure 2-3). It doesn't take a rocket scientist to figure out this situation is both a strategic and economic problem for us.

Forty percent of our energy comes from petroleum, 23 percent from coal, and 23 percent from natural gas. The remaining 14 percent comes from nuclear power, hydroelectric, and renewables. As Bob Brant stated, "Our entire economy is obviously dependent on oil."

The U.S. consumes 20.8 million barrels of petroleum a day, of which 9 million barrels is gas. Automobiles are the single largest consumer of oil, consuming 40 percent, and are also the source of 20 percent of the nation's greenhouse gas emissions.

The U.S. has about 22 billion barrels of oil reserves while consuming about 7.6 billion barrels per year. Problems associated with oil supply include volatile oil prices, increasing world and domestic demand, and falling domestic production.

While it's our own fault for letting it happen, the Organization of Petroleum Exporting Countries (OPEC) price hikes have had a disastrous impact on our economy, our transportation system, and our standard of living. The Arab Oil Crisis of 1973 and subsequent ones were not pleasant experiences. After each crisis, the United States vowed to become less dependent on foreign oil producers—yet exactly the opposite has happened.

Increasing Long-Term Oil Costs

There is a fixed amount of oil/petroleum in the ground around the world, and there isn't going to be any more. We're going to run out of oil at some point. Before that happens, it's going to get expensive.

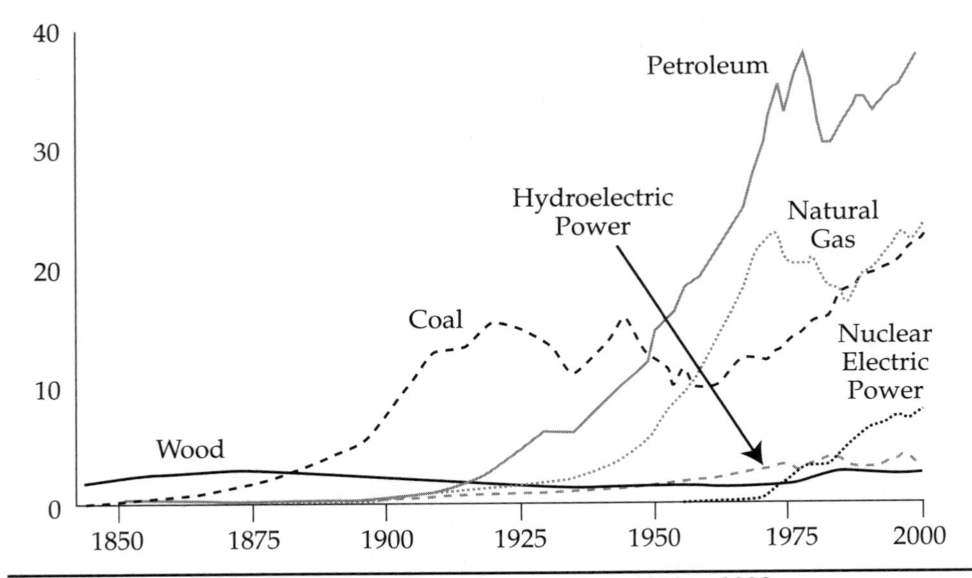

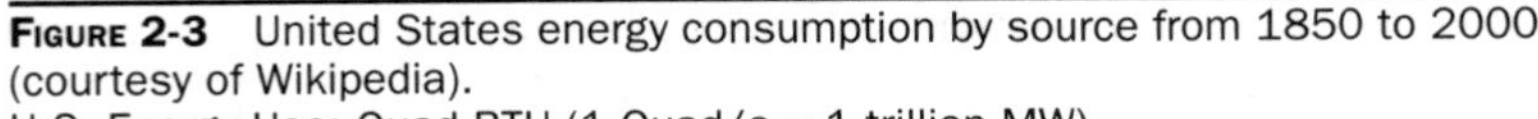

Figure 2-3 United States energy consumption by source from 1850 to 2000 (courtesy of Wikipedia).
U.S. Energy Use: Quad BTU (1 Quad/s = 1 trillion MW)

How did we get into this situation? Today we are already past that amount. No one can accurately predict what fuel prices will be this summer or next year, and whether there will be a shortage or abundance of supplies. Everyone agrees that this is a bad situation. We need to take real steps to correct the problem. Since none of us has the luxury of skipping going to work, many creative "work-around" solutions would come forward if gasoline suddenly cost $4.50 a gallon in the United States or a price for a barrel of oil costs more than $150—for example, electric vehicles.

How Electric Vehicles Can Help

Maintaining stringent toxic air pollution emission levels along with conforming to increasingly higher mandated corporate average fuel economy levels puts an enormous burden on internal combustion engine vehicle technology and on your pocketbook. Automotive manufacturers have to work their technical staffs overtime to accomplish these feats, and the costs will be passed on to the new buyer.

Pollution control equipment is a problem each internal combustion engine vehicle owner has to revisit every year: smog checks, emission certificates, replaced valves, pumps, filters, and parts—all extra cost-of-ownership expenses and inconveniences.

How can electric vehicles reduce toxic air pollution emissions? Easy. All electric vehicles are by definition zero emission vehicles (ZEVs): they emit nothing. That's why California and other states have mandated electric vehicles to solve their air pollution problems. To quote Quanlu Wang, Mark A. DeLuchi, and Dan Sperling, who studied the subject extensively:

"The unequivocal conclusion of this paper is that in California and the United States the substitution of electric vehicles for gasoline-powered vehicles will dramatically reduce carbon dioxide, hydrocarbons and to a lesser extent, nitrogen oxide emissions."

The earlier comments regarding power plant emissions associated with electricity production for electric vehicle use are equally applicable to air pollution as they are greenhouse gas production. In addition, shifting the burden to coal-powered electrical generating plants for electric vehicle electricity production has these effects:

- Focuses smokestack "scrubber" and other mandated controls on stationary sites that are far more controllable than internal combustion engine vehicle tailpipes.
- Shifts automotive emissions from congested, populated areas to remote, less populated areas where many coal-fired power plants are located.
- Shifts automotive emissions to nighttime (when most electric vehicles will be recharged) when fewer people are likely to be exposed and emissions are less likely to react in the atmosphere with sunlight to produce smog and other by-product pollutants.

Electric vehicles generate no emissions whatsoever and reduce our reliance on imported oils. Frankly, until you get an appreciable number of electric cars on the road today (hundreds of thousands to millions), they do not impact emissions from electrical generating plants.

Toxic Solid Waste Pollution

Almost everything going into and coming out of the internal combustion engine is toxic. In addition to the internal combustion engine vehicle's greenhouse gas and toxic air pollution outputs, consider its liquid waste (fuel spills, oil, antifreeze, grease, etc.) and solid waste (oil-air-fuel filters, mufflers, catalytic converters, emission control system parts, radiators, pumps, spark plugs, etc.) byproducts. This does not bode well for our environment, our landfills, or anything else—especially when multiplied by hundreds of millions of vehicles.

How can the electric vehicle help? The only waste elements of an electric vehicle are its batteries. For example, lead-acid batteries—the kind commonly available today—are 99.99 percent recyclable. In processing many tons per day, almost every ounce is accounted for. This means 99.99 percent of all such batteries and the products that go into them (the sulfuric acid, the lead, and even the plastic of their cases) is recoverable.

Toxic Input Fluids Pollution

Remember, almost everything going into and coming out of the internal combustion engine is toxic. The fuel and oil you put into an internal combustion engine, the fuel vapors at the pump (and those associated with extracting, refining, transporting and storing fuel), and the antifreeze you use in its cooling system are all toxic and/or carcinogenic, as a quick study of the pump and container labels will point out. On the output side, when burning coal, oil, gas, or any fossil fuel, you create more problems either by the amount of carbon dioxide or by the type of other toxic emissions produced.

Everything you pour into an internal combustion engine is toxic, but some chemicals are especially nasty. In addition to more than 200 compounds on its initial hazardous list, the Clean Air Act of 1990 amendments said:

> "... study shall focus on those categories of emissions that pose the greatest risk to human health or about which significant uncertainties remain, including emissions of benzene, formaldehyde and 1, 3 butadiene."

Fouling the environment as in the Exxon Valdez oil spill disaster of the 1990s is one thing. Poisoning your own drinking water is another. Those enormous holes in the ground near neighborhood gas stations everywhere (as they rush to be compliant with federal regulations regarding acceptable levels of gasoline storage tank leakage) make the point. So does the recall of millions of bottles of Perrier drinking water where only tiny levels of benzene contamination were involved.

How can the electric vehicle help? The only substance you pour into your electric vehicle occasionally is water (preferably distilled).

Waste Heat Due to Inefficiency

Although its present form represents its highest evolution to date, the gasoline-powered internal combustion engine is classified among the least efficient mechanical devices on the planet. The internal combustion engine is close to 20 percent efficient. The efficiency of an Advance DC motor runs between 80 and 90 percent, sometimes lower.

In gasoline-powered vehicles, only 20 percent of the energy of combustion becomes mechanical energy; the rest becomes heat lost in the engine system. Of the 20 percent mechanical energy:

- One-third overcomes aerodynamic drag (energy ends up as heat in the air).
- One-third overcomes rolling friction (energy ends up as heated tires).
- One-third powers acceleration (energy ends up as heat in the brakes).

In contrast to the hundreds of internal combustion engine moving parts, the electric motor has just one. That's why they're so efficient. Today's EV motor efficiencies are typically 90 percent or more. The same applies to today's solid-state controllers (with no moving parts), and today's lead-acid batteries come in at 75 percent or more. Combine all these and you have an electric vehicle efficiency far greater than anything possible with an internal combustion engine vehicle.

Electric Utilities Love Electric Vehicles

Even the most wildly optimistic electric vehicle projections show only a few million electric vehicles in use by early in the 21st century. Somewhere around that level, EVs will begin making a dent in the strategic oil, greenhouse, and air quality problems. But until you reach the 10 to 20 million or more EV population level, you're not going to require additional electrical generating capacity. This is due to the magic of *load leveling*. Load leveling means that if electric vehicles are used during the day and recharged at night, they perform a great service for their local electrical utility, whose demand curves almost universally look like that shown in Figure 2-4.

How electricity is generated varies widely from one geographic region to another, and even from city to city in a United States region. In 2007, the net electricity mix generated by electric utilities was 48.6 percent coal, 19.4 percent nuclear fission, natural gas 21.5 percent, hydropower 5.8 percent, 1.6 percent and 2.5 percent for geothermal, solar, and wind, with other miscellaneous sources providing the balance (*Source: Edison Electric Institute*).

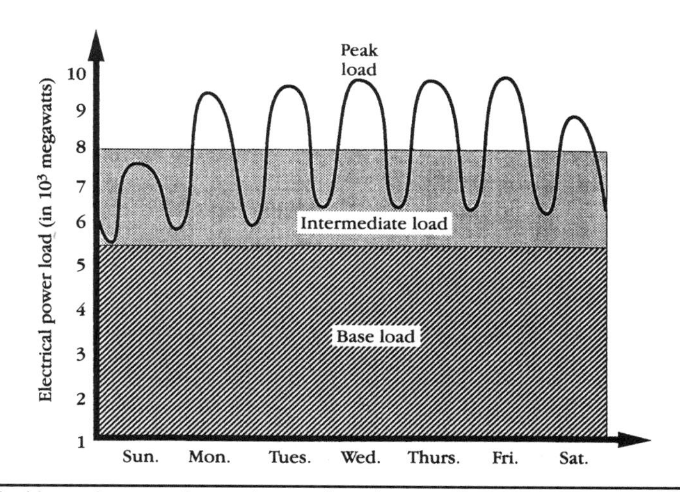

Figure 2-4 Weekly peak power-demand curve for a large utility operating with a weekly load factor of about 80 percent.

Electric utility plants producing electricity at the lowest cost (coal and hydro) are used to supply base-load demands, while peak demands are met by less economical generation facilities (gas and oil). (*Source: Energy Information Administration*).

By owners recharging their electric vehicles in the evening hours (valley periods) they receive the benefit of an off-peak (typically lower) electric rate. By raising the valleys and bringing up its base-load demand, the electric utility is able to more efficiently utilize its existing plant capacity. This is a tremendous near-term economic benefit to our electric utilities because it represents a new market for electricity sales with no additional associated capital asset expense.

Summary

Electric vehicle ownership is the best first step you can take to help save the planet. But there is still more you can do. Do your homework. Write your Senator or Congressperson. Voice your opinion. Get involved with the issues. But don't settle for an answer that says we'll study it and get back to you. Settle only for action—who is going to do what by when and why. I leave you with a restatement of the problem, a possible framework for a solution, and some additional food for thought.

Legacy of Internal Combustion Engine Is Environmental Problems

Internal combustion engine technology and fuel should be priced to reflect its true social cost, not just its economic cost, because of the environmental problems it creates:

- Our dependence on foreign oil and the subsequent security risk problem
- The greenhouse problem
- The air quality problems of our cities
- Toxic waste problem
- Toxic input fluids problem
- Inefficiency problem

Our gasoline's cost should reflect our cost to defend foreign oil fields, reverse the greenhouse effect, and solve the air quality issues. Best of all, our gasoline's cost should include substantial funding to research solar energy generation (and other renewable sources) and electric vehicle technologies—the two most environmentally beneficial and technologically promising gifts we can give to our future generations.

A Proactive Solution

People living in the United States have been extremely fortunate for most of our nations history. However and more now than ever, we have issues with clean air, our natural resources, instable governments, expensive energy costs, and while having a convenient and true standard of living second to no other country on the planet. But nothing guarantees our future generations will enjoy the same birthright. In fact, if we fold our hands behind our backs and walk away from today's environmental problems, we guarantee our children and children's children will not enjoy the same standard of living that we do. For the sake of our children, we cannot walk away, we must do something. We must attack the problem straight on, pull it up by its roots, and replace it with a solution.

Figure 2-5 suggests a possible approach. We need to look at the results wanted in the mid-21st century and work backward—on both the supply and demand sides—to see what we must start doing today. Clearly, it's time for a sweeping change, but we all have to want it and work toward it for it to happen. No one has to be hurt by the change if they become part of the change. Automakers can make more efficient vehicles. Suppliers can provide new parts in place of the old. The petrochemical industry can alter its mix to supply less crude as oil and gas and more as feedstock material used in making vehicles, homes, roads, and millions of other useful items. Long before any of these things happen, you can do your part by building your own electric vehicle.

Today!

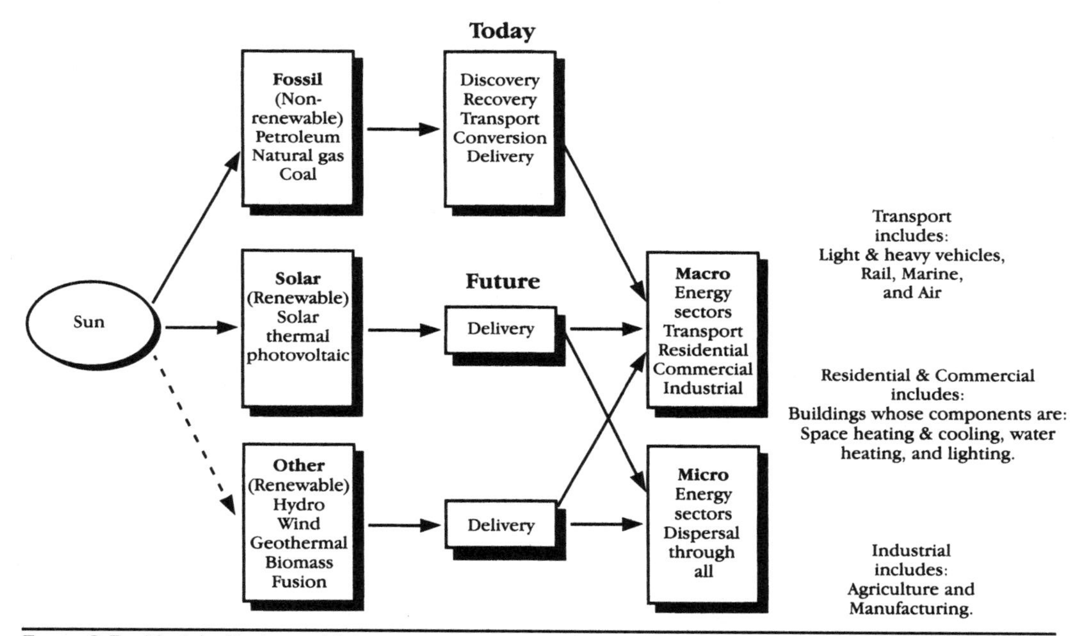

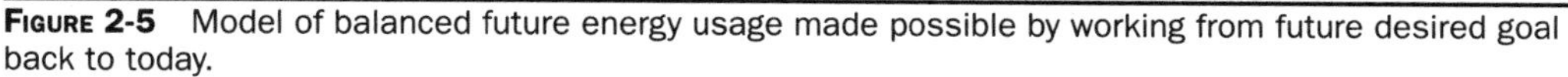

FIGURE 2-5 Model of balanced future energy usage made possible by working from future desired goal back to today.

CHAPTER 3

Electric Vehicle History

"Precedent said, it cannot be done. Experience said, it is done."
—Darwin Gross, Universal Key

Ironically, EVs were around before internal combustion engine vehicles and will also be around after them. The two vehicle types will coexist for some time to come. In this chapter you'll learn about the history of EVs, the forces that shaped their demise, the trends that forced their resurgence, and the tremendous positive role awaiting them in the future.

While modern technology has made electric vehicles better, there is very little new in electric vehicle technology. Today's EV components would be instantly recognizable in those that roamed our streets a century ago. As a potential EV builder or converter, you should be happy to know they have a long and distinguished heritage—you might even get some useful building ideas by looking at the earliest-vintage EVs in an automobile museum.

Before getting into an area that will make EVs sound like something new, let's delve briefly into some basic historical facts about electric vehicles:

- The electric motor came before the internal combustion engine.
- Electric vehicles have been around since the mid-1800s, were manufactured in volume in the late 1800s and early 1900s, and declined only with the emergence and ready availability of cheap gasoline.
- Even so, electric vehicle offshoots—tracked buses, trolleys, subways, and trains—have continued to serve in mass transit capacities right up until the present day because of their greater reliability and efficiency.

One of the greater electric transportation accomplishments I have seen (and became more aware of when I worked for the New York Power Authority [NYPA]) is that most of the Metropolitan Transportation Authority (MTA) subways and trains are electric powered. A minimal number of trains in the MTA fleet are diesel-powered. In addition, each year I worked for the State of New York, the MTA shifted more and more from compressed natural gas to hybrid-electric transit buses. While they are not 100 percent electric, it proves the point that an electric drive is cleaner and more fuel efficient than just an alternative fuel.

The battery and electric motor combination borrowed from the electric vehicles of the "novelty" era up until 1915 and applied as a starter motor for internal combustion engines was responsible for the great upsurge in the internal combustion engine vehicle's popularity. The starter motor systems employed in all of today's internal combustion engine vehicles are virtually unchanged from the original early 1920s concept.

Battery electric vehicles have also been extremely popular in some limited-range applications. Forklifts have been battery electric vehicles (BEVs) since the early 1900s and electric forklifts are still being produced. BEV golf carts have been available for years. Golf carts have led to the emergence of neighborhood electric vehicles (NEVs) or low-speed vehicles (LSVs), which are speed-limited at 25 mph, but are legal for use on public roads. NEVs were primarily offered by car companies during the end of the CA Zero Emission Vehicle (ZEV) Mandate. As of July 2006, there are between 60,000 and 76,000 low-speed, battery-powered vehicles in use in the U.S., up from about 56,000 in 2004 according to Electric Drive Transportation Association estimates. In fact, at the end of my tenure at the New York Power Authority, I managed the LSV donation programs from Ford and Chrysler (GEM) of over 250 vehicles. I believe several thousand vehicles were donated by the car companies to receive ZEV credits for the amount of electric vehicles placed on road in 2003.

By the late 1930s, the electric automobile industry had completely disappeared, with battery-electric traction being limited to niche applications, such as certain industrial vehicles.

The 1947 invention of the point-contact transistor marked the beginning of a new era for BEV technology. Within a decade, Henney Coachworks had joined forces with National Union Electric Company, the makers of Exide batteries, to produce the first modern electric car based on transistor technology, the Henney Kilowatt, produced in 36-volt and 72-volt configurations. The 72-volt models had a top speed approaching 96 km/h (60 mph) and could travel nearly an hour on a single charge. Despite the improved practicality of the Henney Kilowatt over previous electric cars, it was too expensive and production was terminated in 1961. Even though the Henney Kilowatt never reached mass production volume, their transistor-based electric technology paved the way for modern EVs.

Timeline of Vehicle History

Studying vehicle history is similar to looking at any economic phenomenon. iPods are a good example. The first iPod was a novelty; the one hundredth created a strong desire to own one. By the ten thousandth, you own one; by the one millionth, the novelty has worn off; and after the hundred millionth, they're considered ubiquitous. The same with vehicles—past events shift the background climate and affect current consumer wants and needs. The innovative Model T of the 1910s was an outdated clunker in the 1920s. The great finned wonders of the 1950s and muscle cars of the 1960s were an anachronism by the 1970s. A vehicle that was once in great demand is now only junkyard material because consumer wants and needs change.

Figure 3-1 is rather busy, but studying it gives you clues to the rise and fall of the three types of vehicles in one picture—steam, electric, internal combustion—plus the interrelationship between them during the three stages of vehicle history. Figure 3-1 shows that steam has been passed by as a vehicle power source but electric vehicles, dominant in urban areas at the turn of the century, are again returning to favor as the majority of the world's industrial nations become "urbanized," but petroleum-based fuels are becoming more expensive and availability more politically dependent. Nearly

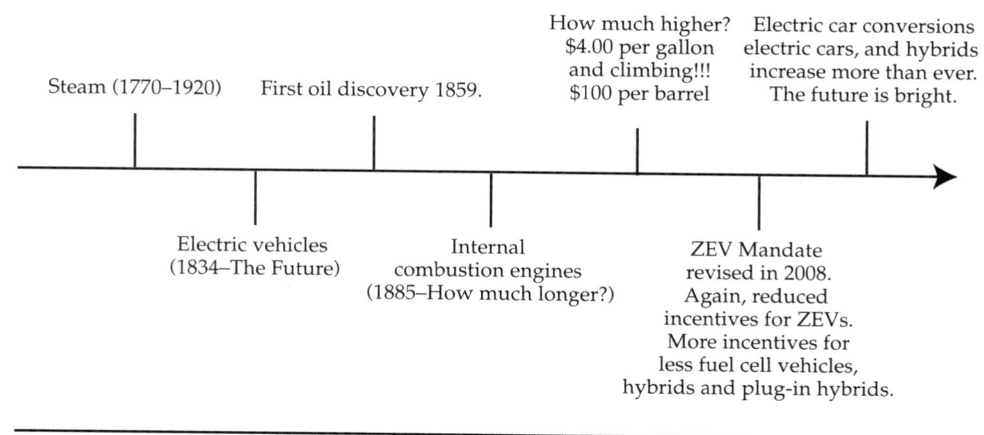

FIGURE 3-1 History of electric vehicles.

100 "golden years" of the internal combustion engine vehicle, which swamped the early steam and electric offerings in a wave of cheap gasoline prices and offered the ability to travel where there were no tracks, are declining. It should be noted, however, that this trend has played out in fits and starts repeatedly over the last several decades, most prominently in the 1990s, when electric vehicles were made in response to California's Zero Emission Vehicle Mandate but were mostly reclaimed and destroyed by automakers a few years later. While we are on the cusp of seeing electric vehicles and plug-in hybrids built, it remains to be seen how these technologies will play out politically and in the marketplace.

A brief look at electric vehicle history is helpful in understanding why electric vehicles came, went away, and are back again.

The Timeline of Electric Cars

Steam engines came first, followed by electric motors, and finally by internal combustion engines. The close proximity of coal and iron deposits in the northern latitudes of what came to be known as the industrialized nations—United States, Europe/England, and Asia—made the steam engine practical. The thriving post-industrial revolution economy provided by the steam engine created the climate for electrical invention. Electrical devices made the internal combustion engine possible. Vehicles powered by them followed the same development sequence.

Up Until 1915

This period marked the transition from the "novelty" era (most bystanders were amazed that these devices actually worked) to the "practical" era, where early buyers just wanted a vehicle to go from one point to another with minimum hassle, and finally to the "production" era. In the production era, after nearly 3,000 vehicle manufacturers had come and gone, the survivors fell into two camps—those able to make a profit catering to custom needs at a high market price, and those offering a standard solution at a low market price. Steam and electric offerings were overwhelmed by the dominance of cheap oil and gasoline and virtually disappeared as competitors to internal combustion engine vehicles after 1915.

Huff and Puff

The same phenomenon that makes the heated tea kettle on your stove whistle, when suitably harnessed, makes a steam engine go. The "huffing and puffing" associated with steam railroad locomotives is absent from the whisper-silent steam automobiles, yet they are equally powerful pound for pound. However, you still need to heat the water, which means burning something, and if you don't continually monitor your boiler steam pressure, everything can blow up.

James Watt's steam engine of 1765—widely acclaimed as responsible for the industrial revolution—was only an improvement on Thomas Newcomen's 1712 machine that, in turn, built on the more primitive 1690-vintage designs of Denis Papin, Christian Huygens, and Robert Boyle, and the initial patent of Thomas Savery in 1698. Steam technology was applied to the first land vehicle—Nicolas Cugnot's tractor—in 1770, to a steamboat by John Fitch in 1787, and to a rail locomotive by Richard Tevithick in 1804.

While the Cugnot steam tractor is a far cry from the Stanley Steamer automobiles of the early 1900s (a streamlined version of the latter set the land speed record at 122 mph in 1906), and still further removed from the high-performance Lear steam cars of a few decades ago, the problem with steam vehicles remains the steam. Water needs a lot of heat to become steam, and it freezes at cold temperatures. To get around these and the basic "time to startup" problems, technical complexity was introduced in the form of exotic liquids to withstand repeated evaporation and condensation, and exotic metals for more sophisticated boilers, valves, piping, and reheaters.

Figure 3-2 shows the steam, electric, and internal combustion vehicle population in the United States from 1900 to 2000. Steam-powered vehicles, popular in the last part of the 1800s, declined in favor of the other two vehicle types after the early 1900s. Electric vehicles enjoyed rapid growth and popularity until about 1910, then a slow decline until their brief resurgence in the 1990s. Internal combustion engine–powered vehicles passed steam and electric early in the 1900s. More than any other factor, cheap and nearly unlimited amounts of domestic (and later foreign) oil, which kept gasoline prices between 10 and 20 cents a gallon from 1900 through 1920, suppressed interest in alternatives to internal combustion engine vehicles until more than 50 years later (the 1970s).

In the early 1900s, steam vehicles unquestionably offered smoothness, silence, and acceleration. But stops for water were typically more frequent than stops for kerosene, and steamer designs required additional complexity and a lengthy startup sequence. While 40 percent of the vehicles sold in 1900 were steam (38 percent were electric), electrics offered simplicity, reliability, and ease of operation, while gasoline vehicles offered greater range and fuel efficiency. Thus steamers declined, and only a handful operate today. (Note: I will debate the issue of fuel efficiency in later chapters.)

The Wheel Goes Round and Round

Electricity is everywhere. In one place it lights a factory, in another it conveys a message, and in a third it drives an electric vehicle. Electricity is transportable—it can be generated at a low-cost location and conveniently shipped hundred of miles to where it is needed. A storage battery, charged from electricity provided by a convenient wall outlet, can reliably carry electricity to start a car anywhere, or power an electric vehicle. Who

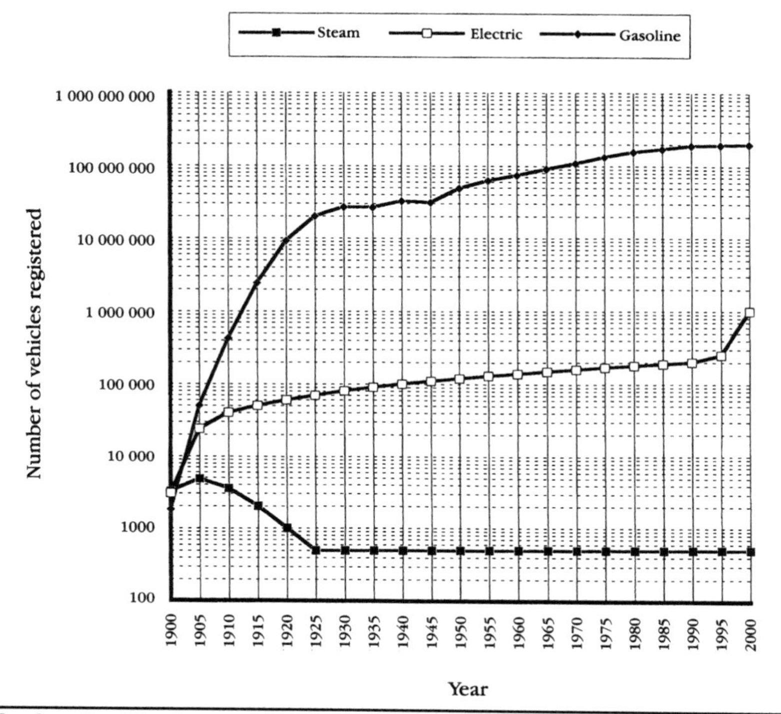

FIGURE 3-2 Growth of the three vehicle types in the United States from 1900 to 2000.

knows (or cares) that the wheels of your vehicle turn because the electricity was generated from water turning a turbine wheel linked to a generator? We take electricity for granted today and are continually developing new uses for it, primarily because of its advantages—it's clean, simple, available, and reliable. But our modern electrical heritage owes a great debt to many pioneers.

Alessandro Volta, building on the experiments of Luigi Galvani in 1782, invented the electric battery—his "Voltaic pile"—in 1800. Joseph Henry, building on the experiments of Han Christian Oersted in 1819 and Andre Ampere in 1820, created the first primitive direct current (DC) electric motor in 1830. Michael Faraday demonstrated the induction principle and the first electric DC generator in 1831. Battery-powered electric technology was applied to the first land vehicle by Thomas Davenport in 1834, to a small boat by M. H. Jacobi in 1834, and to the first battery-powered locomotive—the five-ton "Galvani"—by Robert Davidson in 1838.

Moses Fanner unveiled a two-passenger electric car in 1847, and Charles Page showed off a 20-mph electric car in 1851, but Gaston Plante's lead-acid "rechargeable" battery breakthrough of 1859—improved upon by Camille Favre in 1881 and H. Tudor in 1890—paved the way for extended electric vehicle use. Nikola Tesla's alternating current (AC) induction motor of 1882 and subsequent polyphase patents paved the way for the AC electrical power distribution infrastructure we use today. By the 1890s, DC power distribution via dynamos had been in use for a decade. AC power distribution began with the 1896 Niagara Falls power plant contract award to George Westinghouse

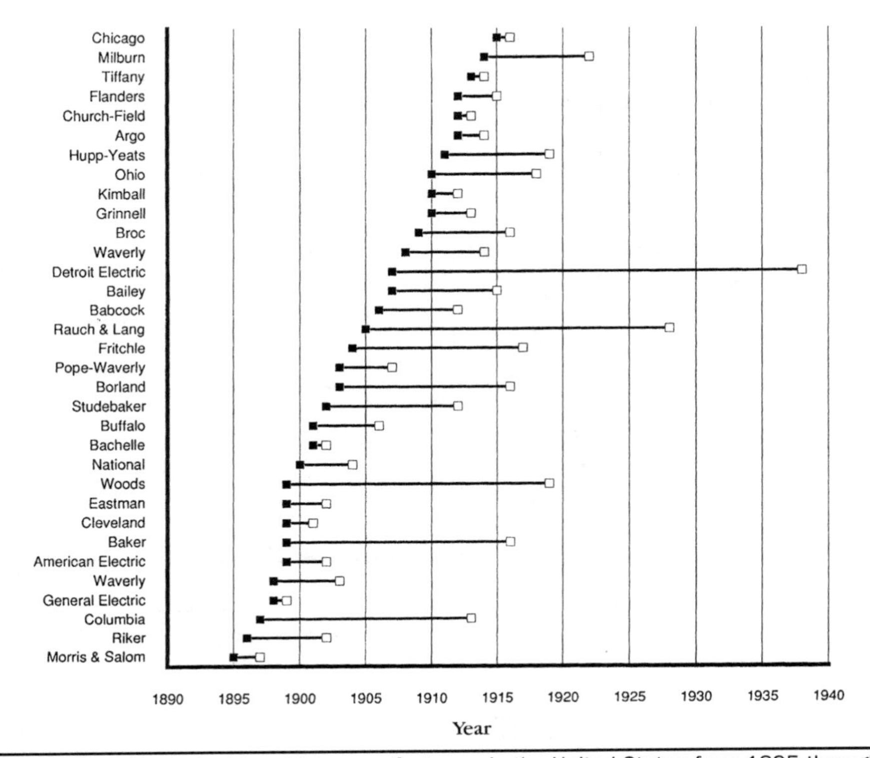

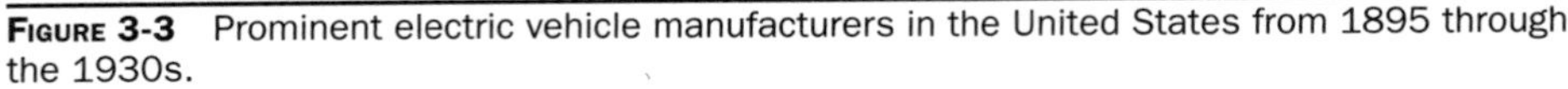

FIGURE 3-3 Prominent electric vehicle manufacturers in the United States from 1895 through the 1930s.

(Nikola Tesla's patents) after a brief but intense battle with Thomas Edison's DC forces.

Figure 3-3 shows a cross-section of the more prominent United States electric vehicle manufacturers in operation from 1895 through the 1930s.

By 1912, the peak production year for early electrics, 34,000 cars were registered. The Reader's Guide to Periodical Literature listings tell the story. The half page of magazine articles listed in the 1890 through 1914 volumes dwindled to a quarter page in 1915–18 and disappeared altogether in the 1925–28 volume.

Early electric vehicle success in urban areas was easy to understand. Most paved roads were in urban areas; power was conveniently available; urban distances were short; speed limits were low; and safety, comfort, and convenience were primary purchase considerations. The quietness, ease of driving, and high reliability made EVs a natural with the wealthy urban set in general and well-to-do women in particular. Clara Bryant Ford (Mrs. Henry Ford) could have any automobile she wanted, but she chose the Detroit Electric now on display at the Henry Ford Museum (shown in Figure 3-4) for getting around the Ford Park Lane estate and running errands. Thomas Edison's 1889-vintage electric vehicle was a test platform for his rechargeable nickel-iron battery experiments. Later, Edison's nickel-iron batteries went into the Bailey Electric and numerous other electrics. Edison had his own personal Studebaker electric vehicle, and both he and Henry Ford were strongly supportive of EVs. At one time these two planned to bring out a lightweight $750 electric auto that was to be called the Edison-Ford.

FIGURE 3-4 Mrs. Henry Ford's Detroit Electric now on display in the Henry Ford Museum.

Electric vehicles also dominated the commercial delivery fleets in urban areas around the world. Department stores, express delivery companies, post offices, utility, and taxicab companies in New York, Chicago, London, Paris, and Berlin used thousands of EVs. High reliability (99 percent of the 300-day work year availability) and low maintenance characterized commercial EVs and made them fleet favorites.

In the EV's performance department, typical 2,500-lb. cars went 20 to 30 mph, and got 50 to 60 miles on a battery charge. Half-ton trucks went 10 to 15 mph and had a 40 to 50 mile range. Ten-ton trucks went 5 to 10 mph and had a 30 to 40 mile range. An electric vehicle set the first land speed record. Camille Jenatzy's "Jamais Contente" (a streamlined vehicle powered by two 12-hp electric motors riding on narrow 25-inch diameter tires) went 66 mph in April, 1899—a record that stood for three years until broken by the Baker Electric "Torpedo" in 1902 at 78 mph, and later by the "Torpedo Kid" in 1904 at 104 mph. In 1900, the French B.G.S. Company's electric car set the world's electric distance record of 180 miles per charge.

A 1915 version of Clara Ford's car, a Detroit Electric powered by a 5.5-hp DC motor driven at 72 volts, when retested 60 years later (with new batteries) by *Machine Design Magazine*, still delivered 25 mph and an 80-mile range. It was still recommended by the magazine as a "best buy," proving the point that we can routinely expect long lifetimes out of electrical machinery. While electric automobiles were a common sight until the mid-1910s, and commercial and industrial EVs have enjoyed continued growth and success on up through today, cheap oil combined with the nonelectrification of rural areas assured victory for internal combustion engine vehicles. Ironically, it was an electric vehicle's motor and battery, adapted as an electric starter for internal combustion engine vehicles by Charles Kettering in 1912, that delivered the crushing coup de grace to early electric autos.

Ashes to Ashes, Dust to Dust

The story of the internal combustion vehicle is inextricably linked to the story of oil itself, but the internal combustion engine's rise in popularity was due more to the great economic advantage of oil rather than any technical advantage of the internal combustion engine. Today, with the United States and other industrialized nations substantially dependent on foreign oil, the strategic economic disadvantage of oil coupled with the environmental disadvantage of the internal combustion engine has created strong arguments for alternative solutions. Let's examine how this situation was created.

Animal oils had been used for centuries to provide illumination. Rock oils (so called to indicate that they derived directly from the ground, and the original name for crude oil or petroleum) were envisioned in the 1850s only as superior alternatives for illumination and lubrication in the upcoming mechanical age. Earlier researchers had discovered that a quality illuminating oil, kerosene, could be extracted from coal or rock oil. Coal existed in plentiful quantities. All that remained was to discover a substantial source of crude oil/petroleum.

The discovery of oil in Western Pennsylvania by Edwin Drake in 1859 was the spark that ignited the oil revolution. Almost overnight, the boom in Pennsylvania oil, with its byproducts exported globally, became vitally important to the United States economy. The promise of fabulous wealth provided the impetus that attracted the best business minds of the age to the quest. Soon the oil business, dominated by kerosene, was controlled by the worldwide monopolies of John Rockefeller's Standard Oil (production/distribution from Pennsylvania in the United States), Ludwig and Robert Nobel (production from Baku on the Russian Caspian Sea), Alphonse and Edmond Rothschild (production from Baku, distribution from Batum on the Russian Black Sea), Shell (production and tanker distribution from Batum/Bomeo to England and the Far East), and Royal Dutch (production from northeast Sumatra in Indonesia).

These monopolies, securely in place before the 1900s, were all based on the markets for oil as kerosene and lubricating products. In the 1890s, gasoline, once thrown away after kerosene was obtained, was lucky to bring two cents a gallon, but that was about to change.

Coal was the foundation for the industrial revolution, and the first internal combustion engine built in 1860 by Etienne Lenoir was fired by coal gas. Nikolaus Otto improved on the design with a four-cycle approach in 1876. But the discovery that gasoline was an even more "combustible" fuel that was also inexpensive, plentiful, and powerful was the spark that ignited the internal combustion engine revolution. All that remained was controlling the explosive gasoline-air mixture—solved by Gottlieb Daimler's carburetor design of 1885—and controlling the timing—solved by Karl Benz's enhanced battery-spark coil-spark plug ignition design of 1885—for the internal combustion engine as we know it today to emerge.

Early internal combustion vehicles were noisy, difficult to learn to drive, difficult to start, and prone to explosions (backfiring) that categorized them as dangerous in competing steam and electric advertisements. Internal combustion vehicle offerings from Daimler (Germany, 1886), Benz (Germany, 1888), Duryea (United States, 1893),

Peugeot (France, 1894), and Bremer (England, 18941) were primitive engineering accomplishments in search of a marketing niche, while contemporary enclosed-body electrics, targeted at the elite urban carriage trade, sold briskly at $5,000 a copy.

This changed quickly, perhaps due to the inspiration of the *Chicago Times-Herald* Thanksgiving Day race of 1895–6, an event held "with the desire to promote, encourage, stimulate invention, development and perfection and general adoption of the motor vehicle in the United States." Won by Frank Duryea, driving a Duryea Brothers motor wagon, it brought instant fame to the brothers but, more importantly, brought most of the United States automotive pioneers together for the first time. Only three years later, more than 200 companies had been organized to manufacture motorcars.

Simultaneously, when internal combustion–powered vehicles were still decades away from dominance, discovery of the Los Angeles field in the 1890s, the "Spindletop" field near Beaumont, Texas in January 1901, and the Oklahoma fields of the early 1900s saw boom and bust times that priced a 42-gallon barrel of crude oil (typically from 15 to 20 percent recoverable as gasoline) as low as three cents a barrel. While DC and AC electrical distribution systems guaranteed that electric lighting would replace the kerosene lamp, cheap domestic oil, which kept gasoline prices between two and ten cents a gallon between 1890 and 1910, guaranteed the success of internal combustion vehicles.

Like Rockefeller with oil, Henry Ford was the individual who was in the right place (Detroit) at the right time (October 1908) with the right idea (Model T Ford) at the right price ($850 FOB Detroit). Ford had attended neither the *Chicago Times-Herald* race nor the earlier World's Columbian Exposition of Chicago that opened May 1, 1893, but written information derived from these events doubtless inspired his first creation—the 1896 Quadricycle 7. By 1908, Henry Ford had produced numerous designs. Hand-built 1899 and 1901 models followed the 1896 Quadricycle; Ford Motor Company models A, B, C, F, N, R, S, and K preceded the T; and Ford had won races with his Grosse Point racer of 1901 and famous "999" Barney Oldfield racer of 1902. But it was the innovation of the mass-produced, one-color-fits-all 1909 Model T at the $850 price that put the internal combustion vehicle on the map. The four-cylinder, 20-hp, 1,200-lb. 1909 Model T's instant success created an enormous demand that lasted nearly 19 years; more than 15,000,000 were manufactured until production ceased in May 1927. By producing nearly the same model, manufacturing economies of scale enabled the price to be dropped year after year until its all-time low of $290 in December 1924. In addition to low purchase price, the Model T's success was also due to its operating economy. Its two forward/one reverse speed planetary transmission and 30-inch wheels (with recommended tire pressures of 60 psi) drove the Model T's engine at 1,000 rpm at 25 mph and 1,800 rpm at 45 mph, producing a typical gas mileage of 20 miles per gallon and up.

Simultaneously, integration (along with the 1905 political problems in Russia) had consolidated the world oil market in the hands of two companies by 1907: Standard Oil and Royal Dutch/Shell. But by 1911, the investigation of Standard Oil launched by president Teddy Roosevelt in 1904 resulted in the United States federal court finding Standard Oil guilty of antitrust violations and ordering its breakup into the companies we recognize today: Standard Oil of New Jersey (Exxon), Standard Oil of New York (Mobil), Standard Oil of California (Chevron), Standard Oil of Ohio (Sohio-BP/America), Standard Oil of Indiana (Amoco), Continental Oil (Conoco), and Atlantic (ARCO/Sun). While this breakup initially led to a decade of peaceful coexistence among the former allied parts, it also paved the way for the oil industry as we know it today, dominated by multiple, large, fiercely competitive, multinational corporations.

Other events of the period also contributed to oil's rise and dominance: the introduction of "thermal cracking" by Standard Oil of Indiana in 1913 (a process that more than doubled the amount of gasoline recoverable from a barrel of crude oil, up to 45 percent); discovery of oil near Tampico, Mexico in 1910; discovery of oil in Persia (Iran) that led to construction of an Abadan refinery in 1912 by Anglo-Persian (a pre–World War I "strategic" decision by Winston Churchill gave the British government, through British Petroleum, 51 percent ownership of Anglo-Persian after it ran into financial difficulties); and World War I itself.

Meanwhile, other internal combustion engine vehicle innovators were busy too: Walter Chrysler, John and Horace Dodge (the brothers who began as captive suppliers to Ford), and numerous others provided innovations that survive to the present day. William Durant incorporated General Motors in September 1908, and by the 1920s its major divisions (Buick, Oldsmobile, Cadillac, Oakland, Chevrolet, GM Truck) and its supporting divisions (Fisher Body, Harrison Radiator, Champion Spark Plug, DELCO, Hyatt Roller Bearing, and others) were household names in the United States. While Durant acquired valuable assets in assembling GM's acquisitions under one holding-company umbrella, it was the talent he obtained (such as Alfred Sloan from Hyatt and Charles Kettering from Cadillac) that paved the way for GM's later rise to dominance. GM innovations such as color, streamlining, smoother, more-powerful six-cylinder engines, and annual model styling changes made Ford's Model T obsolete, despite its $290 price in 1924. The internal combustion vehicles were now on their way.

In 1900, half of the 80 million people in the United States lived in a few large (mostly Eastern) cities with paved roads, and the other half in towns linked by dirt roads or in countryside with no roads at all. Less than 10 percent of the 2 million miles of roads were paved. More than 25 million horses and mules provided mobility for the masses. Electric lighting in the larger cities was dwarfed by the use of kerosene lamps, popularized by the discovery of plentiful amounts of oil, in the countryside. Coal- or wood-burning steam engine locomotives were high tech. Only 200,000 miles of railroad track existed, but some of it provided fast, efficient transportation between major locations—New York to Chicago on the "Twentieth Century Limited" took 20 hours. While a New York-to-California railroad ticket cost $50, vehicles of any kind cost around $5,000 to $50,000 in today's dollars, putting them out of the hands of all but the well-to-do.

Three types of vehicles came into this turn-of-the-century United States environment. Almost 3,000 manufacturers experimented with various combinations of propulsion (steam, electric, or internal combustion engine); fuel (water/kerosene, battery, gasoline/oil); cooling (air or liquid); mounting (front, rear, or middle); drive (front or rear wheels via shaft, gear, chain, or belt); chassis (three- or four-wheel, independent suspension, or leaf/coil springs); and tires (pneumatic or solid). By the time World War I was over, the internal combustion vehicle had emerged as the clear victor.

The Golden Age of Internal Combustion

Internal combustion engine vehicle growth in the United States exploded with World War I. After World War II, world internal combustion engine automotive growth was even more dramatic. What made all this possible was the unprecedented oil availability, and the relative price stability shown in Figure 3-5, which allowed United States gasoline

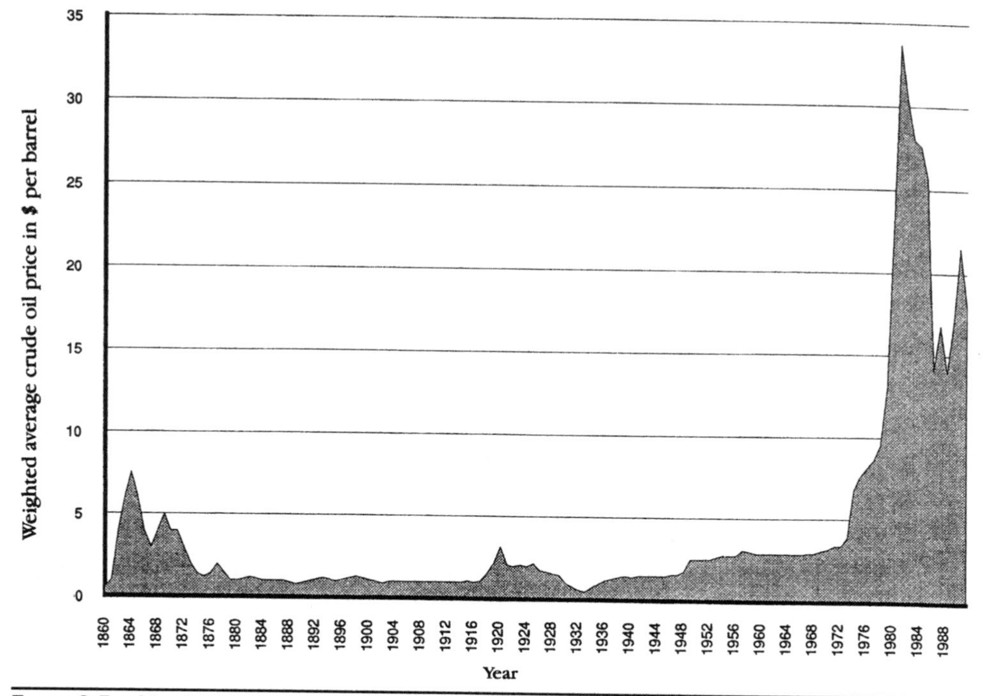

FIGURE 3-5 Oil prices from 1860 to 1990.

prices to move from roughly $0.10 to $0.30 during this 50-year period. Yet in terms of inflation-adjusted "real dollars," the cost of gasoline actually went down. Is it any wonder that no one cared how large the cars were in the 1950s or how much gas they guzzled in the 1960s? Gasoline was cheaper than water.

But there were also some problems: no major discoveries since the Alaskan and North Sea fields of two decades ago; the increasing concerns with oil supplies, beginning during World War II; the introduction of nuclear and natural gas energy alternatives in the 1950s; and the hardening of public opinion with the increasing frequency of smog and air-quality problems, oil and nuclear environmental accidents, and the foreign oil shocks. Already forced to comply with more stringent emission standards by the Clean Air Act of 1968, the first oil shock of 1973 caught the "big three" United States automobile manufacturers with their pants down. Japanese and European auto manufacturers had smaller, more fuel-efficient internal combustion vehicle solutions as a result of years of higher gasoline prices (due to higher taxes earmarked for infrastructure rebuilding). The market share lost by the big three to foreign automakers has never been regained.

By the early 1990s, the wild oil party of the preceding 75 years was over. Environmental problems, the need for energy conservation, and the instability of foreign oil supply all signal that the sun is setting on the internal combustion vehicle. It will not happen overnight. In the near term the industrialized nations of the world and emerging Third World nations will consume ever greater amounts of foreign oil. But it's inevitable that a replacement of the fossil fuel–burning internal combustion vehicle will be found.

Internal Combustion Vehicle Growth Fueled by Cheap and Plentiful Oil

The lessons of World War I were simple. Flexible, oil-powered internal combustion engine cars, trucks, tanks, and airplanes were superior to fixed, coal-powered railway transportation; and those who controlled the supply of oil won the war. The allies had Standard Oil, Royal Dutch/Shell, and Anglo-Persian Oil. The Germans did not have access to vast amounts of oil; the destruction of the Ploesti refinery in Romania and their belated, failed attempts at capturing Baku cost them the war.

Meanwhile, internal combustion engine vehicle registrations in the United States exploded from one-half million in 1910, to 9 million in 1920, to 27 million in 1930, and slowed by the depression, to 33 million in 1940. Gasoline that was sold by the local blacksmith in containers in the early 1900s gave way to 10,000 wooden "filling stations" with gravity-feed tanks at the beginning of the 1920s, to more than 150,000 buried tank/electric-pump-driven "service stations" in the 1930s. More and more paved roads were built; the landscape was changed forever. More oil continued to be found in the 1920s in places like California (Signal Hill), Oklahoma (greater Seminole and Oklahoma City), West Texas (Yates Field), Venezuela (Maracaibo Basin), and Iraq (Kirkuk). Then the biggest oil find of them all was discovered in October 1930—the giant East Texas oil reservoir that later proved to measure 45 miles long and up to 10 miles wide. Crude oil that sold for around $2 a barrel in the mid-1920s dropped to less than 10 cents a barrel in the early 1930s (the low was 4 cents a barrel in May 1933), and gasoline prices that had been chugging along between 10 and 20 cents per gallon from 1910 through the 1920s dropped accordingly. Now the problem was too much oil, and the United States government had to enter the picture to control prices.

1940 to 1989

This period included the "golden age" of the internal combustion engine vehicle and ended up with legislative efforts in states and the federal government regarding oil shocks and a renewed interest in electric cars.

With cheap, available gasoline prevailing as fuel, and basic internal combustion engine vehicle design fixed, manufacturing economies of scale brought the price within reach of every consumer. Expansion away from urban areas made vehicle ownership a necessity. The creation of an enormous highway infrastructure culminated in completion of the interstate highway system. This was accompanied by the destruction of urban non-internal-combustion-powered transit infrastructure by political maneuvering in the United States, and by damage during World War II in Japan and Europe.

World War II Oil Lessons Are Learned by All

Oil was the one resource Japan did not have at all. In retrospect, Japan's war was easy to understand. It needed the oil resources of Indonesia, Malaysia, and Indochina. After an oil embargo against Japan was set up in mid-1941 by blocking the use of Japanese funds held in the United States, Japan was desperate for oil, and did what it had to do to get it. The Pearl Harbor attack was an effort to protect its Eastern flank, but poor timing made it an infamous event (Japan's "declaration of war" didn't get delivered until after the attack). Japan's early loss of planes and ships at Midway meant it was never able to provide adequate protection for its oil tanker convoys from Indonesia. Dwindling oil reserves and a nonfunctioning synthetic fuels program meant new pilots couldn't be trained and ship fleets couldn't maneuver. While Japan "lost" World War II long before 1945, it learned its oil lesson well and converted to the oil standard soon after the war.

On the other hand I.G. Farben, the huge German chemical combine, had mastered synthetic fuel recovery from coal by the early 1920s—hydrogenation was the most popular method—and Germany had plenty of coal. But most of Germany's oil imports came from the West, and increasing demand was causing a foreign exchange hemorrhage. Hitler believed if Russian Baku oil reserves could be added to those of its ally Romania, along with Germany's own 1940 synthetic fuel reserves, the "Thousand-Year Reich" was a cinch via a blitzkrieg-tactics war that didn't consume much fuel. Unfortunately for Hitler, his blitzkrieg advances frequently outran their fuel supply trucks, he never got to Baku, his Romanian oil supply at Ploesti was destroyed early in the war, and the German advance in North Africa ended when its oil tankers couldn't make it across the Mediterranean. In addition, German synthetic fuel aviation gasoline was never quite as "hot" as that produced from real crude oil, and the entire German war machine came to a halt when systematic bombings of its synthetic fuel plants later in the war reduced them to rubble. Germany also lost World War II long before 1945, but learned its oil lesson well and converted to the oil standard soon after the war despite massive reserves of coal.

What did the allies learn from World War II? They relearned the lesson from World War I: *Whoever controls the supply of oil wins the war.*

They also learned the value of a strategic petroleum reserve. Up until 1943, the allies nearly lost the war to the Germans in the North Atlantic—the success of submarine wolf packs made it nearly impossible for allied oil tankers to resupply England, Europe, and Africa. Even in 1945, countless lives were wasted and the Russians moved toward Berlin while Patton's tanks sat without fuel in France, giving the Germans time to regroup and resupply.

While the United States provided six out of every seven barrels of allied oil during World War II, it was recognized by many in government that it would soon become a net importer of oil. More oil had been discovered in Bahrain in 1932, and in Kuwait (Burgan field) and Saudi Arabia (Damman field) in 1938. In 1943, as all eyes turned toward the Middle East with its reserves variously estimated at around 600 billion barrels, the United States government proposed the "solidification" plan to assist the oil companies (that is, share the financial risk) in Saudi Arabian oil development. While this plan and subsequent revisions to it were rejected with much indignation by the oil companies, decades later with 20/20 hindsight, they would all gladly change their votes.

A World Awash in Oil After World War II

After World War II, pent-up consumer demand released another type of blitzkrieg, the internal combustion engine automobile. While Figure 3-6 shows United States automobile registrations were almost quadrupling, the rest of the world's auto population grew more than twentyfold: from 13 million in 1950 to nearly 300 million in 1990. Oil exploration in this period was in high gear. A nearly inexhaustible supply had apparently been found in the Middle East; gasoline prices bounced between 20 and 30 cents per gallon until the early 1970s. Aided by the convenience of the internal combustion automobile, America moved to the suburbs, where distances were measured in commuting minutes, not miles. Fuel-efficient automobiles were the last item on anyone's mind during this period. Gasoline was plentiful and cheap (reflecting underlying oil prices) and regular local retail price wars made it even cheaper.

"Environment" was an infrequently used word with an unclear meaning. Highway construction proceeded at an unprecedented pace, culminating with the Interstate

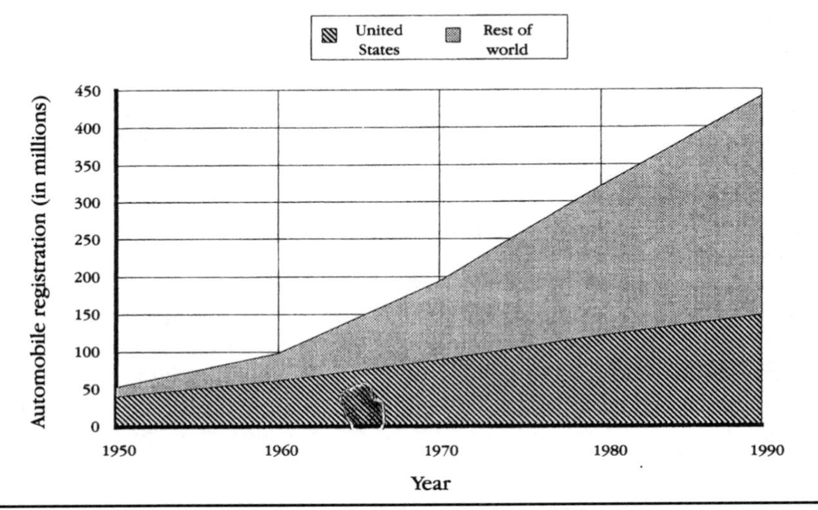

FIGURE 3-6 Growth in automobile registrations from 1950 to 1990.

Highway Bill signed by President Eisenhower in 1956, authorizing a 42,500-mile superhighway system. Public transportation and the railroads—the big losers in Japan and Europe due to World War II damage—also became the big losers in the United States as the U.S. government formally finished the job that major industrial corporations, acting in conspiratorial secrecy and convicted of violating the Sherman Antitrust Act, had started in the 1930s and 1940s: ripping up the tracks, dismantling the infrastructure, and scrapping intercity and intracity light rail and trolley systems that could have saved consumers, cities, and the environment the expenditure of billions of dollars today. (Source: Jonathan Kwitny, "The Great Transportation Conspiracy," *Harper's*, February 1981.)

The post–World War II rebuilding of European and Japanese infrastructures made them more modern than the United States. Germany and Japan (and most of the rest of the industrialized world) rapidly converted from coal to oil economies after World War II, and underwent an unprecedented period of economic and industrial expansion as the surge in automobile registrations outside of the United States, shown in Figure 3-6, attests. All the industrialized economies of the world were now dependent on internal combustion engine vehicles and oil.

The 1960s: The Sleeper Awakens

While electric vehicle automobile development languished since the 1920s (except for Detroit Electric's efforts), commercial and industrial EV activities continued to flourish, perhaps best exemplified by Great Britain's electric milk trucks (called "floats") and its total electric vehicle population of more than 100,000.

The heightened environmental concerns of the 1960s, specifically air pollution, were the first wave upon which electric vehicles rose again. While numerous 1960s visionaries were correctly touting EVs as a solution, the manufacturing technology was, unfortunately, not up to the vision.

Figure 3-7 shows a chronological summary of what was being done by the primary electric vehicle developers in the United States, Europe, and Japan during the four waves.

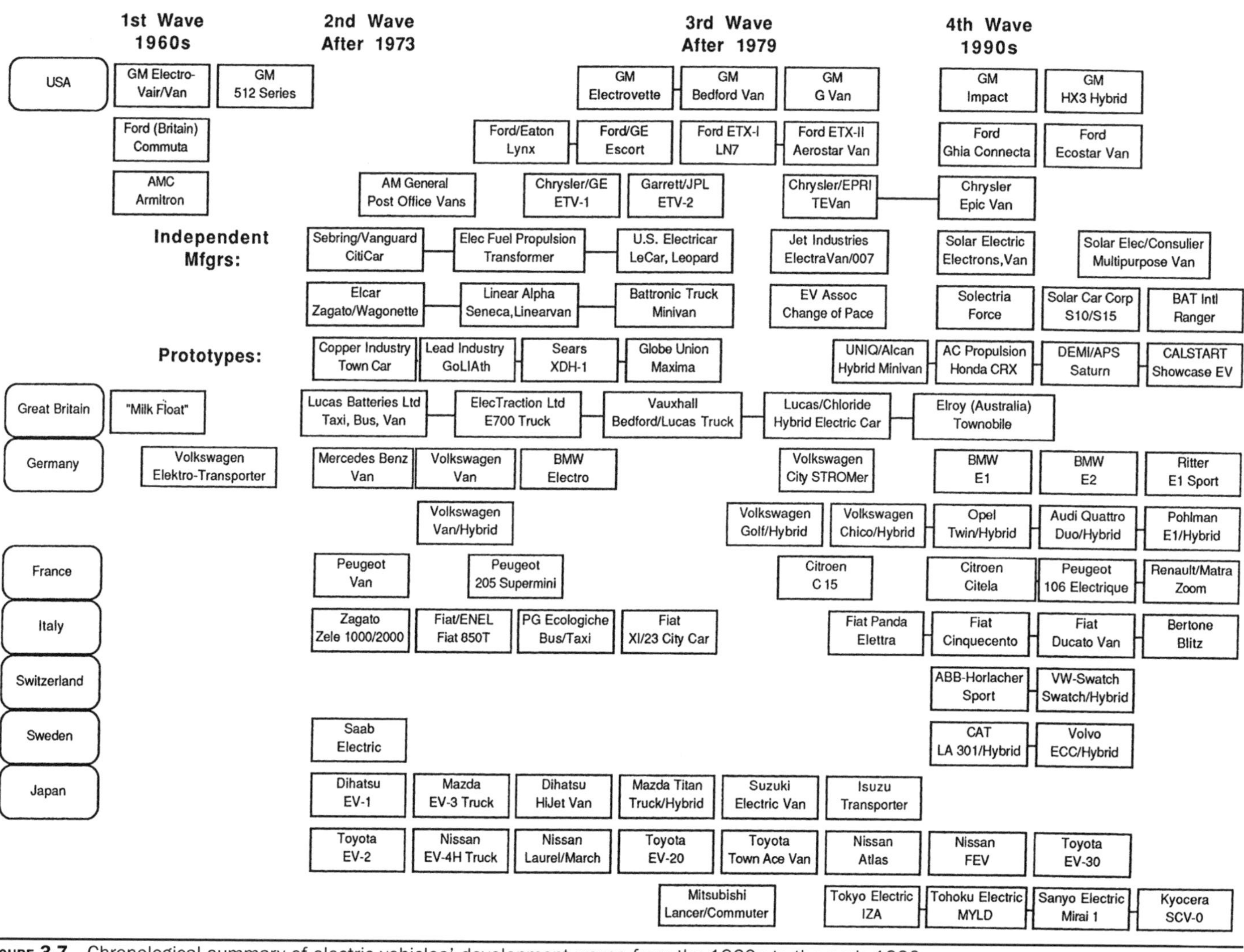

Figure 3-7 Chronological summary of electric vehicles' development waves from the 1960s to the early 1990s.

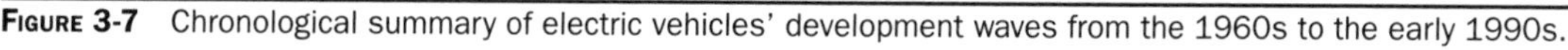

EV interest during the first wave fell into two distinctly different perceptual camps:

- Individuals who successfully converted existing internal combustion vehicles
- Manufacturers who could not figure out how to make the existing technology justify the financial figures, let alone figure out how to market EVs to an American public that wanted quick and large internal combustion engine automobiles

Individuals Lead the Way

Converting was easy enough, and also inexpensive. Then, as today, you picked a vehicle shell—hopefully light in weight and/or easy to modify—added a motor, controller, and batteries, and went. Unfortunately, the most available motors in the appropriate size were decades-old war surplus aircraft starter DC motors; do-it-yourself controllers were barely more sophisticated than their turn-of-the-century counterparts; and battery technology, although cosmetically improved by modern manufacturing and packaging techniques, was virtually unchanged from 1900. As the most readily available controllers came from golf carts that typically used six 6-volt batteries (36 volts), and aircraft starter motors were typically rated at 24 to 48 volts, many first-time do-it-yourself EV attempts suffered from poor performance, and contemporary internal combustion muscle car owners of the 1960s just laughed at them.

Then someone discovered motors were actually underrated to ensure long life, and began driving them at 72 to 96 volts. Some early owners found they could make simple, non-current-limiting controllers, and create vehicles that could easily embarrass any internal combustion muscle car at a stoplight. For a conceptual picture of this, imagine a subway traction motor in a dune buggy. In fact, these owners simply left the starting resistance out of a series DC motor, or equivalently diddled a shunt or compound motor. A series DC motor delivers peak torque at stall, and while starting currents were enormous, these early innovators just made sure they had a load attached when they switched on the juice. The immediate result was a rush. Predictably, the longer-term results were burned-out motors and, occasionally, broken drive shafts or axles. But the sanely driven and controlled 72- to 96-volt EV conversions were not bad at all. This was the 1960s, and the Electric Auto Association was founded in 1967.

Manufacturers Don't

Established internal combustion engine vehicle manufacturers in the late 1960s did not produce much in the way of electric vehicles. The General Motors' XP512E Series (GM's ElectroVairs—converted 1964 and 1966 Corvairs—and ElectroVan—a converted 1966 GMC HandiVan) could have been easily replicated by any individual except for the ElectroVair's high-cost silver-zinc batteries, and ElectroVan's high-cost hydrogen-oxygen fuel cells. Ford of Britain's Comuta (even more easily replicated by anyone) and American Motors' Amitron, like GM's offerings, all resembled souped-up golf carts (though the Amitron was in a class by itself—it featured Gulton's lithium batteries, a solid-state controller, 50-mph speed, and a 150-mile range). It was sad that numerous individuals could develop EV solutions far superior to anything put forth by the giant industrial corporations that had helped to put a man on the moon in the same decade.

The problem was not that these corporations lacked talent, money, or technology. The problem was that corporate thinking of this era was locked into a mode that presumed their current successes would continue forever, and they were committed to maintaining the status quo to assure it.

After 1973: Phoenix Rising, Quickly

The late 1960s policies of the major American automobile manufacturers put them in a poor position to respond to the crisis of the early 1970s—the oil shock of 1973. A huge inventory of stylish but large, gas-guzzling cars, along with four- to five-year new car development cycles, made it an impossible situation. All they could do was wait out the crisis and import smaller, more fuel-efficient cars from their foreign subsidiaries. The higher European and Japanese gasoline prices had, over the years, forced them to develop lighter, more compact automobiles with economical drivetrains. While this helped the Europeans and the Japanese only a little at home after the crisis (their already-high gasoline prices just rose proportionately higher, negating any advantage), the European/Japanese automotive solution was ideal for the United States market of that time. Thus imports, unimportant in the United States until the early 1970s, gained a foothold that was to become a significant factor over time. It was against this backdrop that the electric vehicle rose again, like the phoenix from the ashes. Five trends (shown in Figure 3-7) highlight electric vehicle development during this second wave:

- The inactivity of GM in distinct contrast to Ford, Chrysler, and American Motors
- A period of frantic activity by the independent manufacturers
- A period of strong prototype promotion by industry associations and suppliers
- Resumption of serious overseas development
- Continuation of individuals converting existing internal combustion vehicles

GM Leads, the World Does Not Follow

Under the "2nd Wave" heading in Figure 3-7, GM's inactivity is in marked contrast to what others were doing in electric vehicles during this period. The GM executive's remarks at the June 1975 Congressional hearings, quoted later in this chapter, clearly convey the reason for GM's electric vehicle nonactivity.

While GM's actions are inconsequential today—the times have changed and GM has changed along with them—GM's strong stance against electric vehicles caused much grief among the pro-EV industry forces of the early 1970s.

Late in the decade GM re-entered the electric vehicle world with its ElectroVette (a converted Chevette) and Bedford Van (a converted GM–United Kingdom van). But GM did nothing technically innovative, and both conversion efforts became self-fulfilling prophecies: neither vehicle's performance specs were spectacular, and the economics just didn't make sense for a manufacturer/marketer.

As mentioned in Chapter 1, an 8,000-lb. van would never be my first conversion choice. The Chevette had 20 12-volt maintenance-free batteries, a 53-mph top speed, a 50-mile range at 30 mph, and weighed in at 2,950 pounds—maybe a marginal conversion choice. A contemporary individual electric vehicle converter's outlay might have been $5,000 for the whole package, perhaps only $2,000 with used parts and heavy scrounging, while GM would have been hard pressed to wring a profit out of a $20,000 retail price. The van performance and pricing were even worse. GM could honestly tout their conclusion without mentioning what they hadn't done (such as a total systems design like their 1990 Impact EV).

Ford, Chrysler, and American Motors Move Ahead

Ford's direction was totally different than GM's during this wave. They took a hard look at the problem and decided the two critical areas were battery technology and drivetrain efficiency. This period marked their planting of seeds in sodium-sulfur battery technology (they invented it in 1965) and integrated AC induction motor drivetrains during the Ford/Eaton Lynx and Ford/GE Escort projects that would later bear sweet fruit.

The Chrysler/GE team took advantage of federal government funding of the Electric Test Vehicle One (ETV-1) program in 1977 to couple a lightweight, low–rolling resistance, streamlined body with front-wheel drive to a unique DC motor, transistorized controller, and 108 volts worth of advanced lead-acid batteries. At the same time, a Garrett/JPL team was working on ETV-2 and learning the benefits of a similarly constructed and identically powered vehicle using rear-wheel drive and aided by flywheel energy storage.

In 1971, the United States Post Office studied electric vehicles in a pilot program implemented at the Cupertino Post Office in California, using British Harbilt electric vans. The program was a resounding success. Because of it, the United States Post Office ordered 350 converted "Jeep"-type electric vans from AMC General Corp., a division of American Motors, for the next phase of the program. The AMC General vans were also a resounding success. Both the Harbilt and AMC General vans had enormously high uptimes and low cost per mile while being driven almost continuously during their evaluation periods. The program had strong support inside the Post Office and was cancelled only after the "third oil shock" made it economically unattractive.

Independent EV Manufacturers Rise and Fall

Numerous independent electric vehicle manufacturers came out of the woodwork after the 1973 oil shock; it was a repetition of turn-of-the-century vehicle development with the good, the bad, and the ugly. There were many EVs to choose from, but most were not technically innovative, manufacturing quality was inconsistent, and component quality was occasionally poor.

Prevailing designs used either conversions of existing internal combustion vehicles or unsophisticated new chassis, and many were poorly engineered. In addition, most firms were severely undercapitalized. While the automotive industry marketed its internal combustion–powered vehicles via public relations, lobbying to blunt legislation, measuring public taste through survey and prototype programs, product advertising, and distribution through a dealer network, EV manufacturers of the 1970s used few of these, and haphazardly.

The Sebring/Vanguard CitiCar seen in Figure 3-8 was the most famous and infamous electric vehicle of this period. Sebring/Vanguard was the first manufacturer out of the blocks in 1974, and for a brief time sold all it could make. Eventually, more than 2,000 CitiCars were produced. It was very popular in its day; owners were fiercely loyal, and it received much publicity. Unfortunately, although it was well built (many are still on the road today), efficient, and practical, its design and styling gave it the appearance of a glorified golf cart, similar to the late 1960s engineering prototypes previously mentioned. When they were unjustly crucified by a Consumer's Union report (along with Elcar's golf-cart-sized Zagato imported from Italy), and an unfavorable article went out over the UPI wire, even a letter from the chief attorney of the Department of Transportation saying the criticisms

Figure 3-8 Sebring/Vanguard CitiCar (Source: Wikipedia).

were false failed to undo the damage. The public would always associate EVs with golf carts and some nebulous stigma.

But this painful lesson was well learned by other electric vehicle manufacturers, and later models avoided golf-cart–looking designs like the plague. Another well-known manufacturer of this period was Electric Fuel Propulsion. Their early Renault R 10 and Hornet conversions led to their original and innovative Transformer (featuring 180-volt tripolar lead-cobalt batteries, a 70-mph top speed, and 100-mile range) with its range-extending Mobile Power Plant trailer. Linear Alpha produced the Seneca (Ford) and Linearvan (Dodge) conversions. U.S. Electricar produced the Electricar (Renault LeCar) and Lectric Leopard conversions. Finally, there was Battronic Truck's Minivan, co-produced with the Electric Vehicle Council; more than 60 utilities received production versions of this 6,800-lb. van whose 18 6-volt lead-acid batteries pushed it to 60 mph.

Electric Vehicle Industry Closes Ranks to Show Support

One of the more innovative promotions of this period involved the development of prototypes by industry associations and individual manufacturers who stood to gain from the sale of their product(s) in electric vehicles. The prototypes were used for all sorts of public relations event-style marketing. The result was raising the level of public awareness about EVs—so much so that individuals, thinking these were production products, frequently called these organizations to place orders.

Japan Gets More Serious About EVs

The Japanese rise in and now domination of the electric drive world is remarkable and still continuing. From the Lexus to the Prius to the Camry, hybrid drive is a world-class drive system. From the Toyota RAV4 EV I drove to the plug-in-hybrid cars they will inevitably launch in the future, Japan is an automotive power. Even before they achieved world dominance in the internal combustion engine vehicle in the 1980s, leading visionaries at Japan's state planning agencies had seen the future, and it was electric vehicles. Japan needed little incentive—it was the world's largest importer of oil, had dangerous levels of pollution, and high speeds on its narrow, urban streets were a fact of life.

While the Japanese Electric Vehicle Association and its tight coordination with MITI directives did not arrive on the scene until 1976, Japanese government funding of EV programs began in 1971 with Phase I basic research into batteries, motors, control systems, and components across the spectrum of car, truck, and bus platforms.

The fruits of its labors, augmented by MITI directives to focus on urban acceleration and range, appeared in Phase II. As Table 3-1 attests, Japan's 1970s Phase II offerings from Daihatsu, Toyota, Mazda, and Nissan put it into a world-leadership class. The Nissan EV-4P truck's 188 miles before recharging was a record for lead-acid battery-powered vehicles, and its EV-4H truck's 308 miles was the world record for that period.

Throughout the rest of the 1970s all Japan's big nine automakers—Daihatsu, Honda, Isuzu, Mazda (Toyo Kogyo), Mitsubishi, Nissan, Subaru, Suzuki, and Toyota—were involved in EV activities, although some to a greater extent than others (see Figure 3-7).

Individuals Assisted by More and Better Everything

The best news of the 1970s was for individuals wanting to do EV conversions. More of what was needed was available for conversions, and how-to books even started to appear. Other than the fact that components—particularly the controllers—were still unsophisticated, individual converters enjoyed relating their conversion experiences at regular Electric Auto Association meetings and pushed the outside of the speed and distance envelope at rallies and events. The greatest irony of this period is that at the same time General Motors was providing extremely negative information to the Congressional hearings, the individuals who had actually done a conversion to an electric vehicle were reporting high degrees of satisfaction, with operating costs in the range of two cents per mile, and most had yet to replace their first set of batteries.

	Daihatsu EV-1 car	**Toyota EV-2 car**	**Mazda EV-3 truck**	**Nissan EV-4P truck**	**Nissan EV-4H truck**
Range (miles)	109	283	127	188	308
Top Speed (mph)	55	53	45	54	56
Battery Pack	lead-acid	zinc-air/ lead-acid	lead-acid	lead-acid	zinc-air/ lead-acid
Curb Weight (lbs)	2500	2770	1720	5000	5490

TABLE 3-1 Comparison of Japan's Second Wave Phase II Electric Vehicles

Third Wave After 1979: EVs Enter a Black Hole

While the second "oil shock" of 1979 and the ensuing shortage further spurred electric vehicle development onward, the "oil shock glut" of 1986 and events leading up to it nearly shut development down. While the larger internal combustion automobile manufacturers were "whipsawed"—their crash programs of the late 1970s were now bringing lighter, smaller cars to market that (temporarily at least) no one wanted—the independent electric vehicle manufacturers were simply wiped out.

With oil and gasoline prices again approaching their 1970s levels, everyone lost interest in EVs, and the capital coffers of the smaller EV manufacturers were simply not large enough to weather the storm. Even research programs were affected. From mid-1983 until the early 1990s, it was as if everything having to do with EVs suddenly fell into a black hole—there were no manufacturers, no books, not even many magazine articles. The EV survivors were the prototype builders and converters, the parts suppliers (who typically had other lines of business such as batteries, motors, and electrical components), and EV associations, although their membership ranks thinned somewhat. Four trends (see Figure 3-7) highlight EV development during this third wave:

- Low levels of activity at GM, Ford, and Chrysler
- The best independent manufacturers arrive and then depart
- Low levels of activity overseas
- Continuation of individuals converting existing internal combustion vehicles

Lack of EV Activity at GM, Ford, and Chrysler

In retrospect, given all the other problems the big three had to deal with during this period, it's amazing that electric vehicle programs survived at all. But survive they did, to emerge triumphant in the 1990s.

The GM Bedford van project became the GM Griffon van—the G-Van. With a broad base of participation from the Electric Power Research Institute (EPRI), Chloride EV Systems, and Southern California Edison, the General Motors G-Van, actually an OEM aftermarket conversion by Vehma International of Canada, was widely tested for fleet use. While it was humorous to read numerous complaints about the G-Van's 53–mph top speed, 60-mile range, and 0 to 30 mph in 12 seconds acceleration, one has to wonder how many report readers correctly associated this data with G-Van's 8,120-lb. weight, 36 batteries, and huge frontal area.

Ford's direction was to continue to build on its sodium-sulfur battery and integrated propulsion system technology using government funding. Teamed with General Electric, Ford's ETX-I program adapted sodium-sulfur batteries and an integrated AC induction motor propulsion system to a front-wheel drive LN7 automobile test bed. The follow-on Ford/GE ETX-II program utilized sodium-sulfur batteries and a permanent magnet synchronous motor propulsion system in a rear-wheel-drive Aerostar van. Meanwhile Chrysler, under the sponsorship of the EPRI, used their standard Caravan/Voyager minivan platform, a GE DC solid-state motor, and 30 Eagle Picher NIF 200 6-volt nickel-iron batteries to achieve 65 mph and a 120-mile range in their 6,200-lb. TEVan.

Arrival and Departure of Independent Manufacturers

Numerous independent electric vehicle manufacturers had already come and gone during the previous wave. As an independent EV manufacturer in the third wave, you either were doing something good, or you had come out with something better. Electric Vehicle Associates of Cleveland, Ohio, is the best example of the first type. While their Renault 12 conversion and ElectroVan project with Chloride were interesting diversions, they are best known for their Change of Pace wagons and sedans built on AMC Pacer platforms. The Change of Pace four-passenger sedan weighed in at around 3,990 lbs., and used 20 Globe-Union 6-volt lead-acid batteries driving a DC motor via an SCR chopper to achieve 55 mph and a 53-mile range.

Jet Industries of Austin, Texas, once one of the largest and best of the independent EV manufacturers, was also the last to arrive on the scene. Jet's most popular products were its ElectraVan 600 (based on a Subaru chassis) and 007 Coupe (based on a Dodge Omni chassis). It also offered larger eight-passenger vans and pickups. The 2,690-lb. ElectraVan 600 had a GE 20-hp or Prestolite 22-hp DC series motor, SCR controller, and 17 6-volt lead-acid batteries that could push it to 55 mph with a 100-mile range. Hundreds of ElectraVan 600s and 007 Coupes are prized possessions among Electric Auto Association members today, attesting to their outstanding quality and durability. Jet Industries, alas, is no more.

Needless to say, industry association support of independent EV manufacturers—at its zenith during the previous wave—moved to its nadir during this one. There were no longer any independent electric vehicle manufacturers to support.

Individual Conversions Continue

Individuals assisted by more and better everything during the last wave now had to make do with more modest resource levels. But EV conversions by individuals continued throughout this wave, albeit at a slower pace. The best news of the 1980s was that the resources of the 1970s could still be found and used. During this wave, individual converters still enjoyed relating their conversion experiences at regular Electric Auto Association meetings; they still pushed the outside of the speed and distance envelope at rallies and events; and they still reported high degrees of satisfaction with what they had done.

Mid-1960s to 1990s

This period marked a successively heightened awareness of problems with internal combustion vehicles. Smog problems of the mid-1960s made us aware we were polluting our environment and killing ourselves. Arab oil embargoes, shortages, and gluts of the 1970s, 1980s, and 1990s made us aware of our dependence on foreign oil. Nuclear and oil spill accidents of the 1970s and 1980s made us aware of the long-range consequences of our short-range energy decisions. The internal combustion engine and oil problems that started with a whimper in the mid-1960s turned into a groundswell of public opinion by the 1990s. The net result of new awareness in this period has been the re-emergence of electric vehicles. When legislative action mandating zero emission vehicles in the 1990s forced rethinking of basic vehicle design, current technology applied to the EV concept emerged as the ideal solution.

Twilight of the Oil Gods

By the middle of the 1960s, many in government and industry around the globe became aware that something was very wrong with this picture. Although United States movement towards energy alternatives such as natural gas and nuclear fission (started decades earlier) was rapidly bringing new alternative capacity online, dependence on oil was becoming worse, and smog and environmental issues began coming into the foreground. Passage of the Clean Air Act of 1968 was one result. The 1975 passage of a corporate average fuel economy (CAFE) standards bill was another. The problem was obvious to some, but most of the public chugged merrily along in their internal combustion–powered vehicles.

After the "first shock"—the Arab embargo that followed the October 1973 Yom Kippur War—everyone knew there was a problem. By cutting production 5 percent per month from September levels, and cutting an additional 5 percent each succeeding month until their price objectives were met, the Organization for Petroleum Exporting Countries (OPEC) effectively panicked, strangled, and subverted the industrialized nations of the world to their will. The panic was exacerbated by nations and oil companies scrambling for supplies on the world market, overbuying at any price to make sure they had enough, and consumers doing the same by waiting in lines to "top off their tanks" when weeks before they would have thought nothing of driving around with their gas gauges on empty. When the dust settled, the United States consumer, who had paid about $0.30 a gallon for all the gas he could get a few months earlier, now paid $1 a gallon or more at the height of the crisis, and waited in line to get a rationed amount. Figure 3-9 shows the drastic change.

How could it happen? Easy. The oil crises in 1951 (Iran's shutdown of Abadan), 1956 (Egypt's shutdown of Suez Canal), and 1967 (Arab embargo following June Six-Day War) were effectively managed by joint government and oil industry redirection of surplus United States capacity. But by the early 1970s there was no longer any surplus capacity to redirect—the United States production peak of 11.3 million barrels of oil per day occurred in 1970. Up until the 1970s, the oil industry focused on restraining production to support prices. Collateral to this action, relatively low oil prices forced low investment and discovery rates, and import quotas kept a lid on supplies. But rising demand erased the need for production-restraint tactics and the surplus capacity along with it.

The "second shock" occurred after the Iranian oil strikes began in November 1978. Iran was the second largest oil producer, exporting 4.5 million barrels per day. The strikes reduced this to 1 million in mid-November, and exports ceased entirely by the

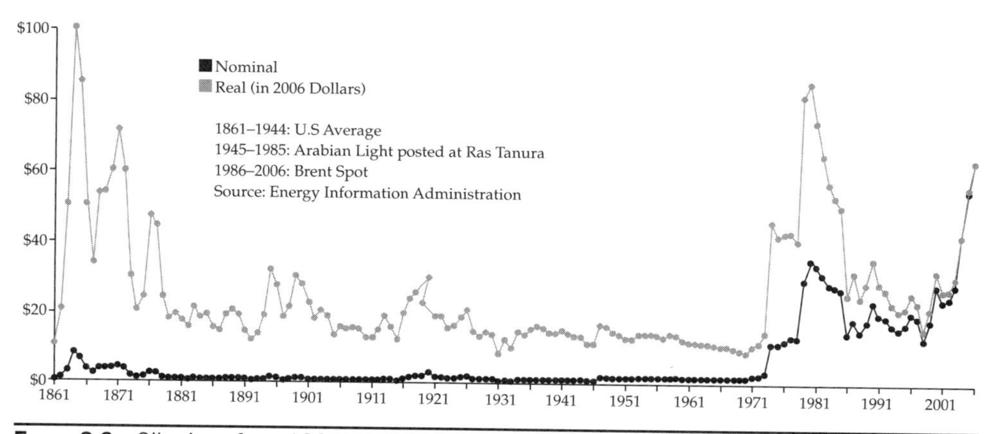

FIGURE 3-9 Oil prices from 1861–2006 (Source: Wikipedia—Created by Michael Ströck, 2006. Released under the GFDL).

FIGURE 3-10 The oil embargo—not fun at all!

end of December. While loss of this supply was partially offset by other suppliers within OPEC, burgeoning TV-network coverage, simultaneously broadcast throughout the world, of the Shah's mid-January 1979 departure and the Ayatollah Khomeini's arrival (along with other internal Iranian events) convinced the world that Iran would never return to its pro-Western ways, and initiated a 1973-style hoarding and panic buying spree. What started as a 2 million barrel per day shortfall became 5 million barrels per day as governments, oil companies, and consumers scrambled for supplies. Hoarding at all levels exacerbated the problem, gas lines appeared again, and oil prices went from $13 to $34 a barrel (see Figure 3-10).

Although Iranian exports returned to the market by March 1979, the ill-timed Three Mile Island nuclear accident of March 28, 1979 further intensified the panic surrounding energy awareness, in addition to forever altering public opinion on nuclear power. Several other factors contributed to making the gasoline crisis that occurred during 1979 in most industrialized nations of the free world more severe than any previous crisis. Many refineries set up to process light Iranian crude could not deliver as much gasoline from the alternate heavier crude oil they were forced to accommodate. Uncooperative (and in some cases, conflicting) policies by federal, state, and local governments and oil companies disrupted the orderly distribution of the gasoline supplies that were available. The appearance of oil commodity traders, who could make huge profits on the play between the long-term contract and spot prices for oil, artificially bid up the price of spot oil in response to the prevailing buy-all-you-can-get-at-any-price mentality. Lengthy gasoline lines and rationing interfered with all levels of business and personal life. By the time the Iran hostage crisis began, a state of anarchy existed in the world oil market that President Carter's subsequent embargo of Iranian oil and freeze on Iranian assets did little to ease.

The "third shock" occurred in the opposite direction—prices went down. The self-correcting market forces that swing into action after any shortage or glut accomplished what the world leadership could not, this time with a vengeance. At the OPEC meeting of June 1980, the "official" price averaged $32 per barrel, but OPEC inventories were high, and approaching economic recession caused price and demand to fall in consuming countries, further swelling their inventory surplus. When Iraq attacked Iran, one of the first steps in its war plan was to bomb the Abadan refinery (September 22, 1980). The net result after reprisals was that Iran oil exports were reduced during the war, but Iraq exports almost ceased. After Saudi Arabia raised prices from $32 to $34 a barrel in October 1981, an unprecedented boom was created, and drillers came out of the woodwork.

Earlier fears of shortage at the beginning of the 1920s and mid-1940s had ended in surplus and glut because rising prices had stimulated new technology and development of new areas. This pattern was repeated with $34 a barrel oil. High OPEC oil prices created a major non-OPEC production build-up in Mexico, Alaska, and the North Sea, as well as in Egypt, Malaysia, Angola, and China. Technological innovations improved exploration, production, and transportation. Meanwhile, demand was reversed by the economic recession, higher prices, government policies, and the growth of nuclear power and natural gas alternatives. Not only did the oil share of the world energy market decrease from 53 percent in 1978 to 43 percent by 1985, but conservation shrank the entire market as well—the 1975 United States CAFE legislation doubled auto fleet mileage to 27.5 miles per gallon by 1985, and this alone removed nearly 2 million barrels per day from the demand side of the world oil ledger. By 1985, the collapse in demand combined with the relentless build-up of non-OPEC supply (and everybody dumping inventory) reduced demand on OPEC by 13 million barrels per day. By the May 1985 Bonn economic summit, excess oil capacity around the world exceeded supply by 10 million barrels per day, the exact opposite of the 1979 situation, but twice as bad!

West Texas Intermediate crude on the futures market plummeted from its all-time high of $31.75 a barrel in November 1985 to $10 a barrel in only a few months. Some Persian Gulf cargoes sold for as little as $6 a barrel. The ensuing "buyers market" saw oil commodity traders sell oil to anyone at any price as spot prices plummeted and sellers scrambled to get out of their long-term contracts. By the time the dust settled and OPEC established a new "official" oil price of $18 in December 1986, consumers were again jubilant, and all fears of permanent oil shortage had been laid to rest. Meanwhile, the Chernobyl nuclear accident in April 1986 gave the now energy-aware public another boost of environmental awareness. The 1989 Exxon Valdez tanker accident in Alaska further heightened this awareness. As oil and internal combustion engine vehicles headed into the 1990s, they faced a tripartite alliance of issues: CAFE standards, environmental standards, and unstable supply dominated by foreign interests.

When the "fourth shock" occurred, the invasion of Kuwait by 100,000 Iraqi troops in August 1990, it removed both the Kuwait and Iraq oil supplies from the market and again sent oil prices climbing. Iraq's Saddam Hussein had problems—his eight-year war with Iran and enormous weapons purchases had run him out of money, his oil production was in disarray, and his political popularity at home was potentially at risk. He needed money, oil, and a new external threat upon which to focus Iraqi citizens' attention. Kuwait had money and oil and was conveniently located at the border. Hussein need only get assurances from the U.S. State Department that it would look the other way, and mock up a few sovereignty claims. (Iraq had claimed ownership of Kuwait before OPEC in 1961, right after Kuwait became independent of Britain, but in fact Kuwait's origins go back to 1756, predating Iraq's origin. Iraq was formed out of the three former provinces of the Turkish empire by the British in 1920.) Having taken care of both these steps, Saddam Hussein's military might had no problem in quickly dispatching Kuwait's defense and occupying it. Unfortunately for Saddam Hussein, he had stepped on the industrialized world's oil jugular, the implied threat being first

Kuwait, then Saudi Arabia and the rest of the peninsula, and total domination of the world's oil supply. Never before had the industrialized world acted so quickly (and in concert), and Iraq was dispatched in short order. Fortunately for Saddam Hussein, he remained to fight another day.

The Iraq-Kuwait war, 9/11, and now the new Iraq War and its aftermath brought the real priorities of all the industrialized world's citizens, now linked together by real-time TV network news coverage, into clear focus. The supply of foreign (mainly Middle Eastern) oil that makes everything work is, and will continue to be, highly vulnerable. We are now in the fifth shock *with oil costing over $140 a barrel!* While the majority of earth's citizenry continue to drive their internal combustion engine vehicles as if the supply of oil was secure and inexhaustible, fouling the air with pollutants and soiling the lands, rivers, and seas with toxic byproducts, more and more individuals are waking up. The stage is set for rising individual responsibility and the reemergence of the electric vehicle.

Electric Vehicles Rise Again

While electric vehicles of all kinds continued to be built in the United States and overseas during the 1920s through the 1950s, the resurgence of interest in EVs directly coincided with the environmental problems of 1960s, as well as the first, second, fourth, and fifth oil shocks. Unfortunately, the lull between these four waves—particularly the third oil shock glut—was equally responsible for retarding further EV development after each interest peak. What guarantees the lasting impact of the fourth wave are the following universal perceptions:

- The security of our oil sources is a serious problem affecting the whole world.
- Our environment is at risk.
- There is a real need to conserve our scarce planetary resources and nonrenewable fossil fuel supplies.

Thanks to the miracle of instant global TV and telecommunications, almost everyone in the industrialized world now knows this. The individual citizens of the advanced industrialized nations, nagged by the increasingly insistent urgings of their conscience, find it harder to conduct "business as usual" if it involves polluting the air, land, or water, or wasting a suddenly precious, precarious, and limited commodity, oil. The handwriting is on the wall. The heydays of gas-guzzling cars are gone forever, as are smoke-belching junkers. Even so, these beg the ultimate question—our proven oil reserves should last us 40 years, as shown in Chapter 2, but what then? Isn't there a better way than oil today?

Of course there is: electric vehicles and renewable energy sources. But before EVs could reappear in quantity on our streets and highways, consumers had to believe in them, and companies had to believe they would be profitable to invest in. All this has taken time. Heading toward 2010, the momentum seems to be supporting electric drive as a potentially dominant technology, but there is still work to be done. This section will explore how the interaction of five diverse areas set the stage for EVs to rise again:

- Interest in electric vehicle speed records
- Interest in electric vehicle distance records
- Development of electric vehicle associations

- Development of U.S. legislation
- Expansion into other electric drive technologies

The remainder of the chapter will explore how the four waves influenced the production history of EVs in the United States, Europe, and Japan.

Manufacturers, in response to growing public interest and awareness, gradually increased their own awareness of electric vehicle feasibility, and established a more open development climate. This fostered technological innovation and, fueled by global competition among United States, European, and Japanese companies, led to the actual rise of today's modern, technically sophisticated electric vehicles.

Setting the Stage for Electric Vehicles

Even those not in the marketing profession know the power of word of mouth. People talk. They tell others about what they like and don't like. In the automotive field, they talk about speed and distance records and automobiles. They go to gatherings to see what's going on and read about the gatherings that they missed. When enough people get together and talk to the government about something, the government listens. Speed and distance records, associations, legislation, and events put EVs back on the map.

The Need for Speed

In Chapter 1 you read about the "speed myth" and learned that you can make electric vehicles go as fast as you want. Each step involved a bigger motor, more batteries, a better power-to-weight ratio, and a more streamlined design.

The very first land speed record of any kind was set by an electric vehicle in 1899 at 66 mph. In the same year the first speeding ticket awarded to any sort of vehicle went to a Manhattan electric cab zipping along at 12 mph. Just after the turn of the century, a Baker Electric went *104 mph*.

After interest in electrics resumed, Autolite reached 138 mph in 1968, EaglePitcher bumped it to 152 mph in 1972, and Roger Hedlund's "Battery Box" pushed it to its present 175 mph in 1974. In 1992, Satoru Sugiyama, in the Kenwood-sponsored "Clean Liner," was going to try for 250 mph at the Bonneville Salt Flats, using a 650-hp motor Fuji Electric motor from a Japanese bullet train and 113 Panasonic lead-acid batteries. Unfortunately, a simple component failure prevented the first day's run, and wind or weather wiped out the next six days of his Speed Week window.

When you consider that GM's Impact dusted off a Mazda Miata and Nissan 300ZX in 0- to 60-mph standing-start races, there should be no question in your mind that today's electric vehicles can go fast and get there quickly. If you have difficulty with the concept, remember that France's T.G.V. electric train routinely goes 186 mph en route (it can go 223 mph) and the Swedish-built X2000 electric train rode the Washington, D.C. to New York corridor at 125 mph (it can go 155 mph) starting in early 1993. An electric train is nothing more than an electric vehicle on tracks (and with a long cord!). Meanwhile, no one is saying electric vehicles are wimpy in the speed department anymore.

The Need for Distance

You also heard the "range myth" debunked in Chapter 1; it's possible to build EVs with ranges comparable to many traditional internal combustion vehicles.

- In 1900, a French B.G.S. Electric went 180 miles on a single battery charge.
- A Nissan EV-4H truck went 308 miles on lead-acid batteries in the 1970s.
- Horlacher Sport pushed that to 340 miles using Asea Brown Boveri sodium-sulfur batteries in 1992 and BAT International bumped it to 450 miles later in the year using proprietary batteries/electrolyte in a converted Geo Metro. The Horlacher averaged nearly 75 mph during its 10-hour test period, an impressive speed. The BAT Geo Metro was driven at a more sedate 35 mph.

There should also be no question in your mind that electric vehicles can go far. Given the fact that most Americans average less than nine miles a trip, and electricity is ubiquitous in most American urban areas, an electric vehicle is no more likely to run out of juice than an internal combustion vehicle would run out of gasoline, certainly less likely than running out of diesel or any of the new alternate fuels that aren't carried in every filling station!

The Need for Association

There were two types of associations started in this period: those for government and commercial interests, and those for individuals. The government/commercial types can be further divided into associations for sharing information (the British formed the first in 1934); associations for lobbying (like the Electric Vehicle Council, Electric Vehicle Development Corporation, and Electric Vehicle Association of the Americas, now known as the Electric Drive Transportation Association); and consortium associations for advancing research like the U.S. Advanced Battery Consortium and CALSTART.

Not all government/commercial associations for sharing information are created equal. The British association was very "loose," but had the world's largest EV population by the 1970s—approximately 150,000 vehicles of all kinds. The German Gesellschaft fur Electrischen Strassenzerkehr (GBR) clearinghouse, a subsidiary of their largest electric utility, was instrumental in focusing German EV efforts; and Japan's Electric Vehicle Association, under the guidance of its Ministry of International Trade and Industry (MITI), gave them a definitive EV mission statement—focus on urban acceleration and range. In contrast to the earlier efforts of the British, and the focused EV efforts of the Germans and the Japanese in the 1970s, the United States didn't have any focused industry effort until the USABC and CALSTART consortia of the 1990s.

Although individual members of the Electric Auto Association (formed in 1967) were a "voice in the wilderness" during this period, they made it increasingly difficult for people to say "it couldn't be done" after they were already doing it.

The Need for U.S. Legislation

It is an enigma that we had the legislation (the U.S. government-funded National Aeronautics and Space Agency–guided Apollo program) to put an electric vehicle on the moon before we had the legislation to develop EVs on earth.

When the very first bills were being introduced in Congress in 1966 to sponsor electric vehicles as a means of reducing air pollution, government-funded contractors had Lunar Rover EV prototypes already working at their facilities. While the pros and cons of EVs were argued during various congressional hearings in the early 1970s, leading to the passage of the landmark Public Law 94-413, the Electric and Hybrid Vehicle Research, Development and Demonstration Act of 1976, Lunar Rover electric

Figure 3-11 Apollo 17 Lunar Rover vehicle on the moon with astronaut Eugene A. Ceman at the controls.

vehicles had already performed flawlessly on the moon during three separate 1971 and 1972 Apollo missions.

The Lunar Rover, shown in Figure 3-11, was optimized for light weight. It had four 0.25-hp series DC motors, one in each wheel, and woven wire wheels and aluminum frame that gave it a mere 462-lb. weight (which could carry a 1,606-lb. payload). Its silver-zinc nonrechargeable batteries gave it an adequate one-time 57-mile range and 8-mph speed. Future moon travellers only have to bring a new set of batteries with them—three Lunar Rovers are already there!

The point is, legislation and subsequent funding, along with the focus and emphasis provided by consolidating the major federal energy functions into one Cabinet-level Department of Energy in 1977, were vital components. Until these pieces were in place, EV development and renewal could not proceed.

But while the government-funded early ETV-I and ETV-II programs in 1977 provided the impetus to jumpstart electric vehicle activity again, it was clear that government legislation and funding amounted to the merest beginning, not the path for introducing electric vehicles in widespread numbers. The testimony of one General Motors executive at the June 1975 Congressional hearings drives home the point: "General Motors does not believe that much can be gained by subsidizing the sale of electric vehicles ...we feel that building more electric vehicles is a waste of resources."

Fifteen years later, the world's largest automaker appeared to have had a turnaround in opinion (although it didn't last for long in the early 2000s). An August 1991 *Motor Trend* article noted, "Among all the automakers, General Motors is by far the most vocal in support of the electric car as a real alternative in the U.S. market."

Much of the enthusiasm on the part of automakers for electric vehicles during this period can be traced to government support. The U.S. Advanced Battery Consortium (whose principals are General Motors, Ford, and Chrysler, with Department of Energy participation), the CALSTART consortium (involving utilities and large and small aerospace/high-tech companies including Hughes, then a subsidiary of General Motors), and research institutions driven by the Los Angeles Clean Air Initiative provided support for research and development of advanced batteries and other electric vehicle technologies.

As we head back into another era of plug-in vehicles, these and other collaborative efforts have experienced a resurgence after several years of seeming to be relatively dormant.

The Need for Events

Some people are thinkers, others are doers, some just like to tinker around—and EVs provide a fertile field for all three types.

- Thinkers can attend symposia; the largest and most well-known is the Electric Vehicle Symposium (EVS), which has been held in cities such as Paris; Washington, DC; Toronto; Hong Kong; Milan; and Anaheim.
- Doers can go to races and road rallies; what started out as the simple MIT versus Cal Tech Great Electric Vehicle Race of 1968 evolved to the huge crowds and multiple classes of the Phoenix 500 race by the 1990s. The Northeast Sustainable Energy Association (NESEA) had the Tour De Sol event, which allows kids in high school and college to race prototype electric vehicles and also has events at each leg of the race to promote electric cars and electric drive.
- In recent years, the National Electric Drag Racing Association has gained prominence, as have races such as the "Power of DC" in Washington, DC, and the "Battery Beach Burnout" in Southern Florida. Chapters of the Electric Auto Association have been holding road rallies since 1968. Some regard the greatest of all challenges (and the greatest of all publicity stunts) to be the World Solar Challenge: a 1,900-mile race across the Australian outback using a motor the size of a coffee can powered only by sunlight.

The 1990s–2000s

Environmental and conservation concerns put real teeth back into EV efforts, and even General Motors got the message. Indeed, GM did a complete about-face and led the parade to electric vehicles. Resumption of interest in EVs during this wave was led by unprecedented legislative, cooperative, and technological developments.

Electric vehicles of the 1990s also benefited from improvements in electronics technology, because the 1980s mileage and emission requirements increasingly forced automotive manufacturers to seek solutions via electronics. Although EV interest was in a lull during the 1980s, that same decade saw a hundredfold improvement in the capabilities of solid-state electronics devices. Tiny integrated circuits replaced a computer that took up a whole room with a computer on your desktop at the beginning of the 1980s, and by one that could be held in the palm of your hand by the beginning of the 1990s. Development at the other end of the spectrum—high-power devices—was just as dramatic. Anything mechanical that could be replaced by electronics was, in order to save weight and power (energy). Solid-state devices grew ever more muscular in response to this onrushing need. Batteries became the focused targets of well-funded government-industry partnerships around the globe; lead-acid, sodium-sulfur, nickel-iron, and nickel-cadmium batteries advanced to new levels of performance. Everyone began quietly looking at the mouth-watering possibilities of lithium-polymer batteries.

Table 3-2 compares electric vehicle specs from the major automotive manufacturers at the beginning 1960s to the 1990s. At first glance, Table 3-2 appears to be a step backward from Japan's accomplishments of the 1970s as shown in Table 3-1. But first impressions are misleading. The 1990s electric vehicles offer substantial technology improvements under the hood, are more energy-efficient, and are closer to being manufacturable products than engineering test platforms. Six trends highlight EV development during this fourth wave:

- High levels of activity at GM, Ford, and Chrysler
- Increased vigor created by new independent manufacturers
- New and improved prototypes
- High levels of activity overseas
- High levels of hybrid activity
- A boom in individual internal combustion vehicle conversions

Leading up to the 1990s, General Motors warmed to EVs through a series of events. First, GM decided to enter the 1987 Australian World Solar Challenge, a project championed from within by Howard Wilson, then vice president of Hughes. They turned for help to a little company called AeroVironment, founded by Paul MacCready. MacCready had already won prizes for the longest human-powered flight with his ultra-lightweight and efficient "Gossamer Condor" in 1977, and for the first human-powered plane to cross the English Channel with his "Gossamer Albatross" in 1979; in addition, his solar-powered "Solar Challenger" flew 163 miles from Paris to the English coast in 1981. AeroVironment was now given the ultimate challenge from GM—creating a vehicle capable of going 1,900 miles across the Australian outback on solar power only. AeroVironment did it. GM's winning "Sunraycer" beat the competition by 2½ days, proving what was possible with electric drive.

	GM Impact car	**Ford Ecostar van**	**Chrysler Epic van**	**BMW E2 car**	**Nissan FEV car**
Range (miles)	120	100	120	150	150
Top speed (mph)	75	65	65	75	80
Power train	2 ea 54 hp ac	75 hp ac	65 hp dc	45 hp dc	2 ea 27 hp ac
Battery pack	Lead-acid	Sodium-sulfur	Nickel-iron	Sodium-sulfur	Nickel-cadmium[1]
Curb weight (lbs)	2200	3100	3200	2020	1980

[1] Augmented by crystal-silicon solar battery.

TABLE 3-2 Leading Electric Vehicles From The 1990s Compared

FIGURE 3-12 The GM Impact electric vehicle, which later became the GM EV1.

MacCready and Wilson wondered what might be possible with an electric passenger vehicle. Wilson was able to present the question personally to Bob Stemple, who sought approval in 1988 by then-chairman Roger Smith, and the rest, as they say, is history.

GM's Impact electric vehicle, shown in Figure 3-12, appropriately debuted at the January 1990 Los Angeles Auto Show, and promptly set the automotive world on its ear. Offering 50- to 70-mile range (a "Gen II" EV1 with NiMH batteries provided 120–140 miles of range), a 0-to-60 time of under 8 seconds, 80-mph freeway capability in a slippery package that had a 0.19 coefficient of drag (still the most aerodynamic production car ever made), the Impact was not your grandparents' electric vehicle. For a while, GM proceeded to show the rest of the world the way.

With his "do more with less" philosophy, MacCready himself provided much of the inspiration: "No one had ever tried to build a super-efficient car from scratch. That's because no one had ever needed to. Energy had always been cheap and pollution controls were relatively recent, so automakers never needed to pay fanatical attention to efficiency."

GM produced 50 Impacts and loaned them to utilities, local governments, and individuals to provide feedback about performance capabilities and user requirements in an innovative program known as "PrEView Drive," which lasted from 1994 to 1996. Meanwhile, GM hedged its bet by working on the hybrid I-IX3 concept minivan that used a gasoline-powered 40-kW generator to extend the range of its two 60-hp AC front-wheel-drive electric motors.

Ford had a different idea; by using its European Escort van as the platform, building on the sodium-sulfur battery technology it invented in 1965, and applying its 1980s ETX-I and ETX-B drivetrain experience, Ford was hoping to leapfrog its competition and be the first to production with 80 Ford Ecostar vans (see Figure 3-13) planned for distribution to fleet customers in 1994. Ford was serious—management, resources, tooling, and facilities had been put in place. Weighing in at 3,100 lbs. and driven by a solid state–controlled 75-hp AC induction motor coupled to an integrated front-wheel drive and powered from sodium-sulfur batteries, the Ford Ecostar's specs were impressive at the time—75 mph and 100-mile range, with the ability to carry an 850-lb. payload. All of which led automotive writer Dennis Simanaitis to comment in the February 1993 issue of *Automobile*, "The first electric vehicle you're likely to see is the most transparent we've driven so far."

Chrysler took yet another approach, building on the TEVan ("T" for T-115 minivan platform, "E" for electric). The already-proven electronics, drivetrain, and batteries lifted from their TEVan were used in their next generation lighter-weight Epic van,

Figure 3-13 Ford Ecostar van electric vehicle.

shown in Figure 3-14, giving them immediate parity in the performance specs (see Table 3-2).

Regulation in California

This period initially marked a brief resurgence of internal combustion vehicles, particularly larger cars, trucks, and sport utility vehicles (SUVs). The ZEV Mandate in California created a backlash against electric vehicles by automakers, who didn't want to be legislated to build anything, let alone something other than their core internal combustion products. After the automakers banded together with the federal government and started legal proceedings against the state of California, the California Air Resources Board gutted the ZEV Mandate (now known as the ZEV Program), effectively releasing automakers from having to build electric vehicles at all. Once automakers were no longer required to market electric vehicles, EV programs were quickly terminated.

During this phase, other vehicle technologies were embraced to various degrees by automakers. The Partnership for a New Generation of Vehicles program, under President Clinton's administration, stimulated the development of gasoline hybrid

Figure 3-14 Chrysler Epic van electric vehicle.

vehicles by domestic automakers. While Detroit never actually deployed hybrid cars during this phase, competitive spirit compelled Japanese automakers to do so. This led to popular vehicles like the Honda Civic hybrid and the Toyota Prius, with most major automakers eventually offering at least one hybrid model. Among domestic automakers, hydrogen became the alternative fuel of choice for new concept cars, which were accompanied by promises to mass-market these vehicles by 2010. As we near the end of this decade, approximately 175 hydrogen fuel cell vehicles have been deployed in test fleets, but none have appeared in showrooms.

As public awareness of the issues surrounding petroleum dependence—climate change, political instability, and public health issues due to poor air quality, to name a few—has increased, the tide seems to be turning back toward plug-in vehicles. This has been stimulated both on a mass level by pop-culture devices such as *An Inconvenient Truth* and *Who Killed the Electric Car?* (the #1 and #3 documentaries of 2006, respectively) as well as on a very personal level with rising gasoline prices. Increasingly, plug-in vehicles, which were once seen as a crunchy, environmental choice, are gathering bipartisan support as those concerned with energy security are beginning to embrace the alternative of using cheap, clean, domestic electricity to power vehicles instead of foreign, expensive, comparatively dirty petroleum. With this broad coalition of support and declining auto sales, automakers have had little choice but to get on board with newer alternatives to internal combustion vehicles.

New technology has also stimulated enthusiasm; in addition to electric vehicles, automakers have started working on low-speed electric vehicles, hybrid electric vehicles, plug-in hybrid electric vehicles (PHEVs) (which combine a certain number of electric miles, with the "safety net" of a hybrid propulsion system), and people building their own electric cars. Depending on the vehicle's configuration, drivers might drive only in electric mode for their weekly commutes, and use gasoline only when driving long distance. This "best of both worlds" concept has renewed enthusiasm for electric vehicles as well, and both types of plug-in vehicles are benefiting from newer lithium-ion battery technology, which stores more energy than previous lead-acid and nickel–metal hydride types, providing longer range.

Legislation provided both the carrot and the stick to jumpstart EV development. California started it all by mandating that 2 percent of each automaker's new-car fleet be comprised of zero emission vehicles (and only electric vehicle technology can meet this rule) beginning in 1998, rising to 10 percent by the year 2003. This would have meant 40,000 electric vehicles in California by 1998, and more than 500,000 by 2003. California was quickly joined in its action by nearly all the Northeast states (ultimately, states representing more than half the market for vehicles in the United States had California-style mandates in place)—quite a stick! In addition, for CAFE purposes, every electric vehicle sold counted as a 200- to 400-mpg car under the 1988 Alternative Fuels Act. But legislation also provided the carrot. California provided various financial incentives, totalling up to $9,000 toward the purchase of an electric vehicle, as well as nonfinancial incentives, such as HOV lane access with only one person. The National Energy Policy Act of 1992 allowed a 10 percent federal tax credit up to $4,000 on the purchase price of an EV. Other countries followed suit. Japan's MITI set a target of 200,000 domestic EVs in use by 2000; and both France and Holland enacted similar tax incentives to encourage electric vehicle purchase.

The California program was designed by the California Air Resources Board (CARB) to reduce air pollution and not specifically to promote electric vehicles. The regulation initially required simple "zero emission vehicles," but didn't specify a required technology. At the time, electric vehicles and hydrogen fuel cell vehicles were the two known types of vehicles that would have complied; because fuel cells were (and remain) frought with technological and economic challenges, electric vehicles emerged as the technology of choice to meet the law. Eventually, under pressure from various manufacturers, and the federal government, CARB replaced the zero emissions requirement with a combined requirement of a very small number of ZEVs to promote research and development, and a much larger number of partial zero emission vehicles (PZEVs), an administrative designation for *super ultra low–emissions vehicles* (SULEVs), which emit about 10 percent of the pollution of ordinary low-emissions vehicles and are also certified for zero evaporative emissions. While effective in reaching the air pollution goals projected for the zero emissions requirement, the market effect was to permit the major manufactures to quickly terminate their public BEV programs.

Since the electric car programs were destroyed, the market has developed an expansive appetite for hybrid electric cars and cleaner gasoline cars. GM EV1 (see Figure 3-15 and 3-16) and EV2, Chrysler's Epic minivan, and Ford Ranger, as well as Honda EVPlus, Nissan Hypermini, and Toyota RAV4 and Toyota electric cars, were recalled and destroyed. Roughly 1,000 of these vehicles remain in private hands, due to public pressure and campaigns waged by a grassroots organizations such as "dontcrush .com" (now known as Plug In America), Rainforest Action Network, and Greenpeace. Contradicting automaker claims of anemic demand for EVs, these vehicles now often sell for more on the secondary market than they did when they were new. The whole episode became known as such a debacle that it spawned a feature-length documentary, directed by former EV1 driver and activist Chris Paine, entitled *Who Killed the Electric Car?*, which premiered at the 2006 Sundance Film Festival and was released in theaters by Sony Pictures Classics.

9/11, Oil, and Our New Understanding for Electric Cars

Understandably, the attack on the World Trade Center, the Pentagon, and the plane that crashed in Pennsylvania on September 11, 2001 clearly shows that our reliance on imported oils is damaging our national and financial security. With oil reaching over

Figure 3-15 The GM EV1 next to a Detroit Electric Car from 1915 (Source: Russ Lemons).

Figure 3-16 The GM EV1 electric vehicles being crushed, as seen in *Who Killed the Electric Car?* (Source: http://ev1-club.power.net/archive/031219/jpg/after2.htm).

$140 a barrel recently created the resurgence, acceptance, and understanding that we need electric cars. September 11 also ended the period of low and stable oil prices. I remember watching the towers in flames and saying to a colleague that things will never be the same again. Now, we are faced with a serious discussion about energy security, reducing our reliance on imported oils and also it's environmental impact on our globe.

General Motors Awakening—The Volt

Figure 3-15 also shows how what comes around goes around. The electric car has come back to GM since the electric car has been with our American automotive history since the beginning of the car.

The ironic twist of events about oil, national security, and climate change shows an interesting side to GM and their understanding of the need for electric cars to help stablize our economy. While the Volt is planned to become a plug-in hybrid, it is starting out as engineers and designers, building GM's own electric vehicle.

In January 2006, Bob Lutz, Vice Chairman of General Motors, one of the main proponents who originally was for crushing the EV1 program is now a pragmatist who believes the electrification of the car is the only way to preserve American car culture. "We are agonizing over what to do to counter the tidal wave of positive PR for Toyota," Lutz says. That month, GM came up with the Volt.

Even Rick Wagoner, President of GM was not an EV proponent for an electric car again. But then Hurricane Katrina sent oil prices soaring. For Wagoner, it was a sign of how volatile the oil markets were becoming and a harbinger. Even the much maligned energy policy of the Bush Administration was changing: In his State of the Union address, the President urged Congress to impose tougher fuel economy rules. By January of 2008, the Volt had become the centerpiece of GM's green strategy.

Does GM's CEO regret not moving faster? You bet he does. Wagoner wishes he hadn't killed the EV1. And he acknowledges underestimating the emergence of consumer societies in China and India would help put the $100 floor under oil prices. Today, all of that is beside the point. The looming question is whether Wagoner can keep his promises. "It's the biggest challenge we've seen since the start of the industry," he says. "It affects everything we thing about." (Source: David Welsh, GM: Live Green or Die, *BusinessWeek*, May 26, 2008, Pages 38–39.)

Wagoner added in a recent interview on PBS, "We had about 100 years of an auto industry in which 98 percent of the energy to power the vehicle has come from oil. We're really going to change that over the next time period, things like battery development and applying batteries to cars, as we're planning on doing with the Volt, is an important step, kind of, in the next 100 years of the auto industry." (Source: www.pbs.org/newshour/bb/transportation/jan-june08/electriccars_06-25.html)

Besides General Motors seeing the light after 9/11, conversion car companies like Steve Clunn's Grass Roots EV saw an increase in electric car conversion requests. More and more people have been buying electric car conversions, electric cars, low speed electric vehicles, hybrid electric cars, and plug-in hybrid electric cars. Car companies, the public, the media, and the free market are all starting to fully accept the fact that electric cars *need* to be part of our automotive future.

Electric Vehicles for the 21st Century

What goes around comes around, as the saying tells us. People have been building commuter-based electric cars that are getting a lot of attention and are now increasing in sales. The Tango by Commuters Cars Corporation (seen in Figure 3-17), the Corbin Sparrow, and Phoenix Motorcars are great examples. While they are different looking than most cars, they are electric, get over 70 mph, and have a significant range (100 plus miles).

As we have seen, people have been successfully converting internal combustion vehicles to EVs for at least the past 35 years. This entire period has been marked by an almost total absence of comments about this activity from the naysayers.

It is ironic that while electric cars were around before the internal combustion engine, and will also be around after them, the common thinking of today is that the development of electric cars has followed a path from gas, to gas hybrid, to plug-in-hybrid, to electric.

Tesla

Tesla Motors is a Silicon Valley automobile startup company, which unveiled the 185 kW (248hp) Tesla Roadster on July 20, 2006. As of March 2008, Tesla has begun regular production of the Roadster.

The Roadster has an amazing AC drive of a new design, a new controller, new motor, and new battery subsystem.

The Tesla Roadster, shown in Figure 3-18, delivers full availability of performance every moment you are in the car, even while at a stoplight. Its peak torque begins at 0 rpm and stays powerful at 13,000 rpm. This makes the Tesla Roadster six times as efficient as the best sports cars while producing one-tenth of the pollution with a range of 220 miles.

FIGURE 3-17 The Tango from Commuter Cars Corporation.

The Tesla Model S is planned for 2010 delivery. The estimated cost is $60,000 with a $30,000 model planned later on.

The Tesla is another great "build your own" sports car EV company for the masses to look at and be excited about the future for electric cars today. The best part is that they want to build more vehicles that can cost $30,000.00.

FIGURE 3-18 The Tesla Roadster.

Figure 3-19 Tom Gage (President of AC Propulsion), with actress Alexandra Paul, explains the eBox specs (Source: AC Propulsion).

eBox

The eBox is the purest and greatest example of building your own electric vehicle using an existing vehicle platform! Take an existing EV, such as the Scion, trick it out with great batteries, a great motor and controller, and watch it work!

This urban utility vehicle, shown in Figure 3-19, has a range of 120–150 miles, 0–60 mph in 7.5 seconds with a top speed of 95 mph. It has a 30-minute charge for 20–50 miles (all you need around town) and a full charge in two hours. It has regenerative braking, lithium-ion batteries, and all the bells and whistles of a regular internal combustion engine vehicle (navigation system and alloy wheels optional).

Hybrid and Plug-In Hybrid Electric Vehicles

Right now there are already electric vehicles available at almost every price point. If you cannot figure out how to fit one of them into your life, you can buy a new HEV, such as a Toyota Prius, for a bit more than $20,000 or a used one starting at about $10,000. Gasoline-powered HEVs are not the ultimate answer to our energy problems, but they do provide an excellent platform for developing EV components such as electric motors, batteries, and transmissions. They also use much less gas than their internal combustion engine–only brethren.

As you can see in Figure 3-20, hybrid sales on a monthly basis from January 2004 to February 2007 show the Prius as the consistent leader (see Figure 3-22), followed by the Honda Civic hybrid and the Toyota Camry hybrid. Even though the Prius is the most

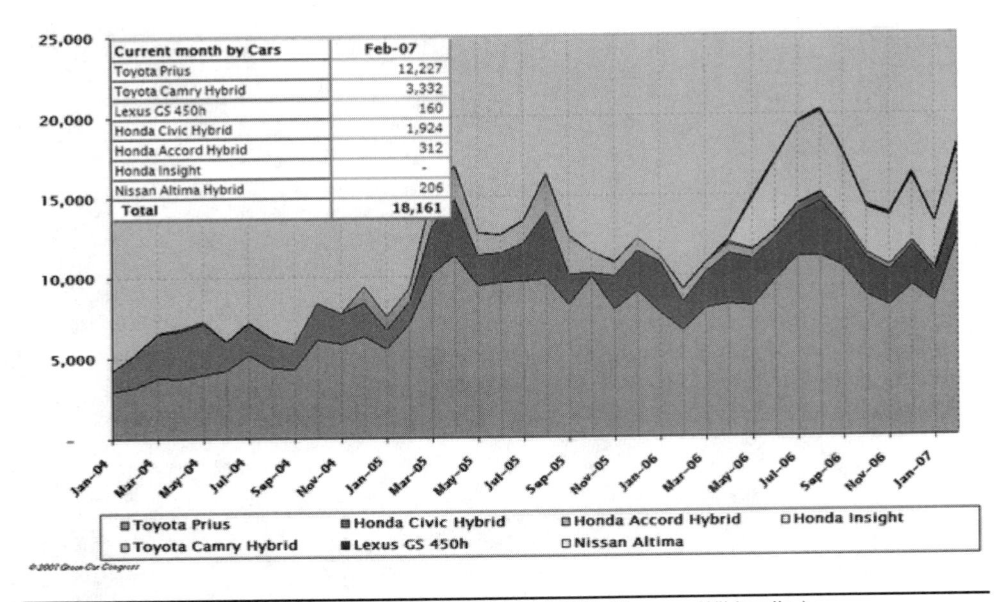

FIGURE 3-20 Hybrid electric car sales from 2004 to 2007. (Source: Wikipedia.)

popular, one of the most fuel efficient hybrids was the Honda Insight (see Figure 3-21), a small compact hybrid.

But there's more good news: there is a wave of EVs and PHEVs under development from companies such as Phoenix Motorcars, AFS Trinity, and a host of others companies. Additionally, Toyota recently announced that it is developing a plug-in version of the Prius with substantial electric-only range.

FIGURE 3-21 The most fuel-efficient hybrid car, the Honda Insight. (Source: Wikipedia.)

FIGURE 3-22 The leading hybrid car, the Toyota Prius (Source: Wikipedia, IFCAR, All Rights Released.)

Many people are not waiting until CAFE standards reach 35 miles per gallon by 2020. They are driving vehicles now that get better than 35 miles per gallon. Some are starting to drive plug-in hybrids that achieve over 100 miles per gallon. In the U.S. 40,000 people drive electric vehicles that use zero gasoline and produce zero emissions.

However, even you do not want to convert your car to electric—the full performance to the low-speed electric vehicles—the total number of electric vehicles has increased significantly within the past 5 to 7 years. Here are just some of the companies out there.

Miles Electric Vehicles, a Los Angeles-based all electric vehicles manufacturer currently produces several low-speed vehicles for military, university, municipal, and private use. The company is also currently developing the XS500 electric vehicle, powered by lithium-ion phosphate batteries. It has a top speed of 80 mph and a range of 120 miles. The company says it plans to crash test the car and release it as a 2009 model with a minimum four star crash test safety rating. The projected price is $30,000.

Phoenix Motorcars (www.phoenixmotorcars.com) based in Ontario, California, plans to build both a mid-sized SUV and an SUT (sports utility truck) with 130 mile range for $45,000 using NanoSafe batteries from Altairnano. Five hundred cars are planned for delivery in early 2008 to fleet customers. A consumer version is planned for release in late 2008. A version with a 250+ mile range is also in development. Their lithium-ion battery is enhanced by nanotechnology that can recharge in under ten minutes, drive for a hundred miles on a charge, and has a top speed of 95 mph.

Tesla Motors, Silicon Valley automobile startup company, unveiled the 185 kW (248 hp) Tesla Roadster on July 20, 2006. As of March 2008, Tesla has begun regular production of the roadster. The Tesla Model S is planned for 2010 delivery. The estimated cost is $60,000 with a $30,000 model planned later on.

ZAP produces several electric vehicles, most notably, the Xebra three-wheel electric vehicle, which is classified as a motorcycle. Top speed is 40 mph and costs about $10,000. According to reports, it has sold about 200 through U.S. dealers. The ZAP-X (www.zapworld.com) is a four door crossover engineered by Lotus with a price tag of around $60,000. No production date has been given.

ZENN (www.zenncars.com) has a fully-enclosed three-door hatchback low-speed vehicle (LSV). I met Ian Clifford, the President of ZENN when I worked for NYPA in 2003. We talked about the possibility of using the vehicle for the Clean Commute Program at NYPA. ZENN is now planning a launch in fall 2009. It will be the first electric car to use the revolutionary EEStor ultra-capacitor.

Smith electric vehicles also produces a 4.5 and 7.5 ton electric truck. The Might-E truck is a heavy duty electric work truck.

Canadian Company Dynasty EV has a neighborhood electric vehicle with a design reminiscent of the Volkswagen Beetle.

REVA is an Indian-built city car. The REVA is sold in the United Kingdom as the G-Wiz as well as in several European countries. In the United States, it is classified as a NEV or LSV and is limited for use as a neighborhood electric vehicle. The CityEl (www.cityel.de) is a three-wheeled EV, produced in Germany. Then there is Modec (www.modec.co.uk/) which is based in the United Kingdom and they build electric delivery vans.

There are many independent companies that are selling electric cars. The free market even shows how internationally, electric cars are increasing in size. There are other companies out there, so research the best one for you!

This is a tell tale sign of the market for electric cars.

Low-Speed Electric Vehicles

People should realize that electric vehicles are not just for the rich. Low-speed vehicles are priced at between $8,000 and $10,000 and only go 25 miles per hour. When I worked for the State of New York, after the full-performance electric cars were recalled, we placed several thousand low-speed electric vehicles in the state. Today, there are over 20,000 low-speed, neighborhood electric vehicles (NEVs) on the road today. While they were not on-road (highway) vehicles, they reduced vehicle emissions. It had an effect. From Battery Park City to the Town Niagara, from the State University of New York at Albany to Staten Island Zoo, these electric vehicles were great in city applications. While the enclosures for those vehicles were not appropriate for the Northeast, many low-speed electric vehicles today have enclosures that are like doors. While they are only allowed on roads that do not exceed 35 miles per hour, there are plenty of areas in the country where they work just fine. They are and were perfect for university campuses, large facilities that need zero-emission indoor vehicles, shuttles in corporate multi-building campuses, and even the military. A low-speed electric car can be a popular second car in two-vehicle households. These low-cost EVs are fine for those who will compromise on speed and range. Reasonably priced new vehicles are coming with few compromises and many exciting features.

Today, there are also low-speed vehicles, such as the ZENN car, that look more like an actual car for many applications. With the ever increasing LSV market, electric cars are increasing significantly worldwide.

Near Future Trends For Electric Drive

There are many options today for electric cars. A Tesla costs about $100,000 for the Roadster, and the company is expecting to launch a vehicle for $30,000 and $60,000. You can order an eBox from AC Propulsion for $73,000 fully loaded. Also, Vectrix is now selling a freeway-legal electric motorcycle for $11,000. Or, you can buy one of the many electric bikes online for less than $1,000 to about $2,000 for really high-performance electric bikes.

Sherry Boschert, author of *Plug-in Hybrids: The Cars That Will Recharge America*, rides on sunlight. She charges her electric vehicle with her home's solar power. Her Toyota RAV4 runs fast on freeways and silent on quiet streets. She uses a zero-emission approach to transportation. Mitsubishi is introducing the iMiEV Sport, which it plans to launch in Japan and possibly other countries in 2009. The car has a range of 93 miles (150 km) and a top speed of 93 mph (150 km/h).

The GM Volt is expected to go on sale by November 2010. GM plans to start selling the Saturn Vue plug-in hybrid in 2009. It is said that it will likely offer the best mileage of any SUV on the market.

TH!NK is starting to sell cars in Europe and has now started TH!NK North America, run by Vicky Northrup. I recently called her to congratulate her on starting back up in the United States. Vicky and I had worked closely together on the NYPA/TH!NK Clean Commute Program™. TH!NK has even developed a full-performance electric car, which can only improve its chances for success.

Toyota, Ford, Volvo, and Saab all have plug-in hybrids in early fleet trails. Other fleets are doing their own custom integration of plug-in hybrids from sedans to heavy vehicles.

Nissan and Renault are planning to mass produce an electric car that will be initially marketed for large cities by 2012. Internationally, from London to Shanghai, there are increased possibilities that only ZEVs will be allowed in major urban areas.

On the international front, Israel has committed to electric vehicle infrastructure across the country. The State of Israel is working directly with Renault-Nissan, who will manufacture an electric car. They are also working with a California start-up founded by former SAP executive Shai Agassi who will build the electric vehicle infrastructure. They are hoping to have over 500,000 charging stations across the country and up to 200 battery-exchange locations. While they plan to start with a pilot program of a few vehicles, they would like to mass market them by 2011. In addition, Israel cut the taxes on electric vehicles by 10% to incentivize consumers to buy electric cars. In the meantime, they can convert to electric today and will have EV infrastructure on the road for their public use. We should applaud the efforts by the State of Israel and Renault-Nissan to work together to create this EV market.

We must realize that electric drive technology is already and will continue to be part of every day automotive technology. Everyone seems to be building their own electric vehicle. Whether it is a hybrid, plug-in-hybrid, or pure EV, people are switching to electric everyday. In addition, more and more people are converting their cars to electric. Every conversion company I talk with has a backlog and more business just keeps coming. However, it is important to have plug-in-hybrids and hybrid cars. They transition the market toward a fully electric drive vehicle. Even if people are not interested in ending our dependence on foreign oil or saving the environment, electric drive technology makes cars with better acceleration and torque that cost less to run than a regular gas car today.

FIGURE 3-23 *Who Killed the Electric Car?* bumpersticker on an eBox. (How appropriate!)

CHAPTER 4

The Best Electric Vehicle for You

Put an electric motor in your chassis and save a bundle.

You learned from earlier chapters that electric vehicles are fun to drive, save you money as you use them, and help save the planet. The benefit of building/converting your own electric vehicle is that you get this capability at the best possible price with the greatest flexibility.

In this chapter you'll learn about electric vehicle purchase trade-offs, conversion trade-offs, and conversion costs. You'll find out how to pick the best electric vehicle for yourself today—whether buying, converting, or building.

Electric Vehicle Purchase Decisions

When you go out to buy an electric vehicle today, you have three choices: buying a ready-to-run EV from a major automaker or an EV conversion shop, or buying a used EV conversion from an individual. Your two most basic considerations are how much money you can spend and how much time you have. Another dimension of your purchase trade-off is where (and from whom) you can obtain an electric vehicle. This section will present your options and highlight why conversion is your best choice at this time.

Conversions Save You Money and Time

Saving money is an important objective for most people. It's also logical to assume you want to arrive at a working electric vehicle in a reasonably short time. When you combine these two considerations, conversion emerges as the best alternative. Figure 4-1 shows you why at a glance. A conversion:

- Costs less money than either buying ready-to-run or building from scratch.
- Takes less time than building from scratch and only a little more time than buying ready-to-run.

As you can see in Figure 4-1, buying "used" clearly lowers the cost for that category. However, when you compare equivalent used pricing across all three choices, conversion is still the best alternative.

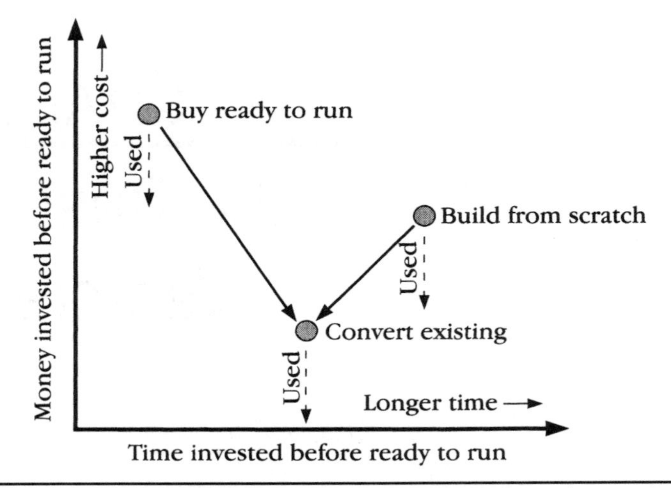

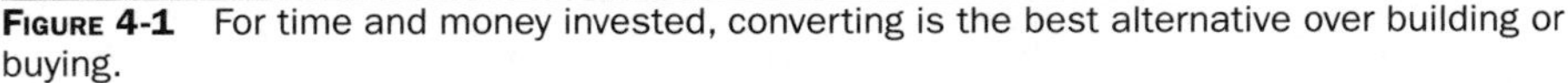

FIGURE 4-1 For time and money invested, converting is the best alternative over building or buying.

A few actual figures make this easy to see. While you might spend upwards of $100,000 to obtain a new electric vehicle from Tesla and could typically spend from $20,000 to $30,000 for a new EV at an independent dealership, in only a month or two, you can be driving around in your own EV conversion for under $10,000. When you build an EV from scratch, on the other hand, your electrical components cost the same as a conversion's, but your frame and body costs start where buying a conversion body leaves off, and you still have to do the work, pushing your time investment way up. Now you know the reason for this book.

Let's look at the buy-build-convert alternatives more closely, then look at the who-where-when trade-off.

Buying Ready-to-Run

Ready-to-run electric vehicles are not available today from any major automaker. Not even ten years ago the major automakers had EVs for lease and sale. Today, most automakers are making hybrids and a few have pure EVs and fuel cell vehicles in various stages of development and testing. This includes GM's Volt, shown in Figure 4-2.

There are a few *new* automakers that are *developing* new EVs. This includes the Tesla and many new NEVs (neighborhood electric vehicles). NEVs (also called low-speed vehicles) are a class of vehicles defined in Vehicle Motor Safety Standards Rule No. 500, which stipulates that the vehicle can be driven on roads only posted at 35 mph or less.

Another category of electric car is the city electric vehicle (such as the TH!NK). The top speed of this vehicle is about 55 mph. It can be driven on roads posted from 40 to 55 mph where a NEV cannot go. This is a popular category overseas where gasoline is very expensive. We also utilized this class of vehicle in the NYPA/TH!NK Clean Commute Program™ since the Ford TH!NK City was perfect for around-town commuting and all of the major roads in these areas did not exceed 55 mph.

Then there is the Tesla.

Figure 4-2 GM's Volt.

Buying Ready-to-Run from an Independent Manufacturer

Ready-to-run electric vehicles are available from independent manufacturers and their dealers today. But these EVs that use an internal combustion engine chassis, such as those from Tesla, cost $40,000 to $100,000. EVs will come down in price as the numbers manufactured increase—their nonrecurring costs can then be amortized over a greater number of units. EVs have the potential to be far lower in cost than internal combustion–powered vehicles in the future when economies of scale from increased production kick in because there are far fewer components that go into an electric vehicle than in an internal combustion–powered vehicle.

Conversion Shops

There are a number of conversion shops around the country. Conversions shops will convert the vehicle of your choice. In many respects, this is an advantage because you get the vehicle you want converted by a professional. Electric Vehicles of America, Inc. is setting up a network of conversion shops across the country.

Converting a Vehicle

Many people have converted their own vehicle to electric. It requires the least amount of money but the most amount of your time. A typical conversion can take 100 hours. There are a number of advantages to doing your own conversion. First, you have control of the project (the converted vehicle and the components selected); second, you have the ability to diagnose, troubleshoot, and repair any problems quickly.

If your desire is to custom-build an EV from the ground up, there is certainly nothing to stop you. But you are certainly going to spend a lot more time and money than in buying ready-to-run (unless you put a zero value on your labor time).

There is also a hidden problem here—the safety aspect. If you buy a vehicle from any manufacturer or convert an existing internal combustion–powered vehicle chassis, the safety aspects have already been handled for you. When you roll your own design, dealing with the safety issues becomes your responsibility. While licensing your one-of-a-kind "Starship Electrocruiser" will consume less time with the local Motor Vehicle authorities than would getting the certifications for a production run from multiple agencies, you are still going to have to convince someone that the wheels aren't going to come off before they permit you to cruise anywhere except in your driveway.

Building to someone's plans is a logical step towards saving time and hassle. You still have to provide the muscle and the bucks, but there's someone who has already blazed a trail before you in terms of licensing and construction shortcuts. In addition, you have someone to write or call if you get hung up on a construction detail.

Putting a custom kit on an existing internal combustion–powered vehicle chassis saves you yet more building time and hassle, but still requires an extensive labor investment on your part and additional out-of-pocket cost. The advantage is that you not only have someone to call or write to for help, but better instructions typically make that help less necessary. Pre-fab parts also go a long way toward ensuring that your finished product looks like a professional job.

There's also a hidden problem with custom-, plan-, or kit-built electric vehicles; it occurs if you decide to sell your creation. With new and do-it-yourself conversions using internal combustion engine vehicles—even if quite radical in the electrical department—the prospective purchaser comes to look at it and the first impression is, "Yeah, it's a Ford" or "OK, it's a Honda" and so forth. The prospective buyer mentally catalogs the make and model of the body, then moves on. With custom-built electrics, you might deal with a lot of questions on the body, such as "How safe is it?" or "Does it leak much when it rains?" or "Will this crack here get much wider?" It can go on and on.

Building an EV from scratch today involves thoroughly planning what you are going to do in advance (Will you make or buy parts, and from what vendors? When will parts be needed? Where will assembly happen? Which subcontractors will help?), then pursuing your plan (leaving room for contingencies, problems, and any bargains that come your way). If you can scrounge, barter, or scavenge used parts, so much the better.

Custom-Built Electric Vehicles

The advantage to custom-building an electric vehicle is that you can "go where no one has gone before." In practical terms, it means you can build something like Don Moriarty's custom sport racer (see Figure 4-3), which is an excellent example of the meticulous attention to design and construction detail that results in winning entries. With the custom-built approach, ideal for high-speed or long-distance race vehicles, you are free to make the design and component trade-offs that optimize your vehicle in the direction of your choice. But this approach also assumes you have the skills, talent, resources, and money to accomplish it.

Electric Vehicles Built from Plans

The best example of building from plans is the Doran three-wheeler electric (shown front and rear in Figure 4-4), whose plans are typically offered by mail-order. With the Doran construction manual and plans you get excellent body, mechanical, and electrical instruction. They result in a less than 1,500-lb. vehicle (motorcycle classification) that goes 80 mph and has an around-town range of 60 miles from its recommended Prestolite 28-hp DC motor and 108-volt (nine 12-volt lead-acid) battery string. But to build even this small, simple vehicle requires mastery of fiberglass-over-foam-core body

Figure 4-3 Don Moriarty's custom sports racer electric vehicle.

construction techniques—or at least making haste slowly while you learn them—and is, again, not for everyone.

Electric Vehicles Built from Kits

The best example of building from kits is the Bradley GT II kit car (see Figure 4-5), of which there are several versions around. This combination of 96-volt electrics on a Volkswagen chassis with a lightweight body delivers impressive performance along with a snazzy, classic body-style, but you're going to have to part with a substantial amount of cash and labor before you make it happen. A kit-built EV can be a highly rewarding and satisfying showpiece project for those who have the experience, enthusiasm, and persistence to pull it off, but (for the third and final time) it's not for everyone.

Converting Existing Vehicles

Conversion is the best alternative because it costs less than either buying ready-made or building from scratch, takes only a little more time than buying ready-made, and is technically within everyone's reach (certainly with the help of a local mechanic, and absolutely with the help of an EV conversion shop).

Conversion is also easiest from the labor standpoint. You buy the existing internal combustion vehicle chassis you like (certain chassis types are easier and better to convert than others), put an electric motor in your chassis, and save a bundle. It's really quite simple; Chapter 10 covers the steps in detail.

To do a smart EV conversion, the first step is to buy a clean, straight, used, internal combustion vehicle chassis. A used model is also to your advantage (as you'll read in Chapter 5) because its already-broken-in parts are smooth and the friction losses are

FIGURE 4-4 Doran three-wheeler electric vehicle.

minimized. A vehicle from a salvage yard or a vehicle with a bad engine may not be the best choice because you do not know if the transmission, brakes, or other components and systems are satisfactory. Once you select the vehicle, then you add well-priced electrical parts or a whole kit from vendors you trust, and do as much of the simple labor as possible, while farming out the tough jobs (machining, bracket-making, etc.). Whether you do the work yourself and just subcontract a few jobs, or elect to have

FIGURE 4-5 Bradley GT II kit car electric vehicle.

someone handle the entire conversion for you, you can convert to an electric vehicle for a very attractive price compared to buying a new EV.

Converting Existing Vans

While large vans, such as Sacramento Municipal Utility District's (SMUD's) converted GMC van (shown in Figure 4-6), make great test-bed vehicles for the utility companies, they are heavy, more expensive to buy, take longer to convert, give less than adequate after-conversion performance, and cost you more to own and operate. For these reasons and others, you will never find an 8,000-lb. van recommended in this book as a potential conversion candidate. On the other hand, minivans—particularly the newer, lighter models—offer intriguing prospects for conversion, and you can look further into them as your needs require. But vans in general, even minivans, are usually more expensive, heavier, or take longer to convert than other chassis styles, so investigate before you invest.

Some Conversion Examples

There are a ton of vehicles that are immediately available for car conversions and it would triple the pages in this book to cover them all. We'll just look (for eye candy) at an electric Rolls Royce, but the principle is simple: *you can convert an existing car to something new!*

Figure 4-7 shows Paul Little's custom-built electric Rolls Royce with a standard (Centurion) body. The vehicle uses a 2,000-amp controller that moves this ship effortlessly around town or on the open road. He even keeps all of the amenities like power brakes, power windows, video screen, automatic transmission, and air

FIGURE 4-6 SMUD converted GMC van electric vehicle.

conditioning. These Rolls Royce–based Centurions are heavy duty and manage a lot of low-slung weight while offering that unsurpassed ride.

Typical Rolls Royce–based models include 26 100-amp hour batteries. By using our custom Royal Centurion body (from the years 1987 to 2001) with 24" wheels, allowances are gained for a complete front to rear battery tray. This creates a very low center of gravity and excellent front to rear weight distribution.

Bill Williams' 1971 Datsun 1200 conversion electric vehicle (shown front and rear in Figure 4-8) is also a 96-volt system, but takes advantage of a little more chassis room to conveniently stow four batteries in the front under-hood area and 12 batteries in the

FIGURE 4-7 Paul Little's electric Centurion built from an older Rolls Royce.

FIGURE 4-8 Bill Williams' 1971 Datsun 1200 conversion electric vehicle.

rear trunk area, keeping them out of the passenger compartment entirely. Notice the neatly laid out components and tied-off wiring runs in the engine compartment.

Paul Little's conversion of a Porsche 959 (shown in Figure 4-9) is today's state-of-the-art conversion. It offers breathtaking performance and range. Paul Little's typical Porsche includes 16–28 batteries depending on range, weight, and speed desired. These electric cars have respectable performance, going 0 to 60 in around five seconds. With a 2,000-amp controller with 269 volts, an 11" warp motor in fourth gear at a .98 ratio can achieve 1,200 ft. lbs. of torque!

FIGURE 4-9 Paul Little's Porsche 959 conversion electric vehicle.

FIGURE 4-10 Lyle Burresci's 1977 Plymouth Arrow rear compartment.

Lyle Burresci's 1977 Plymouth Arrow (Figure 4-10) was the basis for Mike Brown's conversion manual. It also uses a 96-volt system (six batteries in front, ten in the rear) and shows how good chassis selection, construction techniques, and "battery boxes" can transform a potential safety hazard. Figure 4-11 is from Joe Porcelli and Dave Kimmins of Operation Z. It shows their Nissan 280Z rear battery compartment. We'll talk and show you more of their conversion in Chapter 10.

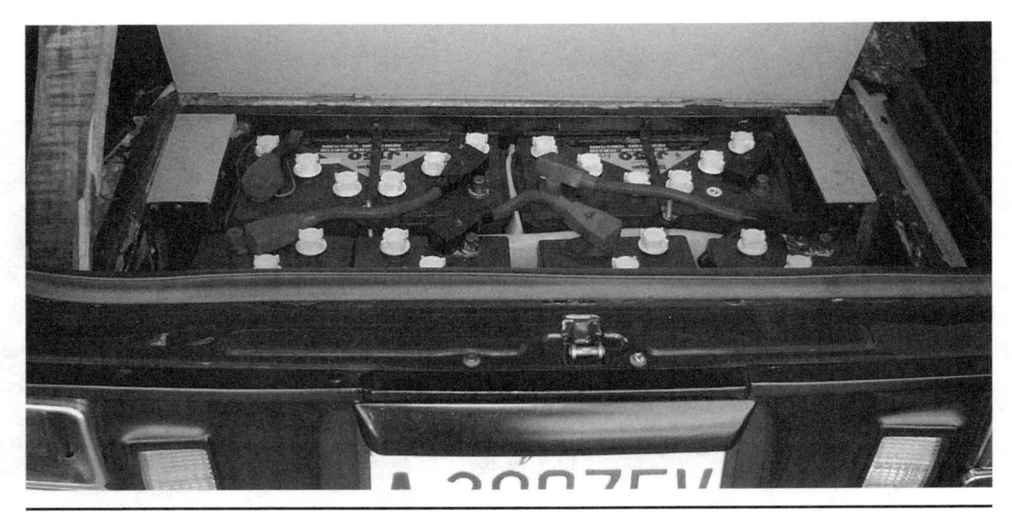

FIGURE 4-11 Joe Porcelli and Dave Kimmins of Operation Z's Nissan 280Z rear battery compartment.

Figure 4-12 Ken Watkin's S-10 conversion (Courtesy of Bob Batson at EV America).

Until recently, not many pickup truck conversions were done. But when EV converters took a closer look at the newer, lighter pickups, this trend reversed itself in a resounding way.

Electric Vehicles of America, Inc. was one of the first to convert small pickup trucks such as the S-10 (shown in Figure 4-12). Their design puts the batteries under the bed for a lower center of gravity and leaves the bed completely free for trips to the dump or local lumberyard. With a greater payload than most cars, trucks can have an even greater range. Pickup trucks are a great EV conversion choice—one you'll get a chance to examine in detail in Chapter 10.

Table 4-1 covers the other dimensions of the buy, convert, or build trade-off: what vehicles or parts are available, who you can get them from, where they are located, and when you will get/assemble them. There are not many places to buy new EVs today. If you don't reside in a place with easy access to an EV manufacturer or dealer, adding the weeks it might take to get your new EV shipped to you puts the conversion option—based on an internal combustion vehicle chassis obtained locally—into an even more attractive light. The weeks in the conversion option "When" boxes are

	Who	Where	When
Buy New	Independent Manufacturers & Dealers	Regional Locations	Buy – Now Shipping – Weeks
Buy Used	EAA Classifieds Manufacturers & Dealers	Anywhere	Buy – Now Shipping – Weeks
Convert New	New Auto Dealers Electric Vehicle Parts Dealers	Local Auto Dealer Mail Order EV Parts	Buy – Weeks Assy – 80-100 Hours
Convert Used	Used Auto Dealers Electric Vehicle Parts Dealers	Local Auto Dealer Main Order EV Parts	Buy – Weeks Assy – 100-200 Hours
Build From Scratch	Manufacturers & Dealers Electric Vehicle Parts Dealers	Local Raw Material Mail Order EV Parts	Buy – Months Assy – 200–??? Hours

Table 4-1 Electric Vehicle Purchase Decisions Compared

primarily consumed by finding (locally) the vehicle chassis you like. Conversion assembly time is then measured in hours or days. Starting with a used chassis will take you longer for cleaning, preparation, and so on. The months it takes to build from scratch are consumed by your finding the vehicle kit, chassis, and parts you like and having them shipped to you. Build-from-scratch assembly times start where conversion times leave off, and can range up into the thousands of hours, depending on how exotic you get.

Electric Vehicle Conversion Decisions

When you do an electric vehicle conversion today, you have many chassis choices: small car, sports car, compact car, crossover, SUV (not recommended), and small trucks. Small cars and sports cars may weigh less, but they also have less room for batteries and minimal payload. Most vehicle models gain 25—50 lbs. each year as the manufacturers add more auxiliaries or sound-deadening materials. Cars and crossovers have an advantage of less aerodynamic drag, something that SUVs and trucks do not have.

Whatever vehicle you choose, select one you personally like. Why spend $10,000 and approximately 100 hours on a vehicle that you do not like? This is a vehicle you want to proudly show.

You have additional choices of an AC (alternating current) or DC (direct current) drive system. Typically, AC systems are higher voltages and give faster performance, but they may cost 2–3 times more than a simpler and proven DC system. Each system requires a motor, controller, batteries, and charger. This section will look at these choices and prepare you for the guidance given by the rest of the book.

To help you make the best decisions, it is necessary to first identify your requirements. Are you looking for a small commuter vehicle for yourself or do you need an in-town vehicle for a small family? The number of people, performance, and range are all basic considerations. The cost of the drive system is directly related to your requirements. Therefore, if you are on a limited budget, it is critical to distinguish between "requirements" and "desires." Many people want the 0–60 mph in four seconds and the 150-mile range available from the Tesla electric vehicle, but not many are willing to pay the $100,000 required.

Your Chassis Makes a Difference

If you're going ahead with the conversion alternative, your most important choice is the chassis you select. Evaluate the vehicle. What is its curb weight? Payload? Can it handle the weight and space required for the batteries required? For example, a Mazda Miata cannot handle the 20 6-volt batteries often used in a pickup truck. Similarly, a 72-volt system used in an in-town commuter (such as the Geo Metro, which used to be popular with Solectria when they were building the Solectria Force) would have been very mediocre performance in a heavier and less aerodynamic pickup truck. Use the Internet to find curb weight and payload capacity of the chassis you're considering. Look at the manufacturer's specs on the driver's door when looking at a vehicle. Once you've identified a few potential candidate vehicles, make cardboard mock-ups of the batteries you might use. Then find a vehicle and see if they will fit.

Minimizing weight is always the number one objective of any EV conversion. When added to the criteria of minimizing conversion time and maximizing the odds for get-it-right-the-first-time success, the trade-off points squarely in the direction of the pickup truck. The van weighs more, and the car is typically the most time-consuming conversion (less room to mount EV options increases the problem of getting parts to fit).

The internal combustion vehicle pickup truck chassis is actually an outstanding electric vehicle conversion choice because:

- A single cab model pickup truck's curb weight, minus its internal combustion engine components, is typically not much more than a car, yet it has considerably more room to add the extra batteries that translate to better performance.
- A manual transmission, no-frills, 4- or 6-cylinder internal combustion engine pickup truck used to be one of the least expensive new or used conversion platforms you could buy.
- The additional battery weight presents no problems for the pickup's structure. Its sturdy frame is specifically designed to carry extra weight and extra or heavier springs are readily available if you need them.
- The pickup isolates the batteries from the passenger compartment very easily, an important safety criteria not found in car or van conversion platforms.
- The pickup is much roomier. The engine compartment and pickup box or bed space offer flexibility for your component design and layout, and a front-wheel-drive model gives you additional flexibility for battery mounting.
- A late-model, compact, or intermediate pickup offers frontal-area comparable to equivalent-sized cars, yet front grill and engine areas can be more easily blocked or covered to reduce wind resistance and engine compartment turbulence.

The pickup truck can be made into an instant hybrid—just load a portable emergency electrical generator and one or more five-gallon gas cans into the back (Figure 4-13).

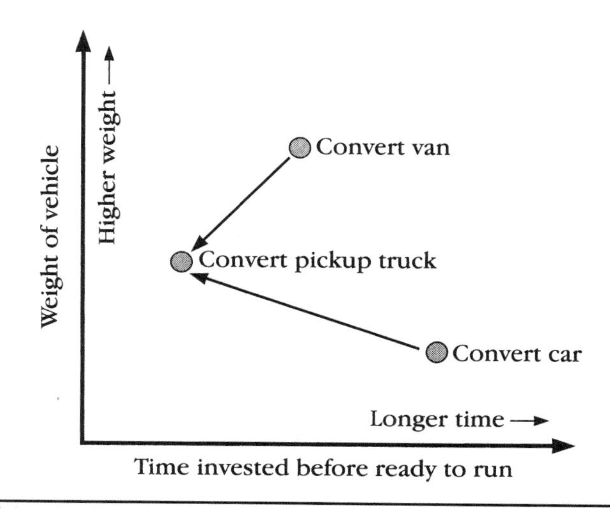

FIGURE 4-13 A pickup truck conversion is the best alternative over a car or van.

Your Batteries Make a Difference

Of all the pickup truck advantages, its extra room makes the biggest difference because it makes more space available to mount batteries. With car conversions, you must either choose a larger chassis or go to 8- or 12-volt batteries. With a pickup truck chassis, you can use more powerful 6-volt batteries to produce 120 volts and more. Both the 6- and 8-volt (more energy storage) and 12-volt (less total weight) options are available with the pickup conversion.

More batteries means higher voltage, which dramatically improves your performance (see Figure 4-14). Vintage EV designs from the 1970s found 72 volts acceptable. Today, you would be unhappy with anything less than 96 volts, and 120 volts—the setup used for Chapter 10's conversion—is even better. Beyond that, a 144-volt battery setup is still better for motors designed to use the extra voltage. Remember, you need to have enough voltage to push the current you need to get the torque you need to go the speed you want. It is usually more efficient to run higher RPMs than higher current to get the power you need. Both maximum RPM and maximum current must be kept within the motor's design spec. When the battery pack is more than 80 percent discharged is when you are most grateful for extra voltage.

A 120-volt battery string comprising 20 6-volt batteries (about 1,200 lbs.) typically delivers a top speed of 60 mph or more and a 60-mile range (at reduced speed) in a 3,000-lb. curb-weight pickup truck with 4-speed transmission. You might get more or less depending on your design and components. Jim Harris' Ford Ranger pickup conversion covered in Chapter 10 goes 75 mph, and his range was still increasing at press time. Increasing voltage will increase maximum motor speed and that equates to being able to drive faster. Increasing battery amp-hr will extend range. Adding batteries will add voltage and weight. But it is possible to increase voltage without changing weight by changing to a different battery.

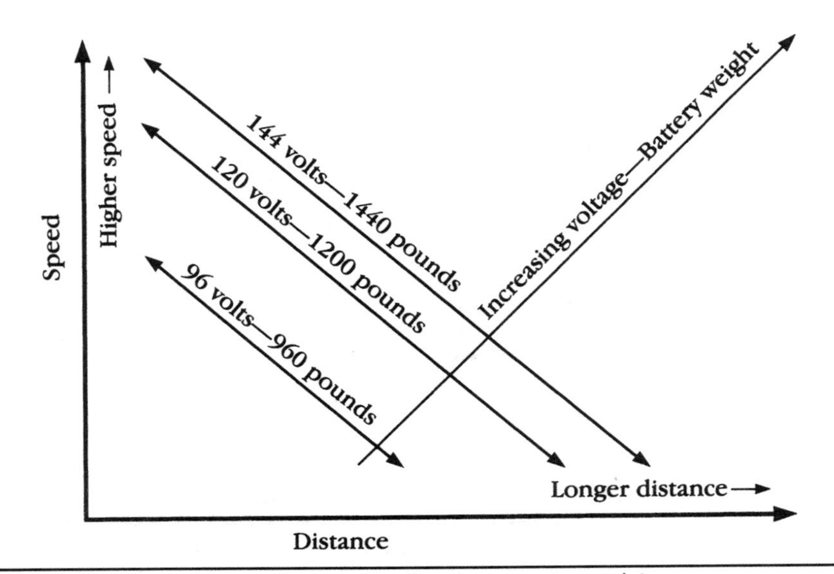

FIGURE 4-14 More batteries are always better than less—up to a point.

Table 4-2 presents all the dimensions of the conversion trade-off: chassis, motor, controller, batteries, and charger versus OK, ideal, best, and unlimited money alternatives. If money is no object, you customize your chassis, add the latest powerful AC induction motor with custom controller, and tool around the countryside powered by nickel metal hydride or lithium polymer batteries towing your power-boost trailer.

The rest of us have to take it a bit slower; our movement toward the best category can only proceed as fast as our pocketbooks allow: a pickup, readily-available series DC motor, Curtis PWM controller, lead-acid batteries, and an onboard 120-volt, 20-amp charger. Actually, anything from the OK or ideal column is acceptable. And while I am obviously prejudiced towards pickups, that doesn't mean you have to be.

The point is, use what's available. DC motors, Curtis controllers, and lead-acid batteries are with us in abundance today and at a reasonable price. While not state of the art, they are proven to work, available from many sources, and you can get lots of help if something goes wrong. By starting with known quantities, your chances of initial success and satisfaction are very high. You can get up and running quickly. When you're ready for your next conversion, you can experiment a bit, "push the outside of the envelope," etc. You know what it's all about, what's working, and where you'd like to make the changes.

The Procedure

Chapters 5 through 9 introduce you to chassis, motors, controllers, batteries, and chargers, and Chapter 12 provides you with some sources to get you started. In Chapter 10, you'll look over my shoulder while I convert a Ford Ranger, following step-by step-instructions you can adapt to nearly any conversion you want. Chapter 11 shows you how to maximize the enjoyment of your EV once it's up and running.

Use the Chapter 12 sources—don't just take my word for it. Join the EAA and subscribe to their newsletter. Read all the books, magazines, and research material you

What	OK	Ideal	Best	Unlimited money
Chassis	Van	Pickup truck	Car	Custom-built Starship Electrocruiser
Motor	DC series 19 hp	DC series 22 hp	DC compound 30 hp	ac induction 50 hp
Controller	Curtis MOSFET PWM	Custom IGBT PWM	Custom IGBT PWH + Regen	Custom ac IGBT PWM + Regen
Battery Pack	96 volts – 16ea 12 volt Lead-Acid	120 volts – 20ea 6 volt Lead-Acid	144 volts – 6 or 12 volt Lead-Acid	120+ volts – NiCd, NaS, Lithium Polymer
Charger	120 volts, 20 amps	120 volts, 20 amps + Onboard	240 volts, 50 amps + Onboard	240 volts, 50 amps + Power boost trailer

TABLE 4-2 Electric Vehicle Conversion Decisions Compared

can. There is also a lot of useful material online (in other words, surf the Net!). Go to meetings, shows, and rallies. Most of all, talk to people who have already done a conversion. If you listen to what they say, you will soon discover there are more opinions on how to do an electric vehicle conversion than there are snowflakes in the known universe. Then integrate all this information and make your own decisions. After you've done your first conversion, you'll notice a new phenomenon—people will start listening to you.

How Much Is This Going to Cost?

Now we come to the part where, as they say, the rubber meets the road—your wallet. Let's look at an actual quote, then add vehicle and battery costs and analyze the results. While you should not consider these costs the last word, you can consider them typical for today's EV conversion efforts. In any event, they will give you a good idea of what to expect for a 144V car system.

Notice also that the professionals such as Electric Vehicles of America, Inc. tell you what performance you can expect, when you're going to get it, how much it's going to cost, and how long the quoted prices are valid. Be sure you get the same information, in writing, out of any supplier you choose. It is important to select the components that will perform as a system. Don't expect to buy random components from the Internet and then get them to function properly as a system. In addition, professionals should be available after the sale to assist you with any problems. Bob Batson of Electric Vehicles of America, Inc. has stated that their services remain with the converted vehicle. So even the second, third, or fourth owner of his conversions is assured of his professional services.

Table 4-3 adds the pickup truck chassis and battery costs to the quote shown in Figure 4-15 (the Typical column); shows what savings you might expect with a used and older chassis (the Economy column); and shows what extra costs to expect when using the latest new chassis and a few extra bells and whistles (the High or custom column). The Typical column summarizes the 1987 Ford Ranger Pickup EV conversion detailed in Chapter 10.

Amounts you might obtain for selling off the internal combustion engine components were omitted from the comparisons; you can expect the vehicle costs to be lower if you do sell them.

Conversion for Fun and Profit

Darwin Gross drew a picture of a two-seater sports car EV on a napkin over lunch one day and said, "You could sell that for $4,995." My own scribbling on the napkin (aluminum tubular frame, plastic body, thin/hard high-pressure tires, no power steering, heater, optional fabric top, motor, controller, batteries, etc.) led me to a more sobering $9,995. But I'm talking about a TR3-sized sports car that could whip a TR3 à la Bertone's Blitz—a mouth-watering idea. In thinking more about it later, if a person applied the Dr. Paul MacCready technique and optimized on cost, such a vehicle is not only energy-efficient and high-performance, but very reachable. Think of it as a "poor man's Impact." And if a MacCready-style team was brought together to accomplish it, you'd have a working model on the streets within a year or so. Somebody's going to make a lot of money on this, or something just like it. You heard it here first.

Item	Economy $1500	Typical $3000	High $10,000	Customer Percent Allocated 28–31–57
Chasis Pickup truck	Earlier model used	Late model used	Last year new	8–8–4
Motor adapter plate Custom	$400 Local or do-it-yourself	$800 Professionally done	$800 Professionally done	15–16–10
Motor Advanced DC 22 hp	$800 Used	$1500 New	$1500 New	8–8–4
Controller Curtis PWM	$400 Used	$750 New	$750 New	12–19–14
Wiring & components Switchers, meters, wire	$600 Used	$1850 New	$2500 New	23–12–7
Battery pack 20ea 6 volt lead-acid	$1200 New	$1200 New	$1200 New	6–6–4
Charger 120V, WA, onboard	$300 Used	$550 New	$800 New	
Total	$5200	$9650	$17,550	

TABLE 4-3 Electric Vehicle Conversion Costs Compared

Motor, Controller, Battery Tests—Your Electric Car Will Run Just Fine!

Current technological capabilities should be sufficient to supply maximum continuous current at freeway speed up a 6 percent grade in a gear with enough torque to not to slow down. The Advanced DC 9" FB1-4001 electric motor can supply more power then an 8" 203-06-4001, but not at 96 volts where power output is about the same and the FB1 is running slower then the 203. At 144 volts you can push the current to get more torque from an FB1 that is beyond the continuous rating of an 203. Advanced DC also makes an 8" XP=1227A that is designed for 240 volts.

The tester of this study, Russ Lemons would not recommend use of the FB1 with battery voltage less then 132 volts or over 156 volts. Motor speed should not exceed 5500 rpm. Russ would also not recommend use of the 203 with battery voltage less than 96 volts or over 132 volts. Motor speed should not exceed 6500 rpm.

Even though the Curtis Controller can handle up to 500 amps, running at high currents is not good for the motor or the battery and will shorten their lives. It is better to slow down and drop down a gear then to overheat the motor. (Source: Russ Lemons)

ELECTRIC VEHICLES OF AMERICA, INC.
11 EAGLE TRACE, P. O. BOX 2037
WOLFEBORO, NH 03894
(603) 569-2100
FAX (603) 569-2900
EVAmerica@aol.com

144V SYSTEM—CAR PACKAGE USING 12V BATTERIES

QTY	DESCRIPTION	UNIT PRICE	TOTAL PRICE
	DRIVE SYSTEM		
1	FB1-4001A Advanced DC Motor with dual shaft 30 hp continuous 100 hp peak	$1,550.00	$1,550.00
1	Curtis 1231C-8601 On-Road EV Controller (96–144V) 500 Amp Limit	$1,495.00	$1,495.00
1	Aluminum Plate/heat sink compound/12V fan	$50.00	$50.00
1	PB-6 Curtis Potbox	$90.00	$90.00
2	Albright Contactors SW-200 (12V coil)	$150.00	$300.00
1	Adaptor Plate with Spacers (2) Manual Transmission–Clutchless	$400.00	$400.00
1	Motor Coupling (Aluminum) Manual Transmission–Clutchless	$325.00	$325.00
	BATTERY SYSTEM		
1	Zivan NG3 Charger 2800 watts 230VAC input 144VDC output	$980.00	$980.00
24	Battery Terminal Protective Covers (Red & Black)	$1.50	$36.00
50	2/0 Cable—25 ft Black, 25 ft Red	$3.25	$162.50
40	Heavy Duty Magna Lugs Plated (36+4)	$2.50	$100.00
5	ft Heat Shrink with sealant	$6.00	$30.00
	INSTRUMENTATION		
1	80–180 Westberg Voltmeter	$65.00	$65.00
1	0–500 Amp Westberg Ammeter	$65.00	$65.00
1	50 mV Shunt	$30.00	$30.00
	POWER BRAKES		
1	Vacuum Pump (12V)	$225.00	$225.00
1	Vacuum Switch	$135.00	$135.00
1	In-line Fuseholders	$5.00	$5.00
	SAFETY		
1	Littelfuse L25S-400	$55.00	$55.00
1	Littelfuse holder	$25.00	$25.00
1	Littelfuse KLK control fuse and holder	$25.00	$20.00

FIGURE 4-15 Sample quotation from Electric Vehicles of America, Inc.

1	Pair Anderson connectors SBX-350	$64.00	$64.00
1	Fuseholder (4)—Control Board	$15.00	$15.00
1	First Inertia Switch–Auto Shutoff (12V Sys) of Power System upon Impact	$45.00	$45.00
	TECHNICAL ASSISTANCE		
A/R	EVA calculations		N/C
1	DVD "Safety First" & S10 Conversion Video		N/C
1	EVA Installation Manual Includes schematics, drawings, etc.		N/C
A/R	On-Line Assistance @ EVAmerica@aol.com		N/C
	1 year Subscription to EVAmerica		N/C
	SUBTOTAL		$6,267.50
	EVAmerica Coupons–Package		–$167.50
	TOTAL (Shipping—not included)		$6,100.00

	OPTIONAL EV COMPONENTS TO REPLACE OR SUPPLEMENT ABOVE		
	DRIVE SYSTEM		
1	203-06-4001A Advanced DC Motor with dual shaft	$1,350.00	$1,350.00
1	1221C-7401 Curtis Controller (72–120V) 400 Amp Limit	$1,075.00	$1,075.00
1	Motor Mount Assembly (RWD)	$180.00	$180.00
	INSTRUMENTATION		
1	0–400 Amp Ammeter Westberg	$65.00	$65.00
	POWER BRAKES		
1	Vacuum Gauge (Initial Set-up)	$15.00	$15.00
	SAFETY		
1	Astrodyne DC-DC Converter w/ relay SB-50 72–132VDC Input 13–14VDC Output Recommended for headlights, etc.	$175.00	$175.00
1	Zivan DC-DC Converter 144VDC Input 14VDC Output Recommended for headlights, etc.	$500.00	$500.00
1	Pair Anderson connectors SB-50	$20.00	$20.00
1	Littelfuse L25S-400–Spare	$55.00	$55.00
14	Ft - 1 1/2 inch clear vinyl hose for 2/0 cable protection	$1.50	$21.00
10	Insulated Metal Clamps for Vinyl Hose	$1.00	$10.00

Figure 4-15 (continued)

CHAPTER 5

Chassis and Design

A 20-hp electric motor will easily push your 4,000-lb. vehicle at 50 mph.

The chassis is the foundation of your electric vehicle conversion. While you might never build your own chassis from scratch, there are fundamental chassis principles that can help you with any EV conversion or purchase—things that never come up when using internal combustion engine vehicles—such as the influence of weight, aerodynamic drag, rolling resistance, and drivetrains.

This chapter will step you through the process of optimizing, designing, and buying your own electric vehicle. You'll become familiar with the chassis trade-offs involved in optimizing your EV conversion. Then you'll design your EV conversion to be sure the components you've selected accomplish what you want to do. When you have figured out what's important to you and verified your design will do what you want, you'll look at the process of buying your chassis.

Knowledge of all these steps will help you immediately (when reading about Chapter 10's pickup truck conversion), and eventually (when picking the best EV chassis for yourself). The principles are universal, and you can apply them whether buying, building, or converting.

Choose the Best Chassis for Your EV

The chassis you pick is the foundation for your EV—choose it wisely. That's the message of this chapter in a nutshell. Since you're likely to be converting rather than building from scratch, there's not a lot you can do after you've made your chassis selection. The secret is to ask yourself the right questions and be clear about what you want to accomplish before you make your selection.

Like a youngster's soapbox derby racer, you want a chassis with an aerodynamic shape and thin wheels, so that you can just give a shove and it runs almost forever. But its frame must also be big enough and strong enough to carry you and your passengers along with the motor, drivetrain/controller, and batteries. In addition, if you want to drive it on the highway, federal and state laws require it to be roadworthy and adhere to certain safety standards.

The first step is to know your options. Your EV should be as light in weight as possible; streamlined, with its body optimized for minimum drag; optimized for minimum rolling resistance from its tires, brakes, and steering; and optimized for minimum drivetrain losses. The motor-drivetrain-battery combination must match the body style you've selected. It must also be capable of accomplishing the task most important to you: high speed, long range, or a utility commuter vehicle midway between the two.

So step two is to design for the capability that you want. Your EV's weight, motor and battery placement, aerodynamics, rolling resistance, handling, gearing, and safety features must also meet your needs. You now have a plan.

Step three is to execute your plan—to buy the chassis that meets your needs. At its heart this is a process no different from any other vehicle purchase you've ever made, except that the best solution to your needs might be a vehicle that the owner or dealer can't wait to get rid of—one with a gas-guzzling, diesel, or otherwise polluting engine—so the tables are completely turned from a normal buying situation.

Used is usually the least expensive, but don't go for something *too* used. You want to feel confident about converting the vehicle you choose before you leave the lot. If it's too small or cramped to fit all the electrical parts, let alone the batteries, you know you have a problem. Or if it's particularly dirty, greasy, or rusty, you need to think twice.

Figure 5-1 gives you the quick picture. The rest of the chapter covers the details. Let's get started.

Optimize Your EV

Optimizing is always step number one. Even if you go out to buy your electric vehicle ready-made, you still want to know what kind of a job has been done so you can decide if you're getting the best model for you. In all other cases, you'll be doing the optimizing—either by the choices you make up front in chassis selection or by your conscious optimizing decisions later on. In this section, you'll be looking to minimize the following resistance factors:

- Weight and climbing and acceleration
- Aerodynamic drag and wind
- Rolling and cornering resistance
- Drivetrain system

You'll look at equations that define each of these factors, and construct a table of real values normalized for a 1,000-lb. vehicle and nine specific vehicle speeds. These values should be handy regardless of what you do later—just multiply by your own EV's weight ratio and use directly, or interpolate between the speed values.

You'll immediately see a number of values reassembled in the design section of this chapter, when a real vehicle's torque requirements are calculated to see if the torque available from the electric motor and drivetrain selected is up to the task. This design process can be infinitely adapted and applied to whatever EV you have.

Conventions and Formulas

This book uses the U.S. automotive convention of miles, miles per hour, feet per second, pounds, pound-feet, etc., rather than the kilometers, newton-meters, etc., in common use overseas. Any formulas borrowed from the Bosch handbook have been converted to U.S. units. Speaking of formulas, you will find the following 13 useful; they have been grouped in one section for your convenience:

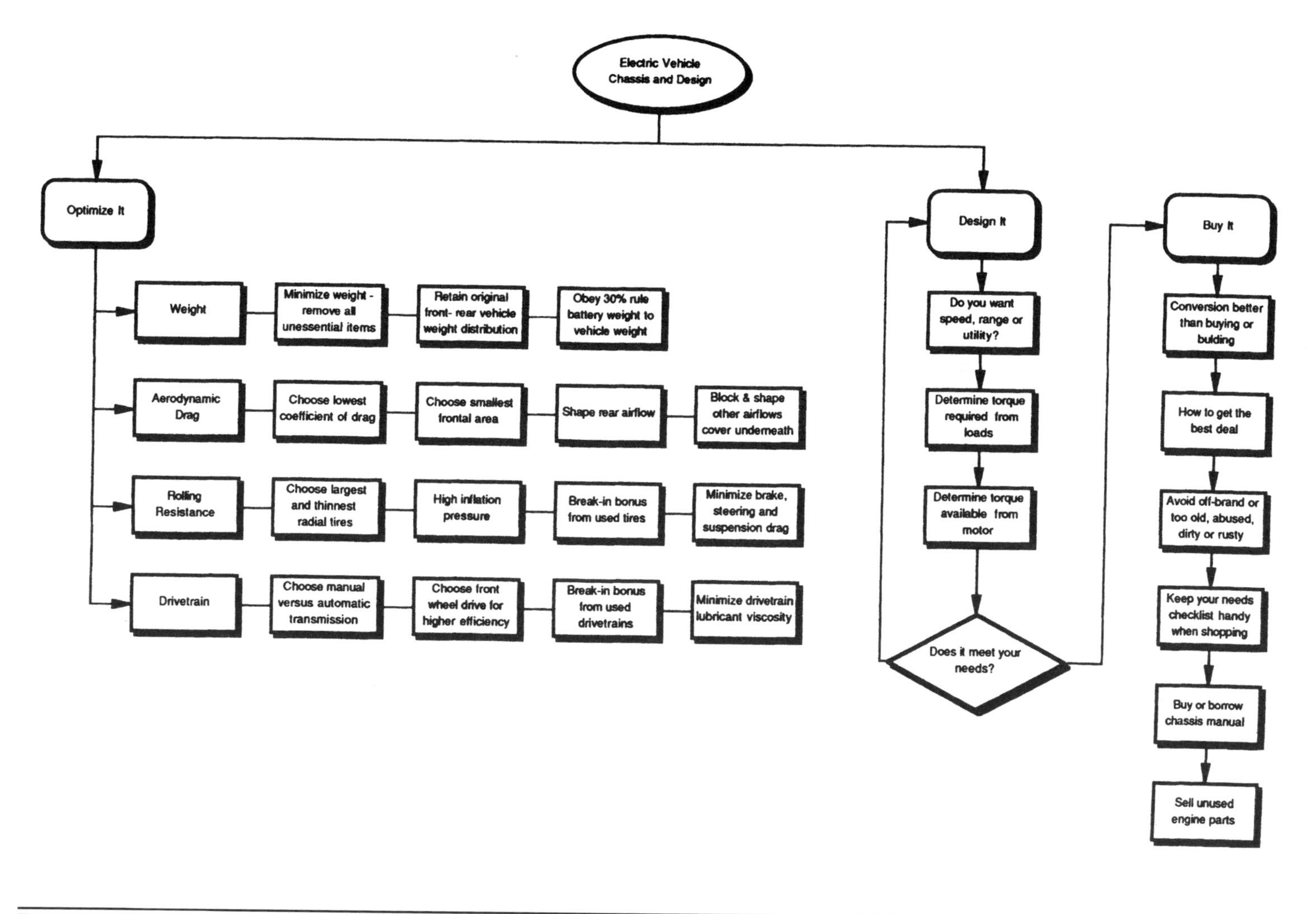

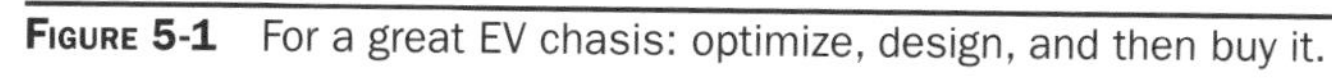
FIGURE 5-1 For a great EV chasis: optimize, design, and then buy it.

1. Power (ft-lb/sec) = Torque (ft-lb) × Speed (radians/sec) = Force in feet per second (FV)
2. 1 Horsepower (hp) = 550 ft-lb/sec

Applying this to equation 1 gives you:

3. Horsepower (hp) =FV/550
   where V is speed expressed in feet/sec.
4. 88 feet/sec = 60 mph
   Multiply feet/sec by (60 × 60)/5280 to get mph.
5. Horsepower (hp) = FV/375
   where V is speed expressed in mph and F is in pounds.
6. Horsepower (hp) = (Torque × RPM)/5252 = 2 π/60 × FV/550
7. Wheel RPM = (mph × Revolutions/mile)/60
8. Power (kW) = 0.7457 × hp
9. The standard gravitational constant (g) = 32.16 ft/sec^2 or almost 22 mph/sec
10. Weight (W) = Mass (M) × g/32.16
    For the rest of this book, we will refer to a vehicle's mass as its "weight."
11. Torque = (F(5280/2π))/(Revolutions/mile) = 840.34 × F / (revolutions/mile)
    Revolutions/mile refers to how many times a tire rotates per mile.
12. $\text{Torque}_{\text{wheel}}$ = $\text{Torque}_{\text{motor}}$ × (overall gear ratio × overall drivetrain efficiency)
13. $\text{Speed}_{\text{vehicle}}$ (in mph) = ($\text{RPM}_{\text{motor}}$ × 60) / (overall gear ratio × revolutions/mile)

It Ain't Heavy, It's My EV

In real estate they say the three most important things are location, location, and location. In electric vehicle conversion the three most important things are weight, weight, and weight. In this section, you'll be taking a closer look at the items in the *Weight* row of Figure 5-1 and the supreme importance of minimizing weight in your EV.

Remove All Unessential Weight

You don't want to carry around a lot of unnecessary weight. But, unless you're starting with a build-from-scratch design, you're inheriting the end result of someone else's weight trade-offs. This means you need to carefully go over everything with regard to its weight versus its value at three different times.

Before You Purchase the Conversion Vehicle

Think about the vehicle's weight-reduction potential before you buy it. Is it going to be easy (pickup) or difficult (van) to get the extra weight out? What about hidden-agenda items? Has a previous accident resulted in a filled fender on your prospective purchase? (Take along a magnet during your exam.) Does its construction lead to ease of weight removal or substitution of lighter-weight parts later? Think about these factors as you look.

During Conversion

As you remove the internal combustion engine parts, it's likely you'll discover additional parts that you hadn't seen or thought of taking out before. Parts snuggled up against the firewall or mounted low on the fenders are sometimes nearly invisible in a crowded and/or dirty engine compartment. Get rid of all unnecessary weight, but do exercise logic and common sense in your weight-reduction quest. Substituting a lighter-weight cosmetic body part is a great idea; drilling holes in a load-bearing structural frame member is not.

After Conversion

Break your nasty internal combustion engine vehicle habits. Toss out all extras that you might have continued to carry, including spare tire and tools.

After all your work, give yourself a pat on the back. You've probably removed from 400 to 800 lbs. or more from a freshly cleaned-up former internal combustion engine vehicle chassis that's soon to become a lean and mean EV machine. The reason for all your work is simple—weight affects every aspect of an EV's performance: acceleration, climbing, speed, and range.

Weight and Acceleration

Let's see exactly how weight affects acceleration. When Sir Issac Newton was bonked on the head with an apple, he was allegedly pondering one of the basic relationships of nature—his Second Law: F = Ma; or force (F) equals mass (M) times acceleration (a). For EV purposes, it can be rewritten as

$$F_a = C_i W a$$

where F_a is acceleration force in pounds, W is vehicle mass in pounds, a is acceleration in mph/second, and C_i is a units conversion factor that also accounts for the added inertia of the vehicle's rotating parts. The force required to get the vehicle going varies directly with the vehicle's weight; twice the weight means twice as much force is required.

C_i, the mass factor that represents the inertia of the vehicle's rotating masses (wheels, drivetrain, flywheel, clutch, motor armature, and other rotating parts), is given by

$$C_i = I + 0.04 + 0.0025(N_c)^2$$

where N_c represents the combined ratio of the transmission and final drive. The mass factor depends upon the gear in which you are operating. For internal combustion engine vehicles, the mass factor is typically: high gear = 1.1; 3rd gear = 1.2; 2nd gear = 1.5; and 1st gear = 2.4. For EVs, where a portion of the drivetrain and weight has typically been removed or lightened, it is typically 1.06 to 1.2.

Table 5-1 shows the acceleration force F_a, for three different values of C_i, for ten different values of acceleration *a*, and for a vehicle weight of 1,000 lbs. The factor *a* is the acceleration expressed in ft/sec^2, rather than in mph/second = 21.95 = 32.2 × (3600/5280)—used only in the formula (because acceleration expressed in mph/second is a much more convenient and familiar figure to work with). Notice that an acceleration of 10 mph/sec, an amount that takes you from zero to 60 mph in 6 seconds nominally requires extra force of 500 lbs.; 5 mph/sec, moving from zero to 50 mph in 10 seconds, requires 250 pounds.

a (in mph/sec)	a'= a/21.95	F_a (in pounds) C_i = 1.06	F_a (in pounds) C_i = 1.1	F_a (in pounds) C_i = 1.2
1	0.046	48.3	50.1	54.7
2	0.091	96.6	100.2	109.3
3	0.137	144.8	150.3	164.0
4	0.182	193.1	200.4	218.6
5	0.228	241.4	250.5	273.3
6	0.273	289.7	300.6	328.0
7	0.319	338.0	350.7	382.6
8	0.364	386.3	400.8	437.3
9	0.410	434.5	450.9	491.9
10	0.455	482.8	501.0	546.6

TABLE 5-1 Acceleration Force, F_a (in pounds), for Different Values of C_i

To use Table 5-1 with your EV, multiply by the ratio of your vehicle weight and use the C_i = 1.06 column for lighter vehicles and C_i = 1.2 column for heavier ones. For example, the 3,800-lb. Ford Ranger pickup truck of Chapter 10 would require 5 mph/sec = 3.8 × 273.3 = 1038.5 lbs.

Weight and Climbing

When you go hill climbing, you add another force:

$$F_h = W\sin\phi$$

where F_h is hill-climbing force, W is vehicle weight in pounds, and ϕ is angle of incline as shown in Figure 5-2. The degree of the *incline*—the way hills or inclines are commonly referred to—is different from the *angle* of the incline, but Figure 5-2 should clear up any confusion for you. Notice that sin ϕ varies from 0 at no incline (no effect) to 1 at 90 degrees; in other words, the full weight of the vehicle is trying to pull it back down the incline. Again, weight is directly involved, acted upon this time by the steepness of the hill.

$$\text{Degree of incline} = 1\% = \frac{1 \text{ foot}}{100 \text{ feet}} = \frac{\text{Rise}}{\text{Run}}$$

$$\text{Angle of incline, } \varnothing = \text{Arc tan}\frac{\text{Rise}}{\text{Run}} = \text{Arc tan } 0.01 = \text{about 0 degrees 34 minutes}$$

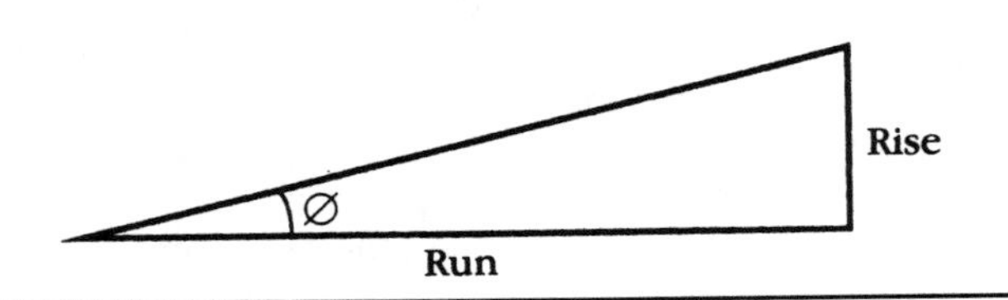

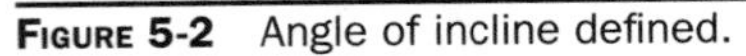

FIGURE 5-2 Angle of incline defined.

Degree of incline	Incline angle θ	sin θ	F_h (in pounds)	a (in mph/sec)
1%	0° 34'	0.00989	9.9	
2%	1° 9'	0.02007	20.1	
3%	1° 43'	0.02996	29.6	
4%	2° 17'	0.04013	40.1	
5%	2° 52'	0.05001	50.0	1
6%	3° 26'	0.05989	59.9	
8%	4° 34'	0.07062	79.6	
10%	5° 43'	0.09961	99.6	2
15%	8° 32'	0.14838	148.4	3
20%	11° 19'	0.19623	196.2	4
25%	14° 2'	0.24249	242.5	5
30%	16° 42'	0.28736	287.4	6
35%	19° 17'	0.33024	330.2	
40%	21° 48'	0.37137	371.4	
45%	24° 14'	0.41045	410.5	

TABLE 5-2 Hill-Climbing Force F_h for 15 Different Values of Incline

Table 5-2 shows the hill-climbing force F_h, for 15 different incline values for a vehicle weight of 1,000 lbs. Notice that the tractive force required for acceleration of 1 mph/sec equals that required for hill-climbing of a 5 percent incline, 2 mph/sec for 10 percent incline, etc., on up through a 30 percent incline. This handy relationship will be used later in the design section.

To use Table 5-2 with your EV, multiply by the ratio of your vehicle weight. For example, the 3,800-lb. Ford Ranger pickup truck of Chapter 10 going up a 10 percent incline would require 3.8 × 99.6 = 378.5 lbs.

Weight Affects Speed

Although speed also involves other factors, it's definitely related to weight. Horsepower and torque are related to speed per equation 3:

$$hp = FV/550$$

where hp is motor horsepower, F is force in pounds, and V is speed in ft/sec. Armed with this information, Newton's Second Law equation can be rearranged as

$$a = (1/M) \times F$$

and because $M = W/g$ (10) and $F = (550 \times hp)/V$, they can be substituted to yield

$$a = 550(g/V)(hp/W)$$

Finally, a and V can be interchanged to give

$$V = 550(g/a)(hp/W)$$

where V is the vehicle speed in ft/sec, W is the vehicle weight in pounds, g is the gravitational constant 32.2 ft/sec², and the other factors you've already met. For any given acceleration, as weight goes up, speed goes down because they are inversely proportional.

Weight Affects Range

Distance is simply speed multiplied by time:

$$D = Vt; \text{ therefore } D = 550(g/a)(hp/W)t$$

So weight again enters the picture. For any fixed amount of energy you are carrying onboard your vehicle, you will go farther if you take longer (drive at a slower speed) or carry less weight. You already encountered the practical results of this trade-off in Figure 4-14 of Chapter 4.

Besides the primary task of eliminating all unnecessary weight, there are two other important weight-related factors to keep in mind when doing EV conversions: front-to-rear weight distribution, and the 30 percent rule.

Remove the Weight but Keep Your Balance

Always focus on keeping your vehicle's front-to-rear weight distribution intact and not exceeding its total chassis and front/rear axle weight loading specifications.

Figure 5-3 shows the magnitude of your problem for a Ford Ranger pickup truck similar to the one used in Chapter 10's conversion. You have pulled out 600 lbs. in engine, fuel, exhaust, emission, ignition starter, and heating/cooling systems. But you're going to be putting 1,400 lbs. back in, including 1,200 lbs. of batteries (20 at about 60 lbs. each). How do you handle it?

Table 5-3 provides the answers. Notice the first row shows the 3,000-lb. curb weight normally distributed 60 percent front (1,800 lbs.) and 40 percent rear (1,200 lbs.) with a 1,200-lb. payload capacity. The second row shows that most of the weight

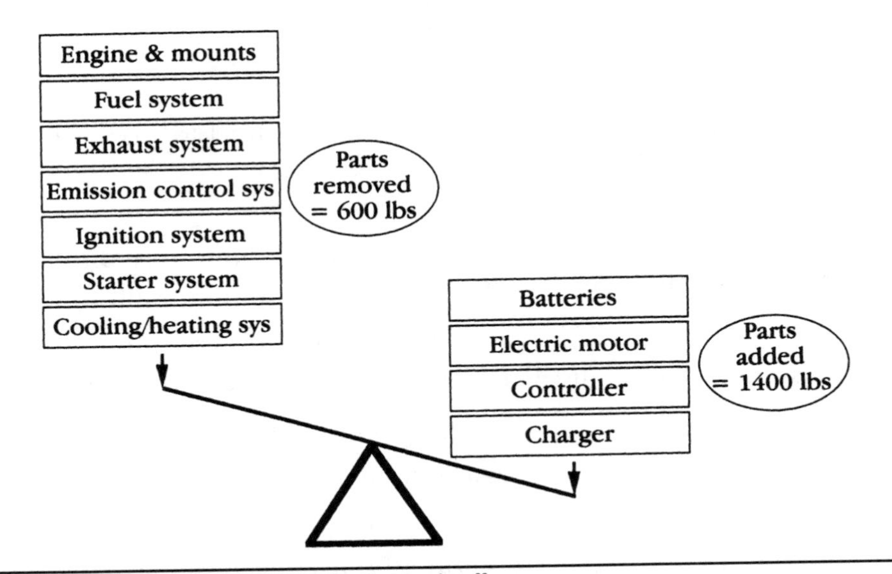

FIGURE 5-3 Front-to-rear weight distribution trade-off.

Item	Curb weight (lbs)	Front axle weight (lbs)	Rear axle weight (lbs)	Payload weight (lbs)
Ford Ranger pickup before conversion	3000	1800	1200	1200
Less IC engine and system parts	<600>	<500>	<100>	
Subtotal before conversion	2400	1300	1100	
Plus electric vehicle batteries, motor, etc.	1400	400	1000	
Ford Ranger pickup after conversion	3800	1700	2100	400
Battery weight 20ea 6-volt @ 60 lbs	1200			
Ratio battery weight to vehicle weight	32%			

TABLE 5-3 Electric Vehicle Conversion Weights Compared

you took out came from in and around the engine compartment—the 600 lbs. you removed took 500 lbs. off the front axle and 100 lbs. off the rear axle. The secret is to put the weight back in a reasonably balanced fashion. This is accomplished by mounting four of the batteries—approximately 240 lbs.—up front in the engine compartment along with the motor, controller, and charger. This puts about 400 lbs. up front and about 1,000 lbs. worth of batteries in the rear. The fifth row shows the results. You're up to a curb weight of 3,800 lbs. with a 1,700- to 2,100-lb. front-to-rear weight distribution, but you're still inside the GVWR, front/rear GAWR, weight distribution, and payload specifications. Furthermore, when you go out and drive the vehicle, it exhibits the same steering, braking, and handling capability that it had before the conversion.

As long as you keep within your vehicle's original internal combustion engine weight loading and distribution specifications, suspension and other support systems will never notice that you've changed what's under the hood. Overloading your EV chassis makes no more sense than overloading your internal combustion engine vehicle chassis. The best and safest solution is to get another larger or more heavy duty chassis.

A postscript: Some owners actually prefer to adjust the shocks and springs of their conversion vehicle at this point to give a slightly firmer ride, or (in the case of pickup truck owners) to return it to its previous firmer ride.

Remember the 30 Percent Rule

The "30 percent or greater" rule of thumb (battery weight should be at least 30 percent of gross vehicle weight when using lead-acid batteries) is a very useful target to shoot for in an EV conversion. You'll want to do even better if you're optimizing for either high-speed or long-range performance goals. Table 5-3 shows that battery weight was 32 percent of gross vehicle weight for this conversion. Notice that if you opt for a

144-volt battery system (adding four more batteries and 240 lbs. additional weight), the ratio goes up to 1440/4040 or 36 percent. Going the other way (taking out four batteries and 240 lbs.), the ratio drops to 960/3560 or 30 percent for a 96-volt system. Taking out four more batteries and going to a 72-volt system, the ratio drops to 720/3320 or 22 percent. The rule of thumb proves to be correct in this case, because you'd be unhappy with the performance of a 72-volt system in this vehicle; even 96 volts is marginal.

Streamline Your EV Thinking

Until fairly recently, most of the automobile industry's wedge-front designs, while attractive, are actually 180 degrees away from aerodynamic streamlining. Look at nature's finest and most common example, the falling raindrop: rounded and bulbous in front, it tapers to a point at its rear—the optimum aerodynamic shape. In addition, new bicycle-racing helmets adhere perfectly to this principle.

While airplanes, submarines, and bullet trains have for decades incorporated the raindrop's example into their designs, automakers' design shops have eschewed this idea as unappealing to the public's taste. With plenty of internal combustion engine horsepower at their disposal, they didn't need aerodynamics, they needed style. Because batteries provide only 1 percent as much power per weight as gasoline, you and your EV do need aerodynamic awareness.

In this section, you'll look at the Aerodynamic Drag row of Figure 5-1 and learn about the factors that come with the turf when you select your conversion vehicle, and the items that you can change to help any EV conversion slip through the air more efficiently.

Aerodynamic Drag Force Defined

Mike Kimbell, an EV consultant, said it best: "Below 30 mph you could put an electric motor on a brick and never notice the difference." The reason is simple: aerodynamic drag force varies with the square of the speed. If you're not moving, there's no drag at all. Once you get rolling it builds up rapidly and soon swamps all other factors. Let's see exactly how.

The aerodynamic drag force can be expressed as

$$F_d = (C_d A V^2)/391$$

where F_d is the aerodynamic drag force in lbs., C_d is the coefficient of drag of your vehicle, A is its frontal area in square feet, and V is the vehicle speed in mph. To minimize drag for any given speed you must minimize C_d, the coefficient of drag, and A, its frontal area.

Choose the Lowest Coefficient of Drag

The coefficient of drag, C_d, has to do with streamlining and air turbulence flows around your vehicle, characteristics that are inherent in the shape and design of the conversion vehicle you choose. C_d is not easily affected or changed later, so if you're optimizing for either high-speed or long-range performance goals, it's important that you keep this critical performance factor foremost in your mind when selecting your conversion vehicle. Figure 5-4 shows the value of C_d for different shapes and types of vehicles. Notice that C_d has declined significantly with the passage of time—the 1920s Ford sedan

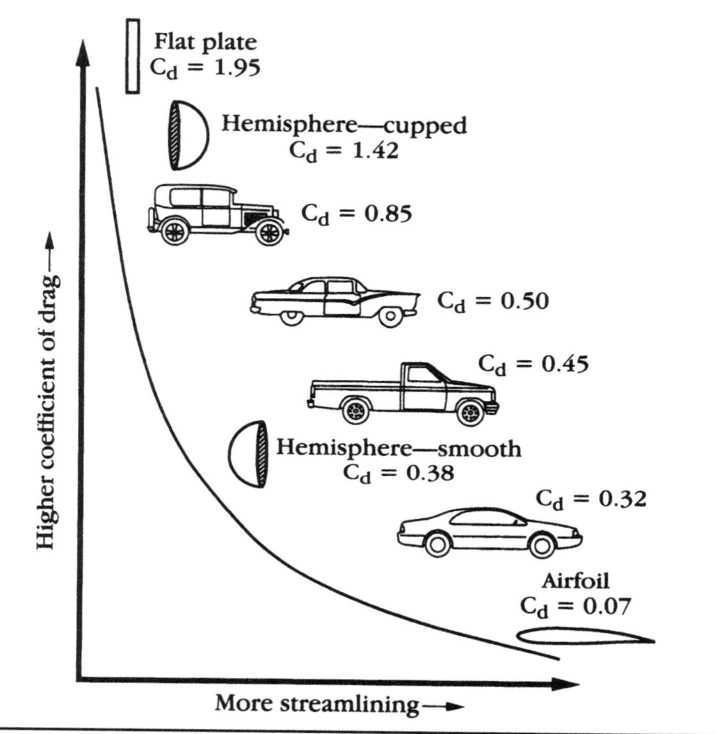

FIGURE 5-4 Coefficient of drag summary for different vehicle types.

had a C_d around 0.85; today's Ford Taurus has a C_d of 0.32. The values of C_d for typical late 1980s cars, trucks, and vans are

- Cars = 0.30 to 0.35
- Vans = 0.33 to 0.35
- Pickup trucks = 0.42 to 0.46

While car C_d values have typically declined from 0.5 in the 1950s, to 0.42 in the 1970s, to 0.32 today, don't be misled. That snazzy contemporary open cockpit roadster can still have a C_d of 0.6. That's because C_d has everything to do with air turbulence caused by open windows, cockpits, and pickup box areas, not with streamlining alone.

Table 5-4 shows how different areas—in this case taken for a 1970s-vintage car—contribute to the C_d of a vehicle. Contrary to what you might think, the body's rear area contributes more than 33 percent of C_d by itself, followed by the wheel wells at 21 percent, the underbody area at 14 percent, the front body area at 12 percent, projections (minors, drip rails, window recesses) at 7 percent, and engine compartment and skin friction at 6 percent each. That's why the C_d of General Motors' highly optimized Impact electric vehicle is 0.19—it has a finely sculptured rear, sculptured or covered wheel wells with thin tires, an enclosed underbody, a low nose with highly sloping windshield

Car Area	C_d Value	Percentage of total
Body–Rear	0.14	33.3
Wheel wells	0.09	21.4
Body–Under	0.06	14.3
Body–Front	0.05	11.9
Projections and indentations	0.03	7.1
Engine compartment	0.025	6.0
Body–Skin friction	0.025	6.0
Total	0.42	100.0

TABLE 5-4 Contribution of Different Car Areas to Overall C_d for a Typical 1970s-Vintage Car

and low ground clearance, no projections, and only two small openings to the engine compartment.

Choose the Smallest Frontal Area

The frontal area (A) for typical late-model cars, trucks, and vans is in the 18- to 24-square-foot range. A 4-foot by 8-foot sheet of plywood held up vertically in front of your vehicle would have a frontal area of 32 square feet. It has to do with the effective area your vehicle presents to the onrushing air stream. Frontal area is also not easily affected or changed later, and there's not too much you can do to significantly minimize it except choose a different vehicle body type. If you're optimizing for either high-speed or long-range performance, keep this critical performance factor in your mind when selecting your conversion vehicle.

Relative Wind Contributes to Aerodynamic Drag

Drag force is measured nominally at 60 degrees Fahrenheit and a barometric pressure of 30 inches of mercury in still air. Normally, those are adequate assumptions for most calculations. But very few locations have still air, so an additional drag component due to relative wind velocity has to be added to your aerodynamic drag force calculation. This is the additional wind drag pushing against the vehicle from the random local winds. The equation defining the relative wind factor, C_w, is

$$C_w = (0.98(w/V)^2 + 0.63(w/V))C_{rw} - 0.40(w/V)$$

where w is the average wind speed of the area in mph, V is the vehicle speed, and C_{rw} is a relative wind coefficient that is approximately 1.4 for typical sedan shapes, 1.2 for more streamlined vehicles, and 1.6 for vehicles displaying more turbulence or sedans driven with their windows open.

Table 5-5 shows C_w calculated for seven different vehicle speeds—assuming the United States average value of 7.5 mph for the average wind speed—for the three different C_{rw} values.

To get your total aerodynamic drag force (still plus relative wind), use the following formula:

$$F_{td} = F_d + C_w F_d \text{ or } F_d(1 + C_w)$$

C_{rw} at average	C_w factor	C_w factor	C_w factor	C_w factor	C_w factor	C_w factor	C_w factor
wind= 7.5 mph	at V= 5 mph	at V= 10 mph	at V= 20 mph	at V= 30 mph	at V= 45 mph	at V= 60 mph	at V= 75 mph
1.2	3.180	0.929	0.299	0.163	0.159	0.063	0.047
1.4 avg sedan	3.810	1.133	0.374	0.206	0.185	0.082	0.062
1.6	4.440	1.338	0.449	0.250	0.212	0.101	0.076

Table 5-5 Relative Wind Factor C_w at Different Vehicle Speeds for Three C_{rw} Values

Aerodynamic Drag Force Data You Can Use

Table 5-6 puts the C_d and A values for actual vehicles together and calculates their drag force for seven different vehicle speeds. Notice that drag force is lowest on a small car and greatest on the small pickup, but the small car might not have the room to mount the batteries to deliver the performance that you need. Notice also that an open cockpit roadster, even though it has a small frontal area A, has drag force identical to the pickup truck.

To use Table 5-5 and Table 5-6 with your EV, pick out your vehicle type in Table 5-6, then multiply its drag force number by the relative wind factor at the identical vehicle speed using the appropriate C_{rw} row for your vehicle type. For example, the 3,800-lb. Ford Ranger pickup truck of Chapter 10 has a drag force of 24.86 lbs. at 30 mph using Table 5-6. Multiplying by the relative wind factor of 0.250 from the bottom row (C_{rw} = 1.6) of Table 5-5 gives you 6.22 pounds. Your total aerodynamic drag forced is then 24.86 + 6.22 or 31.08 pounds.

Shape Rear Airflow

If you've seen a movie of a wind tunnel test with smoke added to make the air currents visible, you've noticed a vortex or turbulence area at the rear of most vehicles tested. Those without access to wind tunnels notice the same effect when a semitrailer truck blows past you on the highway.

As with the falling raindrop shape, a boat tail or rocket ship nose shape is the ideal. While this is difficult to achieve, and no production chassis designs are available to help, you can benefit from rounding your vehicle's rear comers and eliminating all

Vehicle	C_d	A	V= 5 mph	V= 10 mph	V= 20 mph	V= 30 mph	V= 45 mph	V= 60 mph	V= 75 mph
Small car	0.3	18	0.35	1.38	5.52	12.43	27.97	49.72	77.69
Larger car	0.32	22	0.45	1.80	7.20	16.20	36.46	64.82	101.28
Van	0.34	26	0.57	2.26	9.04	20.35	45.78	81.39	127.17
Small pickup	0.45	24	0.69	2.76	11.05	24.86	55.93	99.44	155.37
Roadster	0.6	18	0.69	2.76	11.05	24.86	55.93	99.44	155.37

Table 5-6 Aerodynamic Drag Force F_d at Different Vehicle Speeds for Typical Vehicle C_d and A Values

sharp edges. As you saw in Table 5-4, the body rear by itself makes up approximately 33 percent of the C_d, so any change you can make here should pay big dividends.

In the practical domain of our pickup truck example, putting on a lightweight and streamlined (rounded edges) camper shell should be your first choice. Station wagon shapes actually have better C_d figures than fastback sedan shapes (though the difference in weight negates the advantage). A pickup truck with a streamlined camper shell is very close to a station wagon shape. The next choice is to put a cover over the pickup box. If even this is not feasible, leave the tailgate down and use a cargo net or wire screen instead, or simply remove the tailgate if this presents you with no functional problems.

Shape Wheel Well and Underbody Airflow

Next, we pay attention to the wheels and wheel well area. Table 5-4 showed that the tire and wheel well area by itself contributes approximately 21 percent of the C_d, so small streamlining changes here can have large benefits. Using smooth wheel covers, thinner tires, rear wheel well covers, no mud flaps, or even lowering the vehicle height all come to mind as immediate beneficial steps. Every little bit helps; anything you can do to reduce drag or turbulence is good.

The next obvious area is underneath the vehicle. Table 5-4 showed this contributes approximately 14 percent of your C_d. Automobile designers have traditionally ignored this area because the public doesn't see it. But the onrushing air does. So the built-from-the-ground-up General Motors' Impact EV designers didn't ignore this area. You shouldn't either.

The immediate solution is simple: close the bottom of the engine compartment. There are no longer any bulky internal combustion engine components in the engine compartment, so cover the entire open area with a lightweight sheet of material. You probably still have transmission—and definitely have steering/suspension—components to deal with, so it might take several sheets of material. Whatever material you use, you don't want too thin a material for the body when the onrushing air strikes it, so choose a thickness that eliminates this possibility and fasten the sheet (or sheets) securely to the chassis.

Make the area from behind the underside of your vehicle's nose to underneath the firewall region (or just beyond it) as smooth as possible, using as lightweight a material as you can. If you can go all the way to the rear of your vehicle, so much the better. A fully streamlined underside reduces drag and turbulence, and can only help you.

Block and/or Shape Front Airflow

As you saw in Table 5-4, the body front and the engine compartment combined comprise approximately 18 percent of your C_d. While you cannot make significant changes to the C_d and A values you inherit when you purchase your conversion chassis, you can replace or cover its sharp-cornered front air entry grill and block off the flow of air to the engine compartment. An EV doesn't need the massive air intake required in internal combustion vehicles to feed cooling air to the radiator and engine compartment. A 3-inch diameter duct directing air to your electric motor is perfectly adequate. So anything you can do to round or streamline your vehicle's nose area (the smooth Ford Taurus or Thunderbird shapes of the 1990s as opposed to the shining and sharp grillwork shapes of the 1980s and earlier) is fair game. What you want is the maximum streamlined effect with minimum weight, so use modern kit car plastic and composite materials and techniques—auto-body-filler noses, if you please!

Air entering your EV conversion's now somewhat vacant engine compartment has the negative effect of creating under-the-hood turbulence, so blocking the incoming airflow with a sheet of lightweight material (such as aluminum) placed behind the grill works wonders. Whatever material you choose, just make sure it's heavy enough and fastened securely enough not to buckle, rumple, or vibrate when the air strikes it. Remember to leave a small opening for your electric motor's cooling air duct.

Roll with the Road

As they said in the movie *Days of Thunder*, "Tires is what wins the race." Today tires are fat, have wide tread, and are without low rolling resistance characteristics; they've been optimized for good adhesion instead. As an electric vehicle owner, you need to go against the grain of current tire thinking and learn to roll with the road to win the performance race.

In this section, you'll look at the Rolling Resistance row of Figure 5-1 and learn how to maximize the benefit from those four (or three) tire-road contact patches that are no bigger than your hand.

Rolling Resistance Defined

The rolling resistance force is defined as

$$F_r = C_r W \cos \phi$$

where C_r is the rolling resistance factor, W is your vehicle weight in lbs., and ϕ is the angle of incline as shown in Figure 5-2. Notice that cos ϕ varies from 1 at no incline (maximum effect) to 0 at 90 degrees (no effect). Again, vehicle weight is a factor, this time modulated by the vehicle's tire friction. The rolling resistance factor C_r might at its most elementary level be estimated as a constant. For a typical under-5,000-lb. EV, it is approximately

- 0.015 on a hard surface (concrete)
- 0.08 on a medium-hard surface
- 0.30 on a soft surface (sand)

If your calculations require more accuracy, C_r varies linearly at lower speeds and can be represented by

$$C_r = 0.012\,(1 + V/100)$$

where V is vehicle speed in mph.

Pay Attention to Your Tires

Tires are important to an EV owner. They support the vehicle and battery weight while cushioning against shocks; develop front-to-rear forces for acceleration and braking; and develop side-to-side forces for cornering.

Tires are almost universally of radial-ply construction today. Typically, one or more steel-belted plies run around the circumference (hence, radial) of tire. These deliver vastly superior performance to the bias-ply types (several plies woven crosswise around the tire carcass, hence, *bias* or *on an angle*) of earlier years that were replaced by radials as the standard in the 1960s. A tire is characterized by its rim width, the size wheel rim it fits on, section width (maximum width across the bulge of the tire), section height (distance from the bead to the outer edge of the tread), aspect ratio (ratio of height to width), overall diameter and load, and maximum tire pressure. In addition, the Tire and Rim Association defines the standard tire naming conventions:

- **5.60 × 15 (Typical VW Bug) Bias Tire Size**—*5.60* denotes section width in inches and *15* denotes the rim size in inches.
- **155R13 (Typical Honda Civic) Radial Tire Size**—*155* denotes section width in millimeters and *13* denotes the rim size in inches.
- **P185/75R14 (Typical Ford Ranger Pickup)**—Metric radial tire size. *P* denotes passenger car, *185* denotes section width in millimeters; *75* denotes aspect ratio; *R* denotes radial; *B* belted, or *D* bias; and *14* denotes the rim size in inches.

Although Goodyear has taken a leadership role with their GFE (Greater Fuel Efficiency) tires, Firestone has their Concept EVT series, and other tire-makers are also developing designs that are ideal for EVs. Doing a conversion today means using what's readily available. Table 5-7 gives you a comparison of the published characteristics for the Goodyear Decathlon tire family. This is an economical class of all-weather radials (other tire-makers have similar families) that should be more than adequate for most EV owners' needs. (Note: Tires with low rolling resistance are an excellent choice too.)

Everything in Table 5-7 is from Goodyear's published spec sheets except the Revolutions Per Mile column, which is a nominal value calculated directly from the Overall Diameter column rather than using actual measured data. The calculated value is slightly lower than a measured value when new and, as tread wears down, you are looking at a difference of 0.4 to 0.8 inches less in the tire's diameter, which translates into even more revolutions per mile. The difference might be 30 revolutions out of 900—a difference of 3 percent—but if this figure is important to your calculations,

Tire size	Wheel rim width	Tire rim width	Section width	Tread width	Overall diameter	Revolutions per mile	Maximum load (lbs)
P155/80R13	4.5–5.5	4.5	6.18	3.8	22.76	886	959
P165/80R13	4.5–6.0	4.5	6.5	4.2	23.39	862	1069
P175/80R13	4.5–6.0	5.0	6.97	4.1	24.02	840	1179
P185/80R13	5.0–6.5	5.0	7.24	4.4	24.65	818	1301
P185/75R14	5.0–6.5	5.0	7.24	4.6	24.96	808	1290
P195/75R14	5.0–7.0	5.5	7.72	4.9	25.51	791	1400
P205/75R14	5.5–7.55.5	7.99	5.1	26.14	772	1532	
P205/75R15	5.5–7.55.5	7.99	5.0	27.13	743	1598	
P215/75R15	5.5–7.56.0	8.5	5.3	27.68	729	1742	
P225/75R15	6.0–8.06.0	8.79	5.4	28.31	712	1874	
P235/75R15	6.0–8.06.5	9.25	5.8	28.86	699	2028	

All dimensions in inches.

TABLE 5-7 Comparison of Goodyear's Decathlon Tire Family

measure your tire's actual circumference in your driveway. A chalk mark on the sidewall and a tape measure and one full turn of your tire is all you need to tell if you're in the ballpark.

Ideally, you want soapbox derby or at least motorcycle tires on your EV: thin (little contact area with the road), hard (little friction), and large diameter (fewer revolutions per mile, and thus higher mileage, longer wear). From engineering studies on the rolling loss characteristics of solid rubber tires, we get the following equation:

$$F_t = C_t(W/d)(t_h/t_w)^{1/2}$$

where F_t is the rolling resistance force, C_t is a constant reflecting the tire material's elastic and loss characteristics, W is the weight on the tire, d is the outside diameter of the tire, and t_h and t_w are the tire section height and width, respectively. This is the last you'll see or hear of this equation in the book, but the point is, the rolling resistance force is affected by the material (harder is better for EV owners), the loading (less weight is better), the size (bigger is better), and the aspect ratio (a lower t_h/t_w ratio is better).

The variables in more conventional tire rolling resistance equations are usually tire inflation pressure (resistance decreases with increasing inflation pressure—harder is better), vehicle speed (increases with increasing speed), tire warmup (warmer is better), and load (less weight is better).

Use Radial Tires

Radial tires are nearly universal today, so tire construction is no longer a factor. But you might buy an older chassis that doesn't have radials on it, so check to be sure because bias-ply or bias-belted tires deliver far inferior performance to radials in terms of rolling resistance versus speed, warmup, and inflation.

Use High Tire Inflation Pressures

While you don't want to overinflate and balloon out your tires so that they pop off their rims, there is no reason not to inflate your EV's tires to their limit to suit your purpose. The upper limit is established by your discomfort level from the road vibration transmitted to your body. Rock hard tires are fine; the only real caveat is not to overload your tires.

Brake Drag and Steering Scuff Add to Rolling Resistance

In addition to tires, rolling resistance comes from brake drag and steering, and suspension alignment "scuff." Brake drag is another reason used vehicles are superior to new ones in the rolling resistance department—brake drag usually goes away as the vehicle is broken in. Alignment is another story. At worst it's like dragging your other foot behind you turned 90 degrees to the direction you're walking in. You want to check and make sure front wheel alignment is at manufacturing spec levels, and you haven't accidently bought a chassis whose rear wheels are tracking down the highway in sideways fashion. But neither of these contributes an earth-shattering amount to rolling resistance: the brake drag coefficient can be estimated as a constant 0.002 factor, and steering/suspension drag as a constant 0.001 factor. Taken together, they add only 0.003—or an additional 3 lbs. of force required by a vehicle weighing 1,000 lbs. travelling on a level surface.

Rolling Resistance Force Data You Can Use

For most purposes, the nominal C_r of 0.015 (for concrete) with the nominal brake and steering drag of 0.003 added to it (total 0.018) is all you need. This generates 18.0 lbs. of rolling resistance force for a 1,000-lb. vehicle. The 3,800-lb. Ford Ranger pickup truck of Chapter 10 would have a rolling resistance of 68.4 lbs. (3,800 lbs. of pickup weight × 0.018, or 3.8 × 18 lbs.). At 30 mph, the aerodynamic drag force on the Ford Ranger pickup truck of Chapter 10 is 31.08 lbs.—less than half the contribution of its 68.4 lbs. of rolling resistance drag.

Figure 5-5 shows the aerodynamic drag force and the rolling resistance force on the Ford Ranger pickup truck of Chapter 10 plotted for several vehicle speeds. These two forces, along with the acceleration or hill-climbing forces, constitute the *propulsion* or *road load.*

Notice the 68.4 lbs. of rolling resistance force is the main component of drag until the aerodynamic drag force takes over above 45 mph. Adding the force required to accelerate at a 1 mph/sec rate—nominally equivalent to that required to climb a 4.5 percent incline—merely shifts the combined aerodynamic drag–rolling resistance force curve upwards by 207.9 lbs. (3.8 × 54.7 lbs.) for the pickup. We'll look at these forces once again in the design section.

Less Is More with Drivetrains

In this section, you'll look at the Drivetrain row of Figure 5-1 and see how to get the most out of the internal combustion engine vehicle drivetrain components you adopt for your EV conversion.

The drivetrain in any vehicle comprises those components that transfer its motive power to the wheels and tires. The problem is, two separate vocabularies are used when talking about drivetrains for electric motors as opposed to those for internal combustion

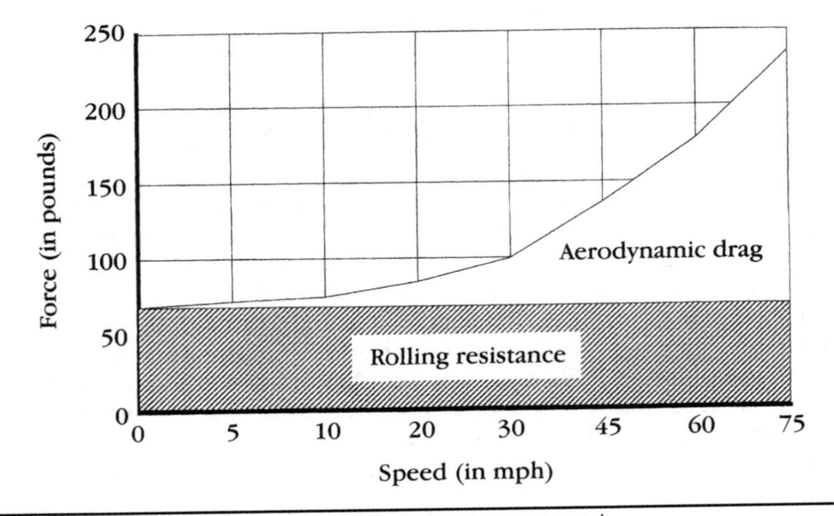

FIGURE 5-5 Rolling resistance and aerodynamic drag versus speed.

engines. This section will discuss the basic components; cover differences in motor versus engine performance specifications; discuss transmission gear selection; and look at the trade-offs of automatic versus manual transmission, new versus used, and heavy versus light fluids on drivetrain efficiencies.

Drivetrains

Let's start with what the drivetrain in a conventional internal combustion engine vehicle must accomplish. In practical terms, the power available from the engine must be equal to the job of overcoming the tractive resistances discussed earlier for any given speed. The obvious mission of the drivetrain is to apply the engine's power to driving the wheels and tires with the least loss (highest efficiency). But overall, the drivetrain must perform a number of tasks:

- Convert torque and speed of the engine to vehicle motion-traction
- Change directions, enabling forward and backward vehicle motion
- Permit different rotational speeds of the drive wheels when cornering
- Overcome hills and grades
- Maximize fuel economy

The drivetrain layout shown in simplified form in Figure 5-6 is most widely used to accomplish these objectives today. The function of each component is as follows:

- **Engine (or Electric Motor)**—Provides the raw power to propel the vehicle.
- **Clutch**—For internal combustion engines, separates or interrupts the power flow from the engine so that transmission gears can be shifted and, once engaged, the vehicle can be driven from standstill to top speed.
- **Manual Transmission**—Provides a number of alternative gear ratios to the engine so that vehicle needs—maximum torque for hill-climbing or minimum speed to economical cruising at maximum speed—can be accommodated.
- **Driveshaft**—Connects the drive wheels to the transmission in rear-wheel-drive vehicles; not needed in front-wheel-drive vehicles.
- **Differential**—Accommodates the fact that outer wheels must cover a greater distance than inner wheels when a vehicle is cornering, and translates drive force 90 degrees in rear-wheel-drive vehicles (might or might not in front-wheel-drive vehicles, depending on how engine is mounted). Most differentials also provide a speed reduction with a corresponding increase in torque.
- **Drive Axles**—Transfer power from the differential to the drive wheels.

Table 5-8 shows that you can typically expect 90 percent or greater efficiencies (slightly better for front-wheel-drive vehicles) from today's drivetrains. Internal combustion engine vehicle drivetrains provide everything necessary to allow an electric motor to be used in place of the removed engine and its related components to propel the vehicle. But the drivetrain components are usually complete overkill for the EV owner. The reason has to do with the different characteristics of internal combustion engines versus electric motors, and the way they are specified.

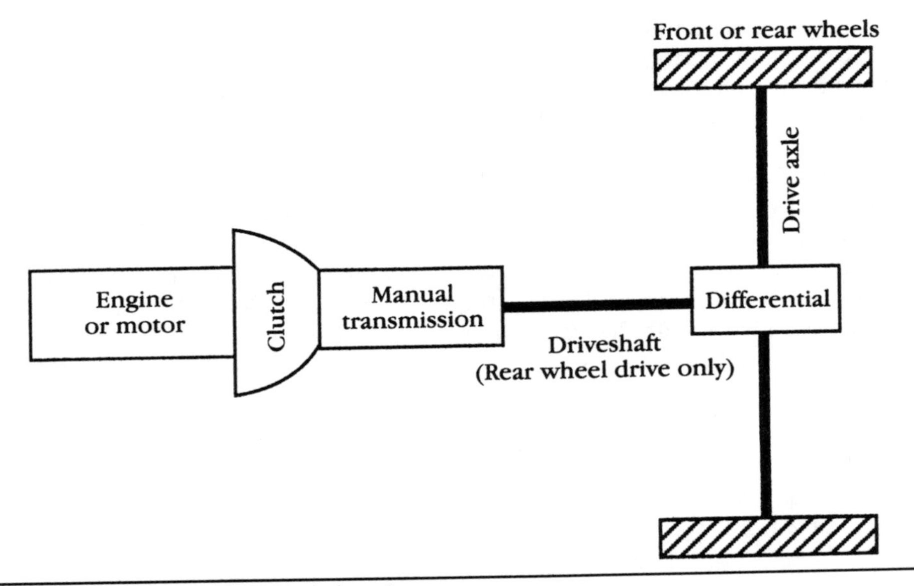

FIGURE 5-6 Simplified EV drivetrain layout.

Difference in Motor vs. Engine Specifications

Comparing electric motors and internal combustion engines is not an "apples to apples" comparison. If someone offers you either an electric motor or an internal combustion engine with the same rated horsepower, take the electric motor—it's far more powerful. Also, a series wound electric motor delivers peak torque upon startup (zero RPM), whereas an internal combustion engine delivers nothing until you wind up its RPMs. An electric motor is so different from an internal combustion engine that a brief discussion of terms is necessary before going further.

There is a substantial difference in the way an electric motor and an internal combustion engine are rated in horsepower. Figure 5-7's purpose is to show at a glance that an electric motor is more powerful than an internal combustion engine of the same rated horsepower. All internal combustion engines are rated at specific RPM levels for maximum torque and maximum horsepower. Internal combustion engine maximum horsepower ratings are typically derived under idealized laboratory conditions (for the bare engine without accessories attached), which is why the rated HP point appears above the maximum peak of the internal combustion engine horsepower curve in Figure 5-7. Electric motors, on the other hand, are typically rated at the continuous

Drivetrain type	Manual transmission	Driveshaft	Differential drive	Drive axle	Overall efficiency
Front wheel drive	0.96	not required	.097	.098	0.91
Rear wheel drive	0.96	0.99	.097	.098	0.90

TABLE 5-8 Comparison of Front and Rear Wheel Drivetrain Efficiences

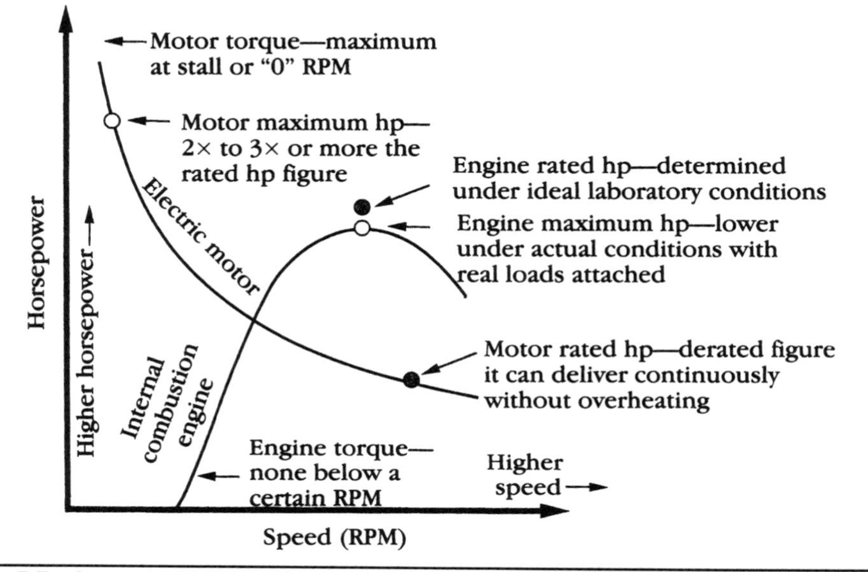

FIGURE 5-7 Comparison of electric motor versus internal combustion engine characteristics.

output level it can maintain without overheating. As you can see from Figure 5-7, the rated HP point for an electric motor is far down from its short-term output, which is typically two to three times higher than its continuous output.

There is another substantial difference. While an electric motor can produce a high torque at zero speed, an internal combustion engine produces negative torque until some speed is reached. An electric motor can therefore be attached directly to the drive wheels and accelerate the vehicle from a standstill without the need for the clutch, transmission, or torque converter required by the internal combustion engine. Everything can be accomplished by controlling the drive current to the electric motor. While an internal combustion engine can only deliver peak torque in a relatively narrow speed range, and requires a transmission and different gear ratios to deliver its power over a wide vehicle speed range, electric motors can be designed to deliver their power over a broad speed range with no need for transmission at all.

All these factors mean that current EV conversions put a lighter load on their borrowed-from-an-internal-combustion-engine-vehicle drivetrains, and future EV conversions eliminate the need for several drivetrain components altogether. Let's briefly summarize:

- **Clutch**—Although basically unused, a clutch is handy to have in today's EV conversions because its front end gives you an easy place to attach the electric motor, and its back end is already conveniently mated to the transmission. In short, it saves the work of building adapters, etc. In the future, when widespread adoption of AC motors and controllers eliminates the need for a complicated mechanical transmission, the electric motor can be directly coupled to a simplified, lightweight, one-direction, one- or two-gear ratio transmission, eliminating the need for a clutch.

- **Transmission**—Another handy item in today's EV conversions, the transmission's gears not only match the vehicle you are converting to a variety of off-the-shelf electrical motors, but also give you a mechanical reversing control that eliminates the need for a two-direction motor and controller—again simplifying your work. In the future, when widespread adoption of AC motors and controllers provides directional control and eliminates the need for a large number of mechanical gears to get the torques and speeds you need, today's transmission will be able to be replaced by a greatly simplified (and even more reliable) mechanical device.
- **Driveshaft, Differential, Drive Axles**—These components are all used intact in today's EV conversions. Because contemporary, built-from-the-ground-up electric vehicles like General Motors' Impact use two AC motors and place them next to the drive wheels, it's not too difficult to envision even simpler solutions for future EVs, because electric motors (with only one moving part) are so easily designed to accommodate different roles.

Going through the Gears

The transmission gear ratios, combined with the ratio available from the differential (or rear end, as it's sometimes called in automotive jargon), adapt the internal combustion engine's power and torque characteristics to maximum torque needs for hill-climbing or maximum economy needs for cruising. Figure 5-8 shows these at a glance for a typical internal combustion engine with four manual forward gears—horsepower/torque characteristics versus vehicle speed appear above the line and RPM versus vehicle speed appear below. The constant engine power line is simply equation 5, hp = FV/375 (V in mph), less any drivetrain losses. The tractive force line for each gear is simply the characteristic internal combustion engine torque curve (similar to the one shown in Figure 5-7) multiplied by the ratios for that gear. The superimposed incline force lanes are the typical propulsion or road-load force components added by acceleration or hill-climbing forces (recall the shape of this curve in Figure 5-5). The intersection of the incline or road-load curves and the tractive-force curves are the maximum speed that can be sustained in that gear. The upper half of Figure 5-8 illustrates how low first gearing for startup and high fourth gearing for high-speed driving apply to engine torque capabilities.

The lower part of Figure 5-8 shows road speed versus engine speed—for each gear appears. The point of this drawing is to illustrate how gear selection applies to engine speed capabilities. Normally, the overall gear ratios are selected to fall in a geometric progression: $1^{st} / 2^{nd} = 2^{nd} / 3^{rd}$, etc. Then individual gears are optimized for starting (1st), passing (2nd or 3rd), and fuel economy (4th or 5th).

Table 5-9 shows how these ratios turn out in an actual production car—in this case a Ford 1989 Taurus SHO. Notice the first two gear pairs are in a 1.5 ratio, whereas the next two move to 1.35. Table 5-9 also calculates the actual transmission, differential, and overall gear ratios (overall equals transmission times differential) for the 1987 Ford Ranger pickup truck that will be later used in the design section. Notice that the Ranger is optimized at both ends of the range but lower in the middle versus the Taurus, reflecting the difference in car versus truck design.

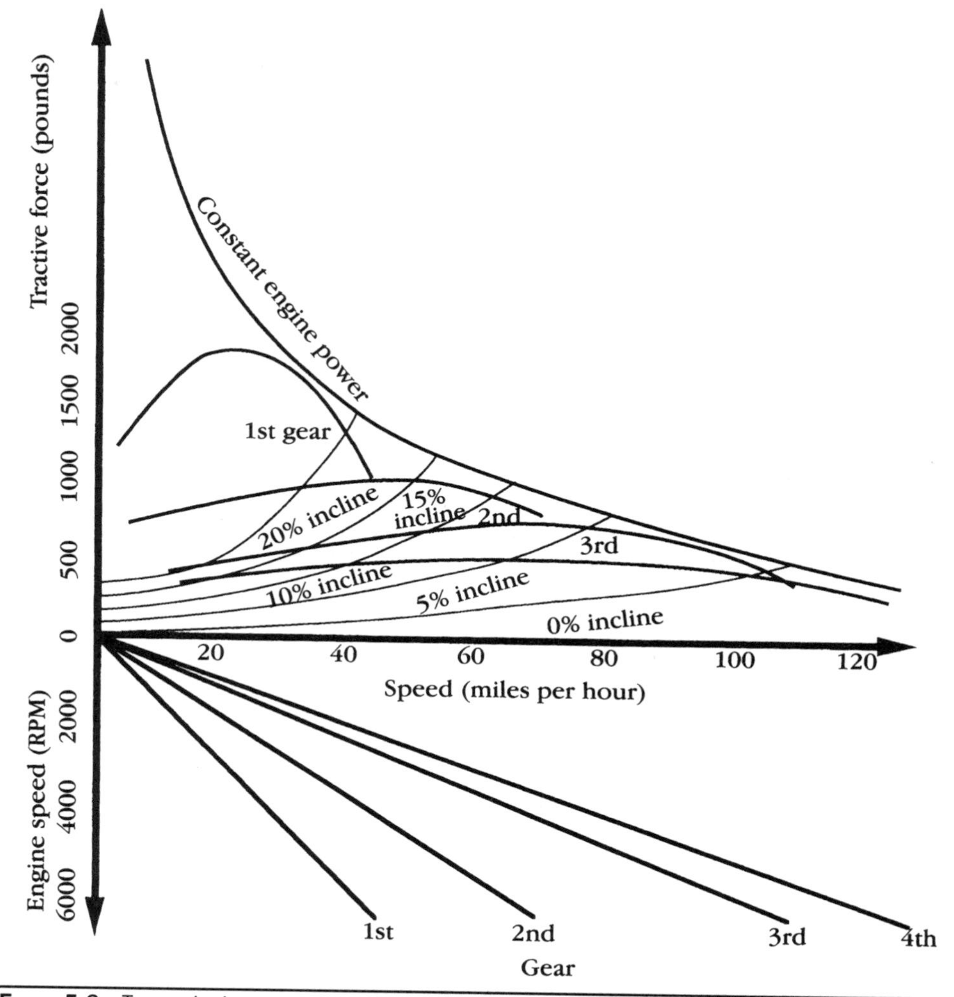

Figure 5-8 Transmission gear ratio versus speed and power summary.

Automatic vs. Manual Transmission

The early transmission discussion purposely avoided automatic transmissions. The reason is simple: EV owners need efficiency, and automatic transmissions are tremendously inefficient. There's another good reason—even with off-the-shelf components, you are not going to shift gears as much with an EV. If you're driving around town, you might only use one or two gears. You just put your EV in the gear you need and go. There's far less need for the clutch too. Remember, when you're sitting at an intersection your electric motor is not even turning.

Two ordinary household fans can demonstrate the torque converter principle that automatic transmissions use. Set them up facing each other about two feet apart and

Transmission Gear	1989 Taurus SHO Overall	Ratio to next gear	1987 Ranger Pickup transmission	1987 Ranger Pickup differential	1987 Ranger Pickup Overall
1st	12.01	1.5	3.96	3.45	13.66
2nd	7.82	1.5	2.08	3.45	7.18
3rd	5.16	1.35	1.37	3.45	4.73
4th	3.81	1.35	1.00	3.45	3.45
5th	2.79	N/A	0.84	3.45	2.90

Table 5-9 Comparison of Overall Transmission Gear Ratios for Ford 1989 Taurus SHO and 1987 Ranger Pickup

turn one on; the other starts to rotate with it. An automatic transmission uses transmission fluid rather than air as the coupling medium, but the principle is the same. An automatic transmission has a maximum efficiency of 80 percent, which drops dramatically at lower engine speed or vehicle torque levels. Automatic transmissions are also typically matched in characteristics to a given family of internal combustion engines with limited peak torque range—exactly opposite in behavior to electric motors. In short, choose a model with a clutch and a manual transmission for your internal combustion engine conversion vehicle.

Use a Used Transmission

There's a bonus here for the efficiency-seeking EV converter. Drivetrains take thousands of miles to wear into their minimum loss condition. When added to the fact that used tires have less rolling resistance, brake drag reduces as drums and linings wear smooth, and oil and grease seals have less drag after a period of break-in wear, you can expect about 20 percent less rolling and drivetrain resistance from a used vehicle than from a brand new one (assuming the used vehicle has not been badly worn or abused). And a used vehicle costs less, so it's a real bonus!

Heavy vs. Light Drivetrains and Fluids

Efficiency-seeking EV converters should not only avoid automatic transmissions, but also oversize axles, transmission, clutches, or anything that adds weight and reduces efficiency. Even a manual-transmission-based drivetrain will exhibit higher losses when operating at a low fraction of the gear's design maximum torque, which is the normal EV mode. The lower EV load will result in a lower efficiency out of a heavy duty part than out of a regular or economy part. Look for a vehicle with a lighter engine and manual transmission (for example, four cylinders rather than six, etc.) in your purchasing search. Earlier vintage models in a series are preferable because manufacturers tend to introduce higher performance options in successive model years.

The corollary of all this is the lubricant you choose. Using a lighter viscosity fluid in your differential lets things turn a lot easier. You're not breaking any rules here. Instead of shoving 500 horsepower through your drivetrain, you're at the opposite extreme—you're putting in an electric motor that lets you cruise at 10 percent of the peak torque load used by the internal combustion engine you just replaced. You're shifting less, using a lower peak torque, and probably using it less often. As a result your electric motor is putting only the lightest of loads on your internal combustion vehicle drivetrain, and you're probably using 50 percent (or less) of your drivetrain's designed capability. So less wear and tear on the gears means you can use a lighter viscosity lubricant and recover the additional benefit of further increased efficiency.

Design Your EV

This is step two. Look at your big picture first. Before you buy, build, or convert, decide what the main mission of your EV will be: a high-speed dragster to quietly blow away unsuspecting opponents at a stoplight; a long-range flyer to be a winning candidate at Electric Auto Association meetings; or a utility commuter vehicle to take you to work or grocery shopping, with capabilities midway between the former two. Your EV's weight is of primary importance to any design, but high acceleration off the line will dictate one type of design approach and gear ratios, while a long-range design will push you in a different direction. If it's a utility commuter EV you seek, then you'll want to preserve a little of both while optimizing your chassis flexibility toward either highway commuting or neighborhood hauling and pickup needs. In this section, you'll learn how to match your motor-drivetrain combination to the body style you've selected by going through the following steps:

- Learn when to use horsepower, torque, or current units, and why
- Look at a calculation overview
- Determine the required torque needs of your selected vehicle's chassis
- Determine the available capabilities of your selected electric motor and drivetrain torque.

The design process described here can be infinitely adapted to any EV you want to buy, build, or convert.

Horsepower, Torque, and Current

Let's start with some basic formulas. Earlier in the chapter, equation 2 casually introduced you to the fact that

$$1 \text{ Horsepower (hp)} = 550 \text{ ft-lb/sec}$$

This was then conveniently bundled into equation 5

$$1 \text{ Horsepower (hp)} = FV/375$$

where V is speed expressed in mph and F in lb force.

Horsepower is a rate of doing work. It takes 1 hp to raise 550 lbs. one foot in one second. But the second equation, which relates force and speed, brings horsepower to you in more familiar terms. It takes 1 hp to move 37.5 lbs. at 10 mph. Great, but you can also move 50 lbs. at 7.5 mph with 1 hp. The first instance might describe the force required to push a vehicle forward on a level slope; the second describes the force required to push this same vehicle up an incline. Horsepower is equal to force times speed, but you need to specify the force and speed you are talking about. For example, since we already know that 146.19 lbs. is the total drag force on the 3,800-pound Ranger pickup at 50 mph, and equation 5 relates the actual power required at a vehicle's wheels as a function of its speed and the required tractive force:

$$hp = (146.19 \times 50)/375 = 19.49 \text{ or } \textit{approximately 20 hp}$$

Only about 20 hp is necessary—at the wheels—to propel this pickup truck along at 50 mph on a level road without wind. In fact, a rated 20-hp electric motor will easily propel a 4,000-lb. vehicle at 50 mph—a fact that might amaze those who think in terms of the typical rated 90-hp or 120-hp internal combustion engine that might just have been removed from the pickup.

The point here is to condition yourself to think in terms of force values, which are relatively easy to determine, rather than in terms of a horsepower figure that is arrived at differently for engines versus electric motors, and that means little until tied to specific force and speed values anyway.

Another point (covered in more detail in Chapter 6's discussion of electric motors and Chapter 9's discussion of the electrical system) is to think in terms of current when working with electric motors. The current is directly related to motor torque. Through the torque-current relationship, you can directly link the mechanical and electrical worlds. (Note: The controller gives current multiplication. In other words, if the motor voltage is one-third the battery voltage, then the motor current is slightly less than three times the battery current. The motor and battery current would be the same only if you used a very inefficient resistive controller.)

Calculation Overview

Notice that the starting point in the calculations was the ending point of the force value required. Once you know the forces acting on your vehicle chassis at a given speed, the rest is easy. For your calculation approach, first determine these values, then plug in your motor and drivetrain values for its *design center* operating point, be it a 100-mph speedster, a 20-mph economy flyer, or a 50-mph utility vehicle. A 50-mph speed will be the design center for our pickup truck utility vehicle example.

In short, you need to select a speed, select an electric motor for that speed, choose the RPM at which the motor delivers that horsepower, choose the target gear ratio based on that RPM, and see if the motor provides the torque over the range of level and hill-climbing conditions you need. Once you go through the equations, worksheets, and graphed results covered in this section, and repeat them with your own values, you'll find the process quite simple.

The entire process is designed to give you graphic results you can quickly use to see how the torque available from your selected motor and drivetrain meets your vehicle's torque requirements at different vehicle speeds. If you have a microcomputer with a spreadsheet program, you can set it up once, and afterwards graph the results of any changed input parameter in seconds. In equation form, what we are saying is

$$\text{Available engine power} = \text{Tractive resistance demand}$$
$$\text{Power} = (\text{Acceleration} + \text{Climbing} + \text{Rolling} + \text{Drag} + \text{Wind})\ \text{Resistance}$$

Plugging into the force equations gives you:

$$\text{Force} = F_a + F_h + F_r + F_d + F_w$$
$$\text{Force} = C_i Wa + W\sin\phi + C_r W\cos\phi + C_d AV^2 + C_w F_d$$

You've determined every one of these earlier in the chapter. Under steady-speed conditions, acceleration is zero, so there is no acceleration force. If you are on a level surface, $\sin\phi = 0$, $\cos\phi = 1$ and the force equation can be rewritten as

$$\text{Force} = C_r\, W\cos\phi + C_d A V^2 + C_w F_d$$

This is the propulsion or road-load force you met at the end of the rolling resistance section and graphed for the Ford pickup in Figure 5-5. You need to determine this force for your vehicle at several candidate vehicle speeds, and add back in the acceleration and hill-climbing forces. This is easy if you recall that the acceleration force equals the hill-climbing force over the range from 1 mph/sec to 6 mph/sec.

You can now calculate your electric motor's required horsepower for your EV's design center.

$$\text{Horsepower (hp)} = (\text{Torque} \times \text{RPM})/5252 = 2\pi/60 \times FV/550$$
$$\text{Wheel RPM} = (\text{mph} \times \text{Revolutions/mile})/60$$

The previous equation can be substituted to give

$$HP_{wheel} = (\text{Torque}_{wheel} \times \text{mph} \times \text{Revolutions/mile})/(5252 \times 60)$$

but,

$$hp_{motor} = hp_{wheel}/n_o$$

where n_o is the overall drivetrain efficiency. Substituting the previous equation into this gives you:

$$hp_{motor} = (\text{Torque}_{wheel} \times \text{mph} \times \text{Revolutions/mile})/(315120 \times n_o))$$

Plugging the values for torque, speed, and revolutions/mile (based on your vehicle's tire diameter) into the equation will give you the required horsepower for your electric motor.

After you have chosen your candidate electric motor, the manufacturer will usually provide you with a graph or table showing its torque and current versus speed performance based on a constant voltage applied to the motor terminals. From these figures or curves, you can derive the RPM at which your electric motor delivers closest to its rated horsepower. Using this motor RPM figure and the wheel RPM figure, which gives you the wheel RPM from your target speed and RPM, you can determine your best gear or gear ratio from

$$\text{Overall gear ratio} = RPM_{motor}/RPM_{wheel}$$

This—or the one closest to it—is the best gear for the transmission in your selected vehicle to use; if you were setting up a one-gear-only EV, you would pick this ratio. With all the other motor torque and RPM values you can then calculate wheel torque and vehicle speed using the following equations for the different overall gear ratios in your drivetrain:

$$\text{Torque}_{wheel} = \text{Torque}_{motor}/(\text{overall gear ratio} \times n_o)$$

$$\text{Speed}_{vehicle}\ (\text{in mph}) = (RPM_{motor} \times 60)/(\text{overall gear ratio} \times \text{Revolutions/mile})$$

You now have the family of torque available curves versus vehicle speed for the different gear ratios in your drivetrain. All that remains is to graph the torque required data and the torque available data on the same grid. A quick look at the graph tells you if you have what you need or if you need to go back to the drawing board.

Torque Required Worksheet

Table 5-10 computes the torque required data for a 1987 Ranger pickup, the vehicle to be converted in Chapter 10. You've met all the values going into the level drag force before, but not in one worksheet. Now they are converted to torque values using equation 11, and new values of force and torque are calculated for incline values from 5 percent to 25 percent. Conveniently, these correspond rather closely to the acceleration values for 1 mph/sec to 5 mph/sec, respectively, and the two can be used interchangeably. The vehicle assumptions all appear in Table 5-10. If you were preparing a computer spreadsheet, all of this type information would be grouped in one section so that you could see the effects of changing chassis weight, C_dA, C_r, and other parameters. You might also want to graph speed values at 5 mph intervals to present a more accurate picture.

Torque Available Worksheet

There are a few preliminaries to go through before you can prepare the torque available worksheet. First, you have to determine the horsepower of an electric motor using the following equation:

$$hp_{motor} = (Torque_{wheel} \times mph \times Revolutions/mile)/(315120 \times n_o)$$

Plugging in real values for the Ranger pickup with P185/75R14 tires and using the torque value from the Torque Required Worksheet at the 50-mph design center speed (recall that overall efficiency for rear-wheel-drive manual transmission was 0.90 from Table 5-8) gives you:

$$hp_{motor} = (152.04 \times 50 \times 808)/(315120 \times 0.9) = 21.66$$

This corresponds quite nicely to the capabilities of the Advanced DC Motors model FB1-4001 rated at 22 hp. From the manufacturer's torque versus speed curves for this motor driven at a constant 120 volts and using this equation:

$$hp = (Torque \times RPM)/5252 = (25 \times 4600)/5252 = 21.89$$

This motor produces approximately 22 hp at 4600 RPM at 25 ft-lbs of torque and 170 amps. Next, calculate Wheel RPM using the following equation:

$$Wheel\ RPM = (mph \times Revolutions/mile)/60 = (50 \times 808)/60 = 673.33$$

You can then calculate the best gear using the next equation:

$$Overall\ gear\ ratio = RPM_{motor}/RPM_{wheel} = 4600/673.33 = 6.83$$

From Table 5-9, the 2nd gear overall ratio of 7.18 for the Ford Ranger pickup comes quite close to this figure, which means it should deliver the best overall performance.

Now you are ready to prepare the torque available worksheet shown in Table 5-11. This worksheet sets up motor values on the far left. Wheel torque and vehicle speed values for the 1st through 5th gears go from left to right across the worksheet, and at higher values of torque and current from top to bottom of the worksheet within each gear. The net result is you now have the wheel torque available at a given vehicle speed for each one of the vehicle's gear ratios. A spreadsheet would make quick work of this; you could see at a glance the effects of varying motor voltage, tires sizes, etc.

Vehicle speed (mph)	10	20	30	40	50	60	70	80	90
Tire rolling resistance C_r	0.015	0.015	0.015	0.015	0.015	0.015	0.015	0.015	0.015
Brake and steering resistance C_r	0.003	0.003	0.003	0.003	0.003	0.003	0.003	0.003	0.003
Total rolling force (lbs)	68.4	68.4	68.4	68.4	68.4	68.4	68.4	68.4	68.4
Still air drag force (lbs)	2.76	11.05	24.60	44.19	69.05	99.43	135.35	176.78	223.73
Relative wind factor C_w	1.338	0.449	0.25	0.169	0.126	0.101	0.083	0.071	0.062
Relative wind drag force (lbs)	3.70	4.96	6.21	7.47	8.73	9.99	11.25	12.51	13.77
Total drag force, level (lbs)	74.86	84.40	99.47	120.07	146.19	177.83	215.00	257.69	305.91
Total drag torque, level (ft-lbs)	77.85	87.78	103.45	124.87	152.04	184.95	223.60	268.00	318.15
Sin ϕ, ϕ = Arc tan 5% incline	0.0500	0.0500	0.0500	0.0500	0.0500	0.0500	0.0500	0.0500	0.0500
Cos ϕ, ϕ = Arc tan 5% incline	0.9988	0.9988	0.9988	0.9988	0.9988	0.9988	0.9988	0.9988	0.9988
Incline force WSin ϕ (lbs)	190.04	190.04	190.04	190.04	190.04	190.04	190.04	190.04	190.04
Rolling drag force C_r WCos ϕ (lbs)	68.31	68.31	68.31	68.31	68.31	68.31	68.31	68.31	68.31
Total drag force, 5% (lbs)	264.81	274.36	289.43	310.02	336.14	367.78	404.95	447.64	495.86
Total drag torque, 5% (ft-lbs)	275.41	285.34	301.01	322.43	349.59	382.50	421.16	465.56	515.71
Sin (Arc tan 10% incline)	0.0996	0.0996	0.0996	0.0996	0.0996	0.0996	0.0996	0.0996	0.0996
Cos (Arc tan 10% incline)	0.9950	0.9950	0.9950	0.9950	0.9950	0.9950	0.9950	0.9950	0.9950
Incline force WSin ϕ (lbs)	378.52	378.52	378.52	378.52	378.52	378.52	378.52	378.52	378.52
Rolling drag force C_r WCos ϕ (lbs)	68.06	68.06	68.06	68.06	68.06	68.06	68.06	68.06	68.06
Total drag force, 10% (lbs)	453.04	462.58	477.65	498.25	524.37	556.01	593.18	635.87	684.08
Total drag torque, 10% (ft-lbs)	471.17	481.10	496.77	518.19	545.35	578.26	616.92	661.32	711.46

(continued on next page)

TABLE 5-10 Torque Required Worksheet for 1987 Ford Ranger Pickup at Different VehicleSpeeds and Inclines*

Vehicle speed (mph)	10	20	30	40	50	60	70	80	90
Sin ϕ, ϕ = Arc tan 15% incline	0.1484	0.1484	0.1484	0.1484	0.1484	0.1484	0.1484	0.1484	0.1484
Cos ϕ, ϕ = Arc tan 15% incline	0.9889	0.9889	0.9889	0.9889	0.9889	0.9889	0.9889	0.9889	0.9889
Incline force WSin ϕ (lbs)	563.84	563.84	563.84	563.84	563.84	563.84	563.84	563.84	563.84
Rolling drag force C_r WCos ϕ (lbs)	67.64	67.64	67.64	67.64	67.64	67.64	67.64	67.64	67.64
Total drag force, 15% (lbs)	637.94	647.49	662.56	683.16	709.27	740.92	778.09	820.78	868.99
Total drag torque, 15% (ft-lbs)	663.48	673.41	689.08	710.50	737.66	770.57	809.23	853.63	903.77
Sin ϕ, ϕ = Arc tan 20% incline	0.1962	0.1962	0.1962	0.1962	0.1962	0.1962	0.1962	0.1962	0.1962
Cos ϕ, ϕ = Arc tan 20% incline	0.9806	0.9806	0.9806	0.9806	0.9806	0.9806	0.9806	0.9806	0.9806
Incline force WSin ϕ (lbs)	745.67	745.67	745.67	745.67	745.67	745.67	745.67	745.67	745.67
Rolling drag force C_r WCos ϕ (lbs)	67.07	67.07	67.07	67.07	67.07	67.07	67.07	67.07	67.07
Total drag force, 20% (lbs)	819.20	828.75	843.82	864.41	890.53	922.18	959.34	1002.0	1050.3
Total drag torque, 20% (ft-lbs)	851.99	861.92	877.59	899.01	926.18	959.08	997.74	1042.1	1092.3
Sin ϕ, ϕ = Arc tan 25% incline	0.2425	0.2425	0.2425	0.2425	0.2425	0.2425	0.2425	0.2425	0.2425
Cos ϕ, ϕ = Arc tan 25% incline	0.9702	0.9702	0.9702	0.9702	0.9702	0.9702	0.9702	0.9702	0.9702
Incline force WSin ϕ (lbs)	921.46	921.46	921.46	921.46	921.46	921.46	921.46	921.46	921.46
Rolling drag force C_r WCos ϕ (lbs)	66.36	66.36	66.36	66.36	66.36	66.36	66.36	66.36	66.36
Total drag force, 25% (lbs)	994.28	1003.8	1018.9	1039.5	1065.6	1097.3	1134.4	1177.1	1225.3
Total drag torque, 25% (ft-lbs)	1034.1	1044.0	1059.7	1081.1	1108.3	1141.2	1179.8	1224.2	1274.4

* Values computed for 1987 Ford Ranger pickup; weight = 3800 lbs; coefficient of drag, C_d = 0.45; frontal area, A = 24 square feet; relative wind factor, C_{rw} = 1.6; relative wind, w = 7.5 mph; tires = P185/75R14; revolutions/mile = 808; torque multiplier = 840.34/(revolutions/mile) = 1.04.

TABLE 5-10 orque Required Worksheet for 1987 Ford Ranger Pickup at Different VehicleSpeeds and Inclines* (continued)

Torque Required and Available Graph

All that remains is to graph the torque required data from Table 5-10 and the torque available data from Table 5-11 on the same grid versus vehicle speed. This is done in Figure 5-9. Notice its similarity to the curve in Figure 5-8 drawn for the internal combustion engine (except that the torque available curves resemble the electric motor characteristics, a comparison made earlier in Figure 5-7).

How do you read Figure 5-9 and what does it tell you? The usable area of each gear is the area to the left and below it, bounded at the bottom by the torque required at the level condition curve. You want to work as far down the torque available curve for each gear as possible for minimum current draw and maximum economy and range. The graph confirms that 2nd gear is probably the best overall selection. You could put the EV into second gear and leave it there, because it gives you 2 mph/sec acceleration at startup, hill-climbing ability up to 15 percent inclines, and provides you with enough torque to take you up to about 52.5 mph. For mountain climbing or quick pops off the line, 1st gear gives you everything you could hope for at the expense of really sucking down the amps, current-wise. But at the other end of 1st gear, if you drive like there's an egg between your foot and the accelerator pedal, it actually draws only 100 amps at 45 mph versus the 210 amps required by 2nd gear. At higher speeds, 3rd gear lets you cruise at 60 mph at 270 amps, and 4th gear lets you cruise at 70 mph at 370 amps. At any speed, 5th gear appears marginal in this particular vehicle; though it can possibly hold 78 mph, it requires 440 amps to do so. At any other speed, other gears do it better with less current draw. While current draw is your first priority, too much for too long overheats your motor. You don't want to exceed your motor's speed rating either, as you would do if you drove much above 45 mph in 1st gear. Is this a usable motor and drivetrain combination for this vehicle? Definitely. If you want to make minor adjustments, just raise or lower the battery voltage. This will shift the torque available curve for each gear to the right (higher voltage) or to the left (lower voltage). A larger motor in this particular vehicle will give you better acceleration and upper-end speed performance; the torque available curves for each gear would be shifted higher. But the penalty would be higher weight and increased current draw and shorter range. A smaller motor would shift the torque available curves lower while returning a small weight and current draw advantage. But beware of underpowering your vehicle. If given the choice, always go for slightly more rather than slightly less horsepower than you need. The result will almost always be higher satisfaction with your finished EV conversion.

Vehicle gear				1st	1st	2nd	2nd	3rd	3rd	4th	4th	5th	5th
Overall gear ratio				13.66	13.66	7.18	7.18	4.73	4.73	3.45	3.45	2.9	2.9
Motor torque multiplier, equation (12)				12.294		6.462		4.257			3.105	2.61	
RPM multiplier, equation (13)					165.56		87.02		57.33		41.81		35.15
Current in amps, torque in ft-lbs, vehicle speed in mph	Motor Current	Motor Torque	Motor RPM	Wheel Torque	Vehicle Speed	Wheel Torque	Vehicle Speed	Wheel Torque	Vehicle Speed	Wheel Torque	Vehicle Speed	Wheel Torque	Vehicle Speed
	100	10	7750	122.94	46.81	64.62	89.06	42.57	135.19	31.05	185.34	26.10	220.50
	125	15	6400	184.41	38.66	96.93	73.54	63.86	111.64	46.58	153.06	39.15	182.09
	150	20	5000	245.88	30.20	129.24	57.46	85.14	87.22	62.10	119.58	52.20	142.26
	170	25	4600	307.35	27.78	161.55	52.86	106.43	80.24	77.63	110.01	65.25	130.88
	190	30	4100	368.82	24.76	193.86	47.11	127.71	71.52	93.15	98.05	78.30	116.65
	210	35	3900	430.29	23.56	226.17	44.82	149.00	68.03	108.68	93.27	91.35	110.96
	230	40	3700	491.76	22.35	258.48	42.52	170.28	64.54	124.20	88.49	104.40	105.27
	250	45	3500	553.23	21.14	290.79	40.22	191.57	61.05	139.73	83.70	117.45	99.58
	270	50	3400	614.70	20.54	323.10	39.07	212.85	59.31	155.25	81.31	130.50	96.73
	290	55	3350	676.17	20.23	355.41	38.50	234.14	58.44	170.78	80.12	143.55	95.31
	305	60	3250	737.64	19.63	387.72	37.35	255.42	56.69	186.30	77.73	156.60	92.47
	320	65	3150	799.11	19.03	420.03	36.20	276.71	54.95	201.83	75.33	169.65	89.62
	335	70	3050	860.58	18.42	452.34	35.05	297.99	53.20	217.35	72.94	182.70	86.78
	355	75	3000	922.05	18.12	484.65	34.47	319.28	52.33	232.88	71.75	195.75	85.35
	370	80	2950	983.52	17.82	516.96	33.90	340.56	51.46	248.40	70.55	208.80	83.93
	390	85	2900	1045.0	17.52	549.27	33.33	361.85	50.59	263.93	69.35	221.85	82.51
	405	90	2850	1106.5	17.21	581.58	32.75	383.13	49.71	279.45	68.16	234.90	81.09
	420	95	2800	1167.9	16.91	613.89	32.18	404.42	48.84	294.98	66.96	247.95	79.66
	440	100	2750	1229.4	16.61	646.20	31.60	425.70	47.97	310.50	65.77	261.00	78.24

* Values computed for 1987 Ford Ranger pickup; tires = P185/75R14; revolutions/mile = 808; overall drivetrain efficiency = 0.90; dc series traction motor is Advanced DC Motors Model FBI-4001; battery pack is 120 volts; equation (12) is $T_{wheel} = T_{motor}$/(overall gear ratio x overall drivetrain efficiency); equation (13) is $Speed_{vehicle} = RPM_{motor}$ x 60)/(overall gear ratio x revolutions/mile).

TABLE 5-11 Torque Available Worksheet for 120-Volt DC Series Motor Powered 1987 Ford Ranger Pickup at Different Motor Speeds and Gear Ratios

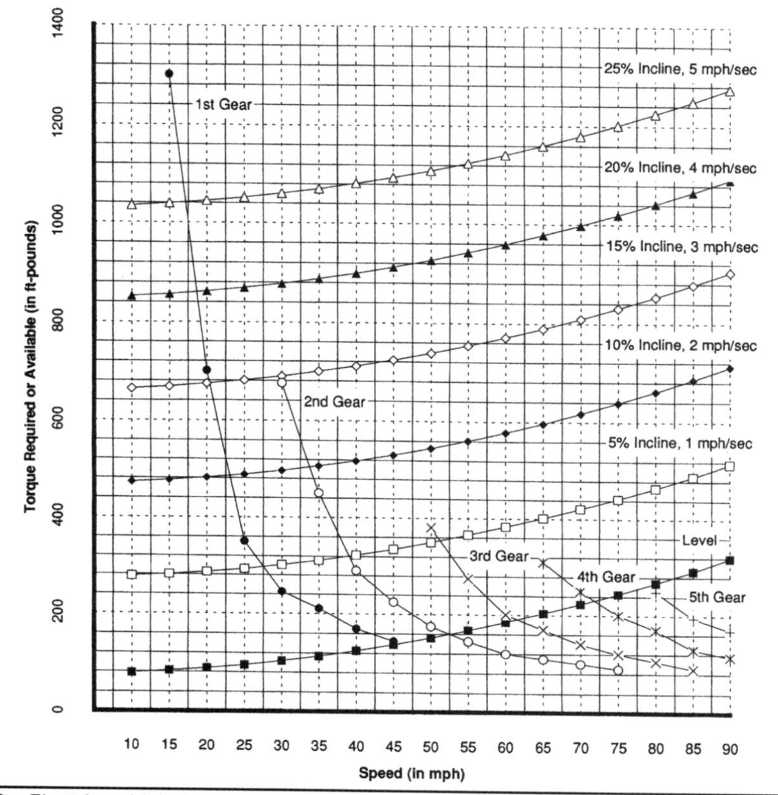

FIGURE 5-9 Electric vehicle torque required versus torque availlable.

Buy Your EV Chassis

This is step three. Even if you go out to buy your EV ready-made, you still want to know what kind of a job the manufacturer has done, so you can decide if you're getting the best model for you. In all other cases, you'll be doing the optimizing, either by the choices you've made up front in chassis selection or by other decisions you make later on during the conversion. In this section, you'll be looking at key points that contribute to buying smart:

- Review why conversion is best—the pro side
- Why conversion might not be for you—the con side
- How to get the best deal
- Avoid off-brand, too old, abused, dirty, or rusty chassis
- Keep your needs list handy
- Buy or borrow the chassis manual
- Sell unused engine parts

Why Conversion Is Best

In the real world, where time is money, converting an existing internal combustion engine vehicle saves money in terms of large capital investment and a large amount of labor. By starting with an existing late-model vehicle, the EV converter's bonus is a structure that comes complete with body, chassis, suspension, steering, and braking systems—all designed, developed, tested, and safety-proven to work together. Provided the converted electric vehicle does not greatly exceed the original vehicle's GVWR overall weight or GAWR weight per axle specifications, all systems will continue to deliver their previous performance, stability, and handling characteristics. And the EV converter inherits another body bonus—its bumpers, lights, safety-glazed windows, etc. are already preapproved and tested to meet all safety requirements.

There's still another benefit—you save more money. Automobile junkyards make money by buying the whole car (truck, van, etc.) and selling off its pieces for more than they paid for the car. When you build (rather than convert) an EV, you are on the other side of the fence. Unless you bought a complete kit, building from scratch means buying chassis tubing, angle braces, and sheet stock plus axles/suspension, brakes, steering, bearings/wheels/tires, body/trim/paint, windshield/glass/wipers, lights/electrical, gauges, instruments, dashboard/interior trim/upholstery, etc.—parts that are bound to cost you more à la carte than buying them already manufactured and installed in a completed vehicle.

The Other Side of Conversion

What's the downside? It's likely that any conversion vehicle you choose will not be streamlined like a soapbox derby racer. It will be a lot heavier than you'd like it, and have tires designed for traction rather than low rolling resistance. You do the best you can in these departments depending on your end-use goals: EV dragster, commuter, or highway flyer.

It's equally likely your conversion vehicle comes with a lot of parts you no longer need: internal combustion engine and mounts, and its fuel, exhaust, emission control, ignition, starter, and cooling/heating systems. These you remove and, if possible, sell.

Then you have additional conversion vehicle components that you might wish to change or upgrade for better performance, such as drivetrain, wheels/tires, brakes, steering, and battery/low voltage accessory electrical system. On these just do what makes sense.

How to Get the Best Deal

There are a number of trends that help you to get into your EV conversion at the best possible price, but you still have to do the shopping. Shopping for your EV chassis is no different from buying any vehicle in general. Put your boots on. Grab a good book on buying new or used cars to help you. Just remember not to divulge your true intentions while bargaining, so you can entertain scenarios like this: When the salesperson says, "Well, to be honest, only two of its four cylinders are working," you say, "No problem. How much will you knock off the price?" Or the ideal situation: "Frankly, that's the cleanest model on the lot but its engine doesn't work," and you say, "No problem. Let me take it off your hands for $100." Your best deal will be when finding exactly the nonworking "lemon" that someone wants to sell—specifically, a lemon in the engine department. If you find a $5,000 vehicle that doesn't run because of an engine problem (but everything else works okay), you can pick it up for a fraction of its value and save nearly the entire cost of conversion. Seek these deals out; they can save you a substantial amount of money.

Specific vehicle characteristics aside, what's the best-vintage vehicle for conversion in terms of cost? As a rule of thumb, used is better than new, but not too used. While older is better in terms of lower cost, you lose a lot of the more recent vehicle safety features if you go too far back. If you go back further than 10 to 15 years, you begin getting into body deterioration and mechanical high mileage problem territory. And if you go back to the classics—the 1960s and earlier vintage models—the price starts going up again. But several important trends are working in the EV converter's benefit.

Late-Model Used Vehicles (Late 1980s and Onward)

These late-model vehicles make ideal EV conversions. Not only are they available at a lower price than new vehicles with all the depreciation worked out, but vehicles with 20,000 to 50,000 miles on the odometer are better for EV converters because vehicle drivetrains, brakes, and wheels/tires generate less friction (burrs and ridges are worn down, shoes no longer drag and seals are seated, etc.) and roll/turn more easily.

Here are some options to look for:

- **Stripped-Down, Late-Model Used Vehicles**—These are an even better deal. Almost everyone wants the deluxe, V-8, automatic transmission, power-everything model. You, on the other hand, are interested in a straight stick, 4-cylinder, no-frills model that nobody wants. The salesperson will fall all over himself/herself trying to help you. Try to keep composed. And don't say what it's for until you've finished the deal.
- **Lightweight, Early 1980s, 4-Cylinder Cars/Trucks**—These represent problem sales for used-auto dealers and are frequently discounted just to move them. Surprise the salesperson who sold you that 4-cylinder lemon—visit next month in your fire-breathing 120-volt EV conversion.
- **Older Lightweight Diesel or Rotary Cars/Trucks**—These also represent problem sales for used-auto dealers because potential buyers are unsure of engine repairability and parts. With current owners, it's more likely they have just gone out of favor. In either case, these represent a buying opportunity for you.
- **Cars/trucks with Blown Engines or No Engines at All**—These are a real problem to move for any owner but a gift for you. It's a marriage made in heaven and you can usually call the terms. Scan the newspapers for these deals.

Avoid the Real Junk

While late-model, nonrunning bargains are great, avoid the problem situations. Avoid buying off-brand, too old, abused, dirty, or rusty chassis. The parts and labor that you have to add to them to bring them up to the level of normal used models makes them no bargain. The salesperson might offer you that used 1957 Mingus for only $50, but where do you get the parts for it?

Dirt and rust are OK in moderation but too much of anything is too much. If you can't tell what kind of engine it is because you can't see it in the engine compartment, pass on this choice. You don't want to spend as much time cleaning as you do converting or pay someone else to do it, only to find that essential parts you thought were present under the dirt are either in poor condition or not present at all.

Rust is rust, and what you can't see is the worst of it. How can you put your best EV foot forward in a rust bucket for a chassis? Don't do it; pass on this choice in favor of a rust-free vehicle or one with minimal rust.

Keep Your Needs List Handy

Regardless of which vehicle you choose for conversion, you want to feel good about your ability to convert it before you leave the lot. If it's too small and/or cramped to fit all the electrical parts—let alone the batteries—you know you have a problem. Or if it's very dirty, greasy, or rusty, you might want to think twice. Here's a short checklist to keep in mind when buying:

- **Weight**—With 120 volts and a 22 hp series DC motor, 4,000 to 5,000 lbs. is about the upper limit. On the other hand, the same components will give you blistering performance and substantially more range in a 2,000- to 3,000-lb. vehicle. Weight is everything in EVs—decide carefully.
- **Aerodynamic Drag**—You can tweak the nose and tail of your vehicle to produce less drag and/or turbulence, but what you see before you buy it is basically what you've got. Choose wisely and aerodynamically.
- **Rolling Resistance**—Special EV tires are still expensive, so look for a nice set of used radials and pump them up hard.
- **Drivetrain**—You don't want an automatic; a 4- or 5-speed manual will do nicely, and front wheel drive typically gives you more room for mounting batteries. Avoid 8- and 6-cylinders in favor of 4-cylinders, and choose the smallest, lightest engine/drivetrain combinations. Avoid heavy duty anything or 4-wheel drive.
- **Electrical System**—Pass on air conditioning, electric windows, and any power accessories.
- **Size**—Will everything you want to put in (batteries, motor, controller, and charger) have room to fit? How easy will it be to do the wiring?
- **Age and Condition**—These determine whether you can get parts for it, and how easy it is to restore it to a condition fit to serve as your car.

Buy or Borrow the Manuals

Manuals are invaluable. If possible, seek them out to read about any hidden problems before you buy the vehicle. After you own it, don't spend hours finding out if the red-striped or the green-striped wire goes to dashboard terminal block number 3; just flip to the appropriate schematic in the manual and locate it in minutes. Component disassembly is easy when you learn that you must always disengage bolt number 2 in a clockwise direction before turning bolt number 1 in a counterclockwise direction, etc. Believe me, these are labor savers.

Sell Unused Engine Parts

Somebody somewhere wants that 4-cylinder engine you are removing from your EV chassis for their car or for scrap metal. Emblazon this above your workbench area before the removed parts begin to accumulate dust or crowd out all other items from your garage. First, make a few phone calls to place want ads; then call dealers and junk yards. If no cash consideration is offered, see if you can trade the parts for something of value. Do all of this early, before you are in the middle of your conversion. Nothing worse than stubbing your toe on an engine block while nonchalantly going for your voltmeter.

CHAPTER 6

Electric Motors

"The superior AC system will replace the entrenched but inferior DC one."
—George Westinghouse (from *Tesla: Man Out of Time)*

The heart of every electric vehicle is its electric motor. Electric motors come in all sizes, shapes, and types and are the most efficient mechanical devices on the planet. Unlike an internal combustion engine, an electric motor emits zero pollutants. Technically, there are three moving parts in an electric motor. Even with three parts, electric motors outlive internal combustion engines every day of the week. The parts are the rotor and two end bearings. This is just one of the main reasons why widespread adoption of EVs or electric drive vehicles are a planet-saving proposition. Recently, I had a discussion with an engineer at my office. At the end of the conversation about fuel cell cars, I reminded him that even a fuel cell car generates electricity and that is why electric cars are the way of the automotive future.

What this means to you is that ownership of a fun-to-drive, high-performance EV will deliver years of low-maintenance driving at minimum cost, mostly because of the inherent characteristics of its electric motor—power and economy.

The objective of this chapter is to guide you towards the best candidate motor for your EV conversion today, and suggest the best electric motor type for your future EV conversion. To accomplish these goals, this chapter will review electric motor basics and provide you with useful equations; introduce you to the different types of electric motors and their advantages and disadvantages for EVs; introduce the best electric motor for your EV conversion today and its characteristics; and introduce you to the electric motor type that you should closely follow and investigate for future EV conversions.

Why an Electric Motor?

In Chapter 1 you learned that the electric motor is ubiquitous because of its simplicity. All electric motors by definition have a fixed *stator* or stationary part, and a *rotor* or moveable part. This simplicity is the secret of their dependability, and why in direct contrast to the internal combustion engine with its hundreds of moving parts, electric motors are a far superior source of propulsion:

- Electric motors are inherently powerful. By selecting a design that delivers peak torque at or near stall, you can move a mountain. Nearly all traction motors deliver near peak torque at zero rpm. That's why electric traction motors have powered our trolley cars, subways, and diesel-electric railroad locomotives for so many years. There is no waiting, as with an internal combustion engine, while it winds up to its peak torque rpm range. Apply electric current to it and you've instantly got torque to spare. If any EV's performance is wimpy, it's due to a poor design or electric motor selection—not the electric motor itself.

- Electric motors are inherently efficient. You can expect to get 90 percent or more electrical energy you put into an electric motor out of it in the form of mechanical torque. Few other mechanical devices even come close to this efficiency.

Horsepower

Since electric motors are efficient, the horsepower behind them in a real electric car can be shocking to the system initially (no pun intended). I just remember the first time I drove an electric car. When I stepped on that accelerator it took off! No questions asked. No engine with excessive parts to get in the way of that.

Here are some technical points to understand when trying to find the right motor for your car.

1. Electric motors are rated at their point of maximum efficiency; they may be capable of 2–4 times their continuous rating but only for a few minutes (acceleration or hill climbing). Internal combustion engines are rated at the peak horsepower. For example, the FB1-4001A motor is rated as 30 hp continuous at 144V and 100 hp peak. The 5-minute rating of the FB1-4001A motor is 48 hp at 144V.
2. Each 1,000 lbs. of vehicle weight after conversion requires 6–8 hp. This is the continuous rating of the motor. So a 3,000-lb. conversion requires a motor that is rated at approximately 20 hp. More horsepower is required for higher speeds, heavier vehicles, and steeper terrains.
3. The available horsepower of a motor increases with voltage; for example, the FB1-4001A motor is rated at 18 hp continuous at 72V but is rated at 30 hp continuous at 144V. As the voltage is increased the rpm increases. Horsepower is a function of rpm × torque.
4. Although electric motors are rated as "continuous," the motor can run at less horsepower. If only 10 hp is required for the speed then the motor runs at that reduced load. This is the function of the motor controller.
5. Operating continuously above the rated horsepower will eventually overheat and damage the motor. A motor that is rated at 150 amps can run at 300 amps for a short time (minutes), but longer periods can easily damage the motor. Do not buy an undersize motor for your vehicle for your application—it will not last long. Current is what overheats components.
6. Highway speeds require greater horsepower. The horsepower required at 70 mph is four times the horsepower required at 35 mph. That means the current required is four times also, which means less range.

Depending on your design and component choices, the electric motor in your EV conversion can smoke its tires and routinely offer 60-mile range on a dollar's worth of electricity. Compare that to 75 cents per mile in a hydrogen FC car, just for the "fuel" alone (of course, that's beside the point, but hydrogen isn't a fuel per se).

DC Electric Motors

An electric motor is a mechanical device that converts electrical energy into motion, and that can be further adapted to do useful work such as pulling, pushing, lifting, stirring, or oscillating. It is an ideal application of the fundamental properties of magnetism and electricity. Before looking at DC motors and their properties, let's review some fundamentals.

Magnetism and Electricity

Magnetism and electricity are opposite sides of the same coin. Electrical and electronics design engineers regularly utilize Maxwell's four laws of electromagnetism based on Faraday's and Ampere's earlier discoveries in their daily work. They might tell you, "magnetism and electricity are inextricably intertwined in nature." In fact, you don't have one without the other. But usually you only look at one or the other unless you are discussing electric motors or other devices that involve both.

If you set a gallon jug of water on the edge of your sink and poke a small pencilpoint-sized hole in the bottom, water squirts out into the sink. When the jug is filled to the top with water, it squirts out all the way across the sink. When only a little water remains in the jug, it squirts out fairly close to the jug. This is due to the weight of the water creating a force which pushes the drops out. That force is analogous to the voltage in a circuit. Voltage is really called an electromotive force. If you refill the jug with the same amount of water but this time only enlarge the hole, water flows out of it faster because there is less resistance to the flow. But it doesn't go out as far, since the same weight doesn't encounter as much resistance. There is better flow, which in this case is analogous to a larger electrical current.

You hook up a light bulb to a battery by completing the wire connections from the battery's positive and negative terminals to the light bulb, and the bulb lights. Hook up two batteries in a series to double the voltage, and the bulb shines even brighter. The water's force or potential—the height of water in the jug—corresponds to the battery's force or potential or voltage; the size of the hole in the bottom of the jug corresponds to the resistance to the flow or the resistance of the light bulb in this case; and the flow of water from the jug corresponds to the flow of electricity or current. To tie things into the electrical realm, there is a mathematical equation that relates these parameters of force, flow, and resistance. The electrical equation, commonly known as Ohm's Law, is

$$V = IR$$

where V is voltage in volts, I is current in amps, and R is resistance in ohms. When you double the voltage, you send twice as much current through the wire and the light bulb becomes brighter. Or, if you reduce the resistance (as was done by enlarging the hole) given the same level of water—you double the flow (increase the current).

The simple bar magnet, which you probably encountered in your school science class, has two ends or poles—north and south. Either end attracts magnetizable objects to it, usually objects containing iron in some form, such as iron filings, paper clips, etc. When two bar magnets are used together, opposite poles (north-south) attract one another and identical poles (north-north or south-south) repel one another.

The compass needle is a magnetized object with its own north and south pole. Lightweight and delicately balanced, it aligns itself with the earth's magnetic field, and tells you which direction is north. But bring the bar-style magnet near it and it will rotate away from the earth's magnetic north pole.

You can create a magnetic field with electricity. Take an iron nail from your toolbox, wrap a few turns of insulated copper wire around it, and hook the ends up to a battery. The plain nail is transformed into a bar-style magnet that can behave just like the compass.

If you were to write a magnetic equation, it would look like this:

$$\text{mmf} = \phi R = \text{flux} \times \text{reluctance}$$

In this equation, commonly known as Rowland's Law, mmf is the magnetomotive force, ϕ is the flux (the flow of magnetism), and R is reluctance, the opposition to the flow. Because designers of electrical transformers and motors work in terms of magnetic field strength H, flux density B, and permeability (u) of a given transmission medium, and because

$$\text{mmf} = \text{H} \times 1;\ \phi = \text{B} \times \text{A};\ \text{and } R = 1/(\text{u} \times \text{A})$$

(where 1 is the length and A is the area), Rowland's Law usually appears in its more common form (for a linear range) as

$$\text{B} = \text{Hu}$$

In this equation, B is the flux density or induction forms in lines per square inch, H is the magnetic field strength or flow of magnetism in the material in ampere-turns per inch, and u is the permeability or resistance of the material to magnetizing force in henrys per inch.

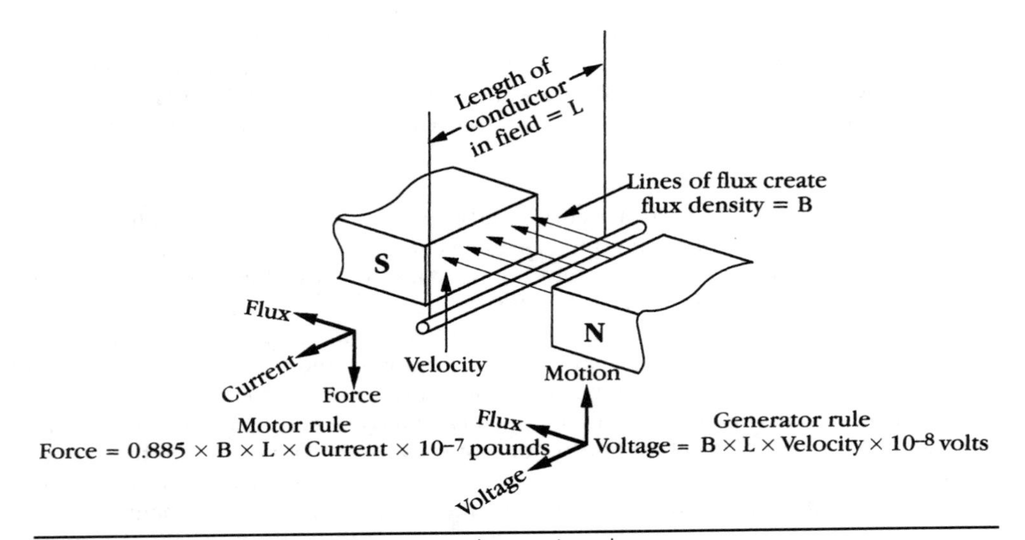

FIGURE 6-1 DC motor basics—the motor and generator rules.

Conductors and Magnetic Fields

If you had a horseshoe-shaped magnet with the ends close together and you moved a wire through its pole are at a right angle to the flux as shown in Figure 6-1, the equation describing it would be

$$V = Blv \times 10^{-8}$$

In this equation, a simplified form of Faraday's Law, V is the induced voltage in volts, B is the flux density in lines per square inch, 1 is the length of that part of the conductor actually cutting the flux, and v is the velocity in inches per second. For example, if you moved a 1-inch wire or conductor at a right angle to a field of 50,000 lines per square inch at 50 feet per second,

$$V = 50000 \times 1 \times 50 \times 12 \times 10^{-8} = 0.3 \text{ volts}$$

This relationship holds true whether the field is stationary and the wire is moving or vice versa. The faster you cut the lines of flux, the greater the voltage, but you must do so at right angles. (At any angle ϕ other than the 90-degree right angle, that equation becomes $V = Blv \times \sin \phi \times 10^{-8}$.) What you have just demonstrated is Faraday's Law or the generator rule: Motion of a conductor at a right angle through a magnetic field induces a voltage across the conductor. If the circuit is closed, the induced voltage will cause a current to flow. A handy way to remember the relationships is the right-hand rule: the thumb of your right hand points upward in the direction of the motion of the conductor, your index finger extends at right angles to it in the direction of the flux (from north to south pole), and your third finger extends in a direction at right angles to the other two, indicating the polarity of the induced voltage or the direction in which the current will flow as shown in Figure 6-1.

The flip-side or corollary to the generator rule is the *motor rule* whose equation is

$$V = 0.885 \times BI \times I \times 10^{-8}$$

where F is the force generated on the conductor in pounds, B is the flux density in lines per square inch, 1 is the length of that part of the conductor actually cutting the flux, and I is the current in amperes. For example, if you had a 1-inch wire at right angles to a field of 50,000 lines per square inch with a current of 100 amperes flowing through it:

$$V = 0.885 \times 50000 \times 1 \times 100 \times 10^{-7} = 0.44 \text{ pounds}$$

What we have just demonstrated here is Ampere's Law or the motor rule: Current flowing through a conductor at a right angle to a magnetic field produces a force upon the conductor. The right-hand rule is again a handy way to remember the relationships, but this time the thumb of your right hand points toward you in the direction of the current through the conductor, your extended index finger at right angles to it points in the direction of the flux (from north to south pole), and your extended third finger points downwards in a direction at right angles to the other two, indicating the direction of the generated force as shown in Figure 6-1. Now you're ready to talk seriously about motors.

DC Motors in General

If you could support the conductor shown in Figure 6-1 so that it could rotate in the magnetic field, you would create the condition shown in the upper part of Figure 6-2.

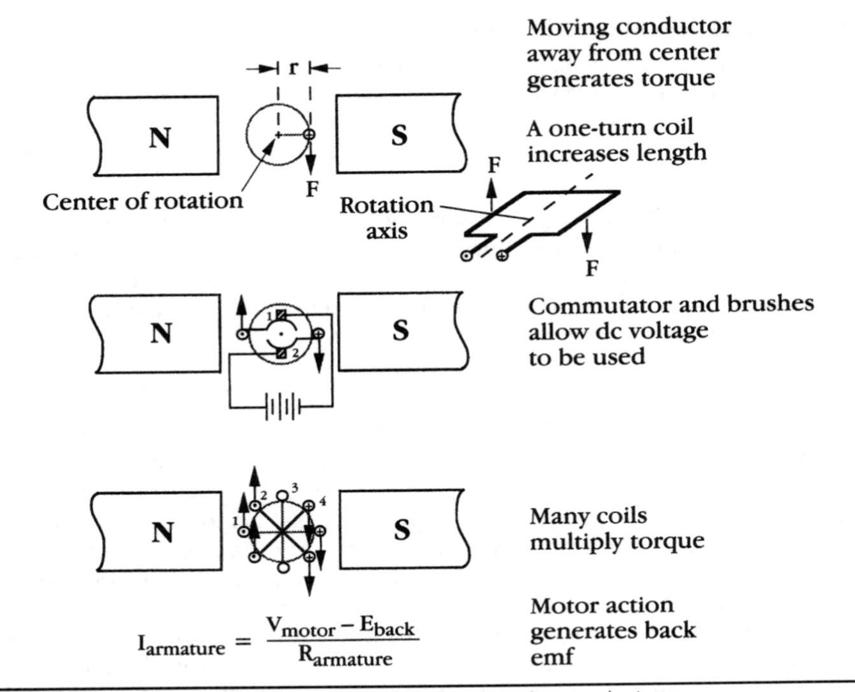

FIGURE 6-2 DC motors basics—obtaining torque from moving conductors.

Now the current through the conductor (which flows into the page—you're seeing the back of the arrow) would exert a force that would tend to rotate it in the clockwise direction. The magnitude of the torque would be given by

$$T = Fr$$

where T is torque in foot-pounds, F is force in pounds, and r is the distance measured perpendicularly from the direction of F to the center of rotation in feet. Now you have a motor design—on paper. You need to take some steps to make it real.

First you need to make it more "force-full." Since the force varies with the length of the conductor, if you make a coil of wire, as shown in the upper right of Figure 6-2, twice as much length is cutting the lines of flux. The force generated on the right-hand wire is downward and the force generated on the left-hand wire is upward; they would assist one another in producing rotation and result in twice the torque.

To further assist rotation, you add the commutator and brushes, shown in the middle of Figure 6-2. This arrangement allows you to power your motor from a constant supply of direct current (DC) voltage. Switching the polarity of the coils when they reach the 12 o'clock or 6 o'clock position (minimum flux point) guarantees that current will always flow in through brush number 2 and out through brush number 1, thus always producing upward force on the left-hand conductor and downward force on the right-hand conductor, and creating a constant rotation.

To further increase the motor's torque abilities, you add additional coils as shown at the bottom of Figure 6-2. In reality, each coil can have many windings, and you can arrange commutator segments to match the number of coils so that you have the force on each of these coils acting in unison with the force on all the other coils.

DC Motors in the Real World

Now it's time you met real-world DC motors—their construction, definitions, and efficiency. Let's start by looking at their components.

Armature

The armature is the main current-carrying part of a motor that normally rotates (brushless motors tend to blur this distinction) and produces torque via the action of current flow in its coils. It also holds the coils in place, and provides a low reluctance path to the flux. (Reluctance is defined as $(H \times 1)/4$ and measured in ampere-turns per lines of flux.) The armature usually consists of a shaft surrounded by laminated sheet steel pieces called the *armature core*. The laminations reduce eddy current losses; steel is replaced by more efficient metals in newer designs. There are grooves or slots parallel to the shaft around the outside of the core; the sides of the coils are placed into these slots. The coils (each with many turns of wire) are placed so that one side is under the north pole and the other is under the south pole; adjacent coils are placed in adjacent slots, as shown at the bottom of Figure 6-2. The end of one coil is connected to the beginning of the next coil so that the total force then becomes the sum of the forces generated on each coil.

Commutator

The commutator is the smart part of the motor that permits constant rotation by reversing the direction of current in the windings each time they reach the minimum flux point. This piece is basically a switch. It commutates the voltage from one polarity to the opposite. Since the motor rotor is spinning and has momentum, the switching process repeats itself in a pre-ordained manner. The alternating magnetic poles continue to provide the push to overcome losses (friction, windage, and heating) to reach a terminal speed. Under load, the motor behaves a bit differently, but the load causes more current to be drawn.

Physically, it's a part of the armature (typically located near one end of the shaft) that appears as a ring split into segments surrounding the shaft. These segments are insulated from one another and the shaft.

Field Poles

In the real world, electromagnets (recall your toolbox nail with a few turns of insulated copper wire wrapped around it) are customarily used instead of the permanent magnets you saw in Figure 6-1 and Figure 6-2. (Permanent magnet motors are, in fact, used, and you'll be formally introduced to them and their advantages later in this section.) In a real motor the lines of flux are produced by an electromagnet created by winding turns of wire around its poles or pole pieces. A pole is normally built up of laminated sheet steel pieces, which reduce eddy current losses; as with armatures, steel has been replaced by more efficient metals in the newer models. The pole pieces are usually curved where they surround the armature to produce a more uniform magnetic field. The turns of copper wire around the poles are called the field windings.

Series Motors

How these windings are made and connected determines the motor type. A coil of a few turns of heavy wire connected in series with the armature is called a *series motor*. A coil of many turns of fine wire connected in parallel with the armature is called a *shunt motor*.

Brushes

Typically consisting of rectangular-shaped pieces of graphite or carbon, brushes are held in place by springs whose tension can be adjusted. The brush holder is an insulated material that electrically isolates the brush itself from the motor frame. A small flexible copper wire embedded in the brush (called a *pigtail*) provides current to the brush. Smaller brushes can be connected together internally to support greater current flows.

Motor Case, Frame, or Yoke

Whatever you wish to call it, the function of this part is not only to provide support for the mechanical elements, but also to provide a magnetic path for the lines of flux to complete their circuit—just like the lines of flux around a bar magnet. In the motor's case, the magnetic path goes from the north pole through the air gap, the magnetic material of the armature, and the second air gap, to the south pole and back to the north pole again via the case, frame, or yoke.

Motors operating in the real world are subject to losses from three sources:

- **Mechanical**—All torque available inside the motor is not available outside because torque is consumed in overcoming friction of the bearings, moving air inside the motor (known as windage), and because of brush drag.
- **Electrical**—Power is consumed as current flows through the combined resistance of the armature, field windings, and brushes.
- **Magnetic**—Additional losses are caused by eddy current and hysteresis losses in the armature and the field pole cores.

In summary:

- Efficiency is simply the power out of a device, relative to the power applied to the device.
- When you apply 100 watts, and get only the work equivalent to 90 watts out of it, you have a 10-watt loss. That is a 90 percent efficient device.
- That rule of efficiency applies to motors, motor speed controllers, battery chargers, etc.

DC Motor Types

Now that you've been introduced to DC motors in theory and in the real world, it's time to compare the different motor types. DC motors appear in the following forms:

- Series
- Shunt
- Compound
- Permanent magnet

- Brushless
- Universal

The last three motor types are just variations of the first three, fabricated with different construction techniques. The compound motor is a combination of the series and shunt motors. For the first three motor types we'll look at the circuit showing how the motor circuit field windings are connected, and then look at the characteristics of the torque and speed versus armature current and shaft horsepower curves that describe their operation (Figure 6-3).

Each of the motor types will be examined for its torque, speed, reversal, and regenerative braking capabilities—the factors important to EV users. The motor types will all be compared at full load shaft horsepower—the only way to compare different motor types of equal rating. Efficiency is a little harder to determine since it also depends on the external resistance of the circuit to which the motor is connected. So efficiency has to be calculated for each individual case.

Series DC Motors

The most well-known of the DC motors, and the one which comes to mind for traction applications (like propelling EVs), is the series DC motor. It's so named because its field winding is connected in series with the armature (Figure 6-3). Because the same current

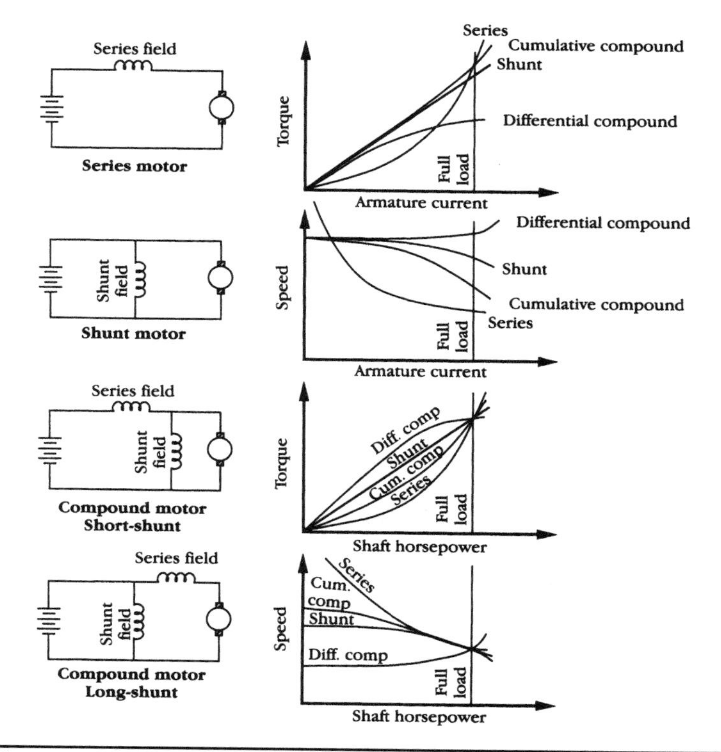

FIGURE 6-3 Summary of DC motor types—windings and characteristics.

must flow through this field winding as through the motor armature itself, it is wound with a few turns of heavy gauge wire.

Torque

As equation (5) showed, Torque = $K \times \phi \times I_a$, and the current in the series field is I_a. I_a can be substituted for ϕ and the series motor torque equation can be written as

$$\text{Torque} = K' \times I_a^2$$

This shows that in a series motor, torque varies with the square of the current—a fact substantiated by Figure 6-3's actual torque versus armature current graph. The graph is actually a little misleading because at startup there is no counter electromotive force (EMF) to impede the flow of current in the armature of a series motor, and startup torque can be enormous; with no current limiting, you can start up at torque values far to the right of the full load line for a series motor. In actual use, the armature reaction and magnetic saturation of the series motor at high currents sets upper limits on both torque and current, although you might prefer to limit your circuit and components to far lower values. High starting torque makes series motors highly desirable for traction applications.

Speed

As the following equation shows,

$$\text{Speed} = (V_t - I_a R_a)/K4)$$

Once again, I_a can be substituted for ϕ, and the series motor speed equation can be rewritten as

$$\text{Speed} = (V_t - I_a(R_a + R_f))/KI_a$$

where R_a and R_f are the resistances of the armature and field, respectively. This shows that speed becomes very large as the current becomes small in a series motor—a fact substantiated by Figure 6-3's actual speed versus armature current graph. High rpms at no load is the series motor's Achilles heel. You need to make sure you are always in gear, have the clutch in, have a load attached, etc., because the series motor's tendency is to run away at no load. Just be aware of this and back off immediately if you hear a series motor rev up too fast.

Field Weakening

This technique is an interesting way to control series motor speed. You place an external resistor in parallel with the series motor field winding, in effect diverting part of its current through the resistor. Keep it to 50 percent or less of the total current (resistor values equal to or greater than 1.5 times the motor's field resistance). The byproduct is a speed increase of 20 to 25 percent at moderate torques without any instability in operation (hunting in rpm, etc.). Used in moderation, it's like getting something for nothing.

Frank G. Willey, a member of the EAA, created the first pulse width modulated motor speed controller using silicon power transistors. He was recognized by the IEEE in the early 1980s for doing so.

Offered in kit form or completely assembled and tested, the Model 9 had ratings of 450 amps for traction pack voltages from 24 to 120 volts. Today, most motor controllers use this approach, or a variation of it. Field weakening had its limitations, which drivers soon didn't want to contend with, given the other better options available to them.

Reversing

The same current that flows through the armature flows through the field in a series motor, so reversing the applied voltage polarity does not reverse the motor direction. To reverse motor direction, you have to reverse or transpose the direction of the field winding with respect to the armature. This characteristic also makes it possible to run series DC motors from AC (more on this later in the section).

Regenerative Braking

All motors simultaneously exhibit generator action—motors generate counter EMF—as you read earlier in this section. The reverse also holds true—generators produce counter torque. Regenerative braking allows you to slow down the speed of your EV (and save its brakes) and put energy back into its battery (thereby extending its drivable range) by harnessing its motor to work as a generator after it is up and running at speed. Put another way, the vehicle, once running at speed, has kinetic energy. Regenerative braking allows you to electronically switch the motor and turn it into a generator, thereby capturing the energy that would normally be dissipated (read: lost) as heat in the brake pads while slowing down. The motor does the braking, not your brake pads. While all motors can be used as generators, the series motor has rarely been used as a generator in practice because of its unique and relatively unstable generator properties.

Shunt DC Motors

The second most well-known of the DC motors is the shunt DC motor, so named because its field winding is connected in parallel with the armature (Figure 6-3). Because it doesn't have to handle the high motor armature currents, a shunt motor field coil is typically wound with many turns of fine gauge wire and has a much higher resistance than the armature.

Torque Characteristics

Because the shunt field is connected directly across the voltage source, the flux in the shunt motor remains relatively constant. Its torque is directly dependent upon the armature current, as described by torque equation:

$$\text{Torque} = K \times \phi \times I_a$$

This shows that, in a shunt motor, torque varies directly with the current, and the straight-line relationship shown in Figure 6-3 results. Although there is initially no counter emf to impede the flow of startup current in the armature of a shunt motor, the shunt motor's linear relationship is quickly established. As a result, the shunt motor does not produce nearly as much startup torque as the series motor. This translates to reduced acceleration performance for owners of shunt motor-powered EVs.

Speed

When a load is applied to any (but here specifically a shunt) motor, it will tend to slow down and, in turn, reduce the counter emf produced.

The increase in armature current results in an increase in torque to take care of the added load. Basically, constant, counter emf tends to remain constant over a wide range of armature current values, and produces a fixed speed curve that only droops slightly at high armature currents.

The shunt motor's linear torque and fixed speed versus armature current characteristics have two undesirable side effects for traction applications when controlled manually.

- First, when a heavy load (hill climbing, extended acceleration) is applied, a shunt motor does not slow down appreciably as does a series motor, and the excessive current drawn through its armature by the continuous high torque requirement makes it more susceptible to damage by overheating.
- Next, in contrast to the "knee" bend in the series motor torque-speed equations (Figure 5-9), shunt motor torque-speed curves are nearly straight lines. This means more speed control or shifting is necessary to achieve any given operating point. Series motors are therefore used where there is a wide variation in both torque and speed and/or heavy starting loads.
- Other than having much lower startup torque, shunt motors can perform as well as series motors in EVs when electronically controlled.
- The downside is that a shunt motor controller can be more complicated to design than a series motor controller.

Field Weakening

You can also achieve a higher-than-rated shunt motor speed by reducing the shunt coil current—in this case, you place an external control resistance in series with the shunt motor field winding. But here, unlike the series motor's runaway rpm region at no load, you are playing with fire, because the loaded shunt motor armature has an inertia that does not permit it to respond instantly to field control changes. If you do this while your shunt motor is accelerating, you might cook your motor or have motor parts all over the highway by the time you adjust the resistance back down to where you started. Be careful with field weakening in shunt motors. Again, these motors soon lost favor due to the inherent simplicity of a series DC motor arrangement. Heating was less of an issue, and certainly not something that drivers want to have to worry about; the series motor was better suited to the application.

A shunt motor is instantly adaptable as a shunt generator. Most generators are in fact shunt wound, or variations on this theme. The linear or nearly linear torque and speed versus current characteristics of the shunt motor manifest as nearly linear voltage versus current characteristics when used as a generator. This also translates to a high degree of stability that makes a shunt motor both useful for and adaptable to regenerative braking applications, either manually or electronically controlled.

Reversing

Current flows through the field in a shunt motor in the same direction as it flows through the armature, so reversing the applied voltage polarity reverses both the direction of the current in the armature and the direction of the field-generated flux, and does not reverse the motor rotation direction. To reverse motor direction, you have to reverse or transpose the direction of the shunt field winding with respect to the armature.

Regenerative Braking

A shunt motor is instantly adaptable as a shunt generator. Most generators are in fact shunt wound, or variations on this theme. The linear or nearly linear torque and speed versus current characteristics of the shunt motor manifest as nearly linear voltage versus current characteristics when used as a generator. This also translates to a high degree of stability that makes a shunt motor both useful for and adaptable to regenerative braking applications—either manually or electronically controlled.

Compound DC Motors

A compound DC motor is a combination of the series and shunt DC motors. The way its windings are connected, and whether they are connected to boost (assist) or buck (oppose) one another in action, determine its type. Its basic characterization comes from whether current flowing into the motor first encounters a series field coil-short-shunt compound motor or a parallel shunt field coil-long-shunt compound motor as shown in Figure 6-3. If, in either one of these configurations, the coil windings are hooked up to oppose one another in action, you have a differential compound motor. If the coil windings are hooked up to assist one another in action, you have a cumulative compound motor. The beauty of the compound motor is its ability to bring the best of both the series and the shunt DC motors to the user.

- **Torque**—The torque in a compound motor has to reflect the actions of both the series and the shunt field coils. Depending on whether you are hooked up in the differential or cumulative position, the shunt torque ϕ and the series torque ϕ either subtract to a difference figure or add together. The effect of these hookup arrangements on torque is illustrated in Figure 6-3, where the differential compound motor builds more slowly to a lower torque value than the shunt curve and the cumulative compound motor builds to a slightly higher torque value than the shunt curve at a slightly higher rate.
- **Speed**—Similar to torque, the speed action in a compound motor will also reflect the dual variables of series and shunt field coil action. Figure 6-3 shows the speed curves. One of the initial benefits of the compound configuration is that runaway conditions at low field current levels for the shunt motor and at lightly loaded levels for the series motor can be eliminated. While the differential compound configuration is of questionable value—its curve shows a tendency to runaway speeds at high armature current values—the cumulative compound motor appears to offer benefits to EV operation. You can tailor a cumulative compound motor to your EV needs by picking one whose series winding delivers good starting torque and whose shunt winding delivers lower current draw and regenerative braking capabilities once up to speed. When you look, you might find these characteristics already exist in an off-the-shelf model.
- **Reversing**—A short-shunt compound motor resembles a series motor, so reversing its applied voltage polarity normally does not reverse the motor direction. A long-shunt compound motor resembles a shunt motor, so reversing the voltage supply leads normally reverses the motor. As with speed and torque, compound motors can be tailored to do whatever you want to do in the reversing department.

- **Regenerative Braking**—A compound motor is as easily adaptable as a shunt motor to regenerative braking. Its series winding gives it additional starting torque, but can be bypassed during regenerative braking, and its shunt winding allowed to give it the more desirable shunt generator characteristic. Manual or solid-state electronics controlled, compound motors are usable for and adaptable to regenerative braking applications.

Permanent Magnet DC Motors

When you were first introduced to the DC motor topic, permanent magnets were used as an example because of their simplicity. Permanent magnet motors are, in fact, being increasingly used today because new technology—various alloys of Alnico magnet material, ferrite-ceramic magnets, rare-earth element magnets, etc.—enables them to be made smaller and lighter in weight than equivalent wound field coil motors of the same horsepower rating. Rare-earth element magnets surpass the strength of Alnico magnets significantly (by 10–20 times), and have been used with great success in other areas such as computer disc drives, thereby helping drive down the production costs. (See Figure 6-4.) While commutator and brushes are still required, you save the complexity and expense of fabricating a field winding, and gain in efficiency because no current is needed for the field.

Permanent magnet motors approximately resemble the shunt motor in their torque, speed, reversing, and regenerative braking characteristics; either motor type can usually be substituted for the other in control circuit designs. But because modem materials can support higher levels of magnetizing force—the H factor in equation 2—the much smaller armature reaction of permanent magnet motors greatly extends the linear characteristics of conventional shunt motor speed/torque curves down to zero speed. This means that permanent magnet motors have starting torques several times that of shunt motors, and their speed versus load characteristics are more linear and easier to predict.

Brushless DC Motors

With no brushes to replace or commutator parts to maintain, brushless motors promise to be the most long-lived and maintenance-free of all motors. You can now custom-tailor the motor's characteristics with electronics (because electronics now represent

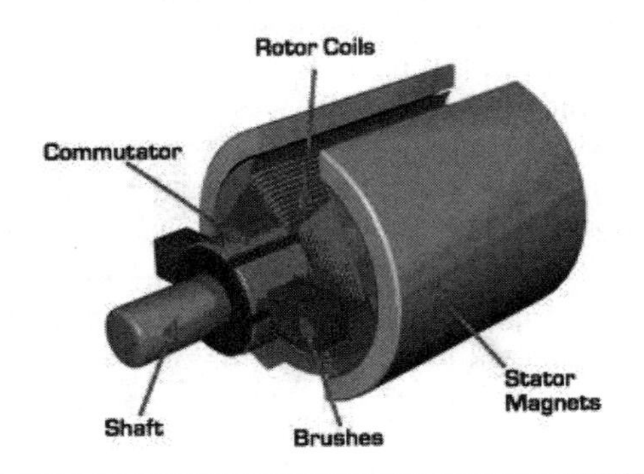

Figure 6-4 Permanent magnet DC motor (Courtesy of Zero Emission Vehicles of Australia and www.electric-cars-are-for-girls.com).

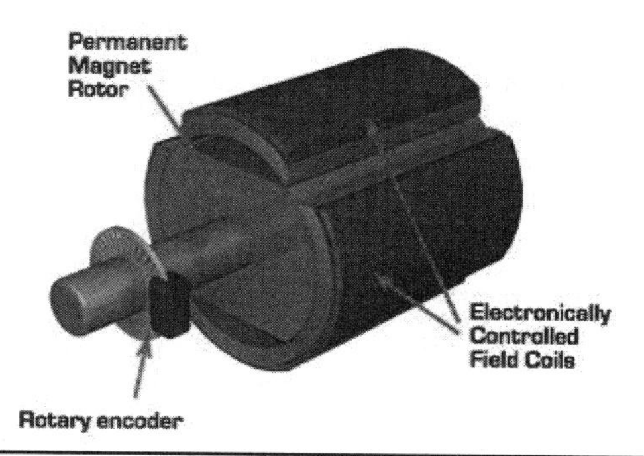

FIGURE 6-5 Brushless DC motor (Courtesy of Zero Emission Vehicles of Australia and www.electric-cars-are-for-girls.com).

half the motor), and the distinction between DC motor types blurs. In fact, as seen in Figure 6-5, the brushless motor more closely resembles an AC motor (which you'll meet in the next section) in construction. Assume that brushless DC motors resemble their permanent magnet DC motor cousins in characteristics—shunt motor plus high starting torque plus linear speed/torque—with the added kicker of even higher efficiency due to no commutator or brushes.

There are other manufacturers of DC motors. They are UQM and AVEOX. For more information, please refer to Chapter 12.

Universal DC Motors

Although any DC motor can be operated on AC, not all DC motor types run as well on AC and some might not start at all (but will run once started). If you want to run a DC motor on AC, you have to design for it. A series DC motor type is usually chosen as the starting point for universal motors that are to be run on either DC or AC. DC motors designed to run on AC typically have improved lamination field and armature cores to minimize hysteresis and current losses (see Figure 6-6). Additional compensating or interpole windings can be added to the armature to further reduce commutation problems by reducing the flux at commutator segment transitions. In general, series DC motors operating on AC perform almost the same (high starting torque, etc.), but are less efficient at any given voltage point.

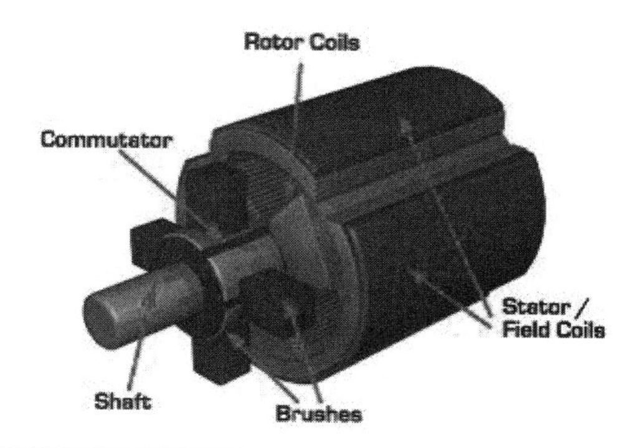

FIGURE 6-6 DC motor with round stator (Courtesy of Zero Emission Vehicles of Australia and www.electric-cars-are-for-girls.com).

The application of a motor in a vehicle doesn't need to contend with multiple sources of power. Universal motors are typically used in appliances, and not much at that, given what happened to the world during the oil crises in 1973 and 1979. Efficiency became a watchword, and a NASA patent was eventually put into the public domain. This resulted in the "Green Plug," which is now standard equipment in all new motor-driven appliances. These motors shouldn't be put in EVs of any sort. There are better solutions.

AC Electric Motors

Now that you've met DC electric motors, it's time to meet the motor you encounter most often in your everyday life—the AC electric motor. The great majority of our homes, offices, and factories are fed by alternating current (AC). Because it can easily be transformed from high voltage for transmission into low voltage for use, more AC motors are in use than all the other motor types put together. Before looking at AC motors and their properties, let's look at transformers.

Transformers

In its simplest and most familiar form, a transformer consists of two copper coils wound on a ferromagnetic core (Figure 6-7). The primary is normally connected to a source of alternating electric current. The secondary is normally connected to the load. When a changing current is applied to the primary coil, a changing magnetic field common to both coils results in the transfer of electrical energy to the second coil.

The other aspect of transformers useful to you (highlighted by Figure 6-7) is that an equivalent circuit of a transformer can be drawn for any frequency, and you can study what is going on. This is useful and directly applicable to AC induction motors.

AC Induction Motors

The AC induction motor, patented by Nikola Tesla back in 1888, is basically a rotating transformer. Think of it as a transformer whose secondary load has been replaced by a rotating part. In simplest form, this rotating part (rotor) only requires its conductors to be rigidly held in place by some conducting end plates attached to the motor's shaft. When a changing current is applied to the primary coil (the stationary part or stator), the changing magnetic field results in the transfer of electrical energy to the rotor via induction. As energy is received by the rotor via induction without any direct connection, there is no longer a need for any commutator or brushes. Because the rotor itself is simple to make yet extremely rugged in construction (typically, a copper bar or conductor embedded in an iron frame), induction motors are far more economical than their equally rated DC motor counterparts in both initial cost and ongoing maintenance.

While AC motors come in all shapes and varieties, the AC induction motor—the most widely-used variety—holds the greatest promise for EV owners because of its significant advantages over DC motors.

These solid-state components have resulted in AC induction motors appearing in variable speed drives that meet or exceed DC motor performance—a trend that will surely accelerate in the future as more-efficient solid-state components are introduced at ever-lower costs. This section will examine the AC induction motor for its torque, speed, reversal, and regenerative braking capabilities—the factors important to EV users. Other AC motor types not as suitable for EV propulsion will not be covered.

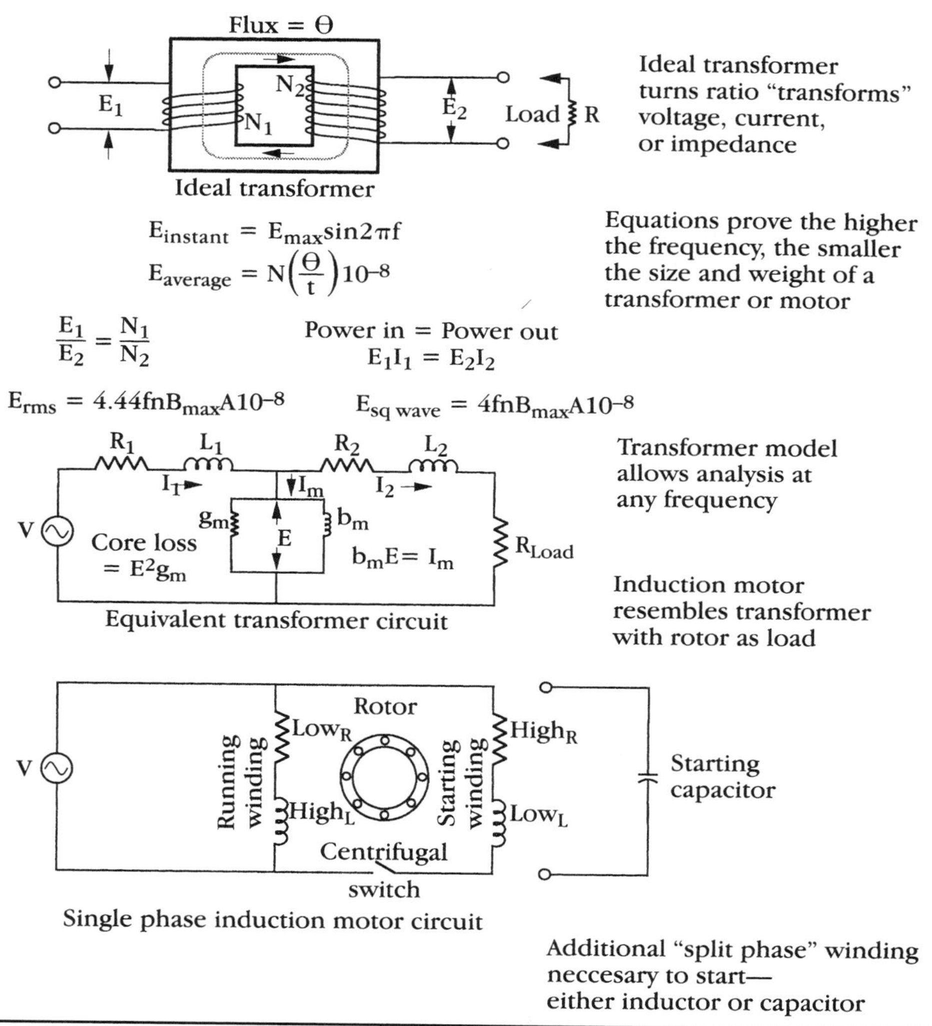

FIGURE 6-7 AC motors resemble transformers.

Single-Phase AC Induction Motors

Recall the universal DC motor discussed earlier. When you connect it to a single-phase AC source, you have little difference in its motor action because changing the polarity of the line voltage reverses both the current in the armature and the direction of the flux, and the motor starts up normally and continues to rotate in the same direction. Not so in an induction motor driven from a single-phase AC source. At startup you have no net torque (or more correctly, balanced opposing torques) operating on its motionless rotor conductors. Once you manually twist or spin the shaft, however, the rotating flux created by the stator currents now cuts past the moving conductors of the rotor, creates a voltage in them via induction, and builds up current in the rotor that follows the rotating flux of the stator.

How do you overcome the problem? The bottom of Figure 6-7 shows one key. If you introduce a second winding that is physically at right angles to the main stator winding, you induce a rotor current out *of phase* with the main rotor current that is sufficient to start the motor. This *split phase* induction motor design—or some variation of it—is the one you are most likely to encounter on typical smaller motors that power fans, pumps, shop motors, etc. To maximize the electrical phase difference between the two windings, the resistance of the starting winding is much higher and its induction is much lower than the running winding. To minimize excessive power dissipation and possible temperature rise after the motor is up and running, a shaft-mounted centrifugal switch is connected in series with the starting winding that opens at about three-fourths of synchronous speed. Figure 6-7 looks like a representation of a shunt motor: the smaller split-phase induction motor speed characteristics look like those of DC shunt motors, but their starting torque is much greater.

The most common split-phase induction motor is one that uses a capacitor-start, also shown in Figure 6-7. The capacitor automatically provides a greater electrical phase difference than inductive windings. This greater phase difference—nearly 90 electrical degrees—also gives capacitor-start split-phase induction motors a much higher starting torque (three to five times rated torque is common). The principle was discovered quite early by Charles Steinmetz and others, but capacitor technology had to catch up before it could be widely introduced on production motors. Capacitor-start design variations include two types: separate starting and running capacitors; and permanent capacitor with no centrifugal cutout switch. The two-capacitor approach brings you the best of both the starting and the running worlds; the permanent capacitor type gives you superior speed control during operation at the expense of lower starting torque. The other common split-phase induction motor design, called *shaded pole,* applies mostly to smaller motors; you are more likely to find it in your electric alarm clock than in your EV, so we'll skip it here.

Polyphase AC Induction Motors

Polyphase means more than one phase. AC is the prevailing mode of electrical distribution. Single-phase 208V to neutral from a three-phase transformer on the pole is the most prevalent form found in your home and office. The phase voltage that comes from the pole is 240V. These are widely available in nearly every city in the industrialized world. If one phase is good, then three phases are better, right? Well, usually. Stationary three-phase electric induction motors are inherently self-starting and highly efficient, and electricity is conveniently available.

Three-phase AC connected to the stator windings of a three-phase AC induction motor produces currents that look like those shown at the top of Figure 6-8—they are of the same amplitude, but 120 degrees out of phase with one another.

As in a DC motor, power and torque are also a function of current in an induction motor. Because the current is equal to the voltage divided by the motor reactance, at any given voltage, current is a function of stator, rotor, and magnetizing reactances that change as a function of frequency. The top of Figure 6-9 shows this at a glance.

The characteristic induction motor torque to slip graph, shown in Figure 6-9 for both its motor and generator operating regions, offers insight into induction motor operation. If an induction motor is started at no load, it quickly comes up to a speed that might only be a fraction of 1 percent less than its synchronous speed. When a load is applied, speed decreases, thereby increasing slip; an increased torque is generated to

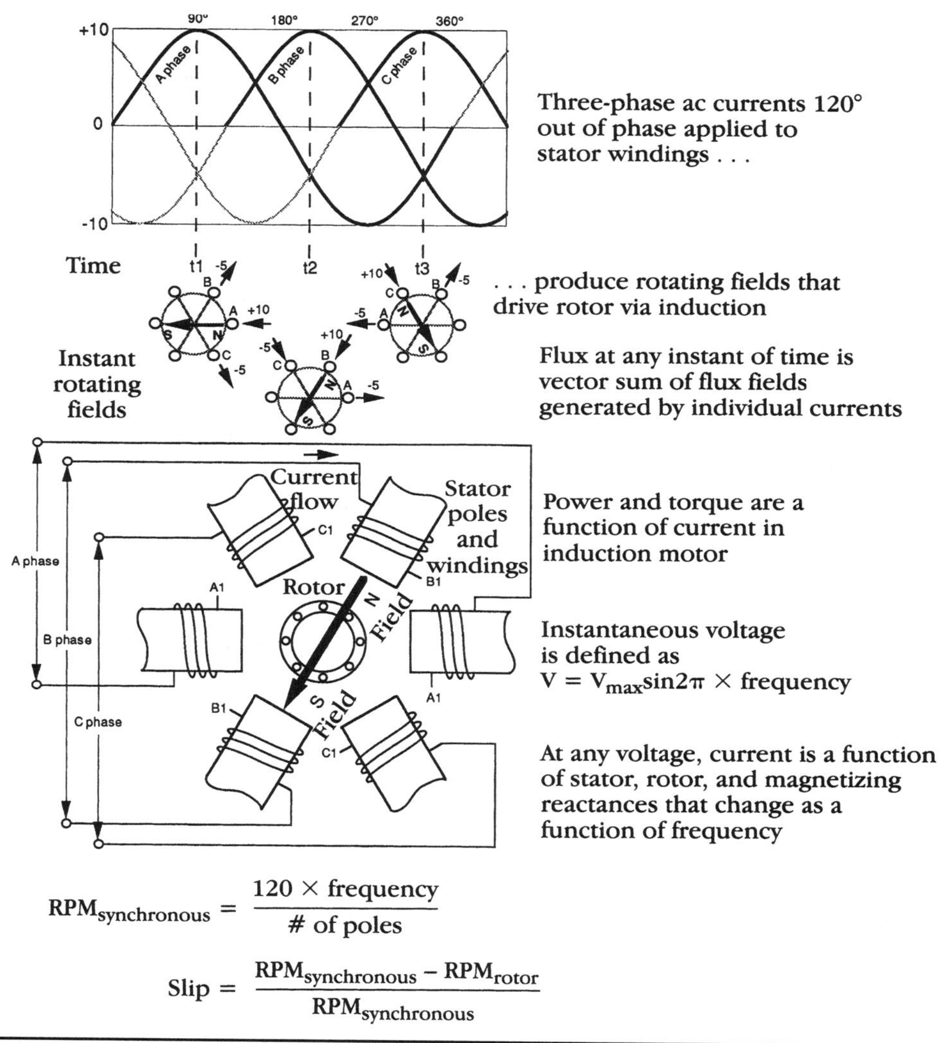

$$RPM_{synchronous} = \frac{120 \times \text{frequency}}{\text{\# of poles}}$$

$$\text{Slip} = \frac{RPM_{synchronous} - RPM_{rotor}}{RPM_{synchronous}}$$

FIGURE 6-8 Polyphase AC motor operation summarized.

meet the load up to the area of full load torque, and far beyond it up to the maximum torque point (a maximum torque of 350 percent–rated torque is typical).

Speed and torque are relatively easy to handle and determine in an induction motor. So are reversing and regenerative braking. If you reverse the phase sequence of its stator supply (that is, reverse one of the windings), the rotating magnetic field of the stator is reversed, and the motor develops negative torque and goes into generator action, quickly bringing the motor to a stop and reversing direction (see Figure 6-10). Regenerative braking action—pumping power back into the source—is readily accomplished with induction motors. How much regenerative braking you apply creates braking (moves the steady state induction motor operating point down the

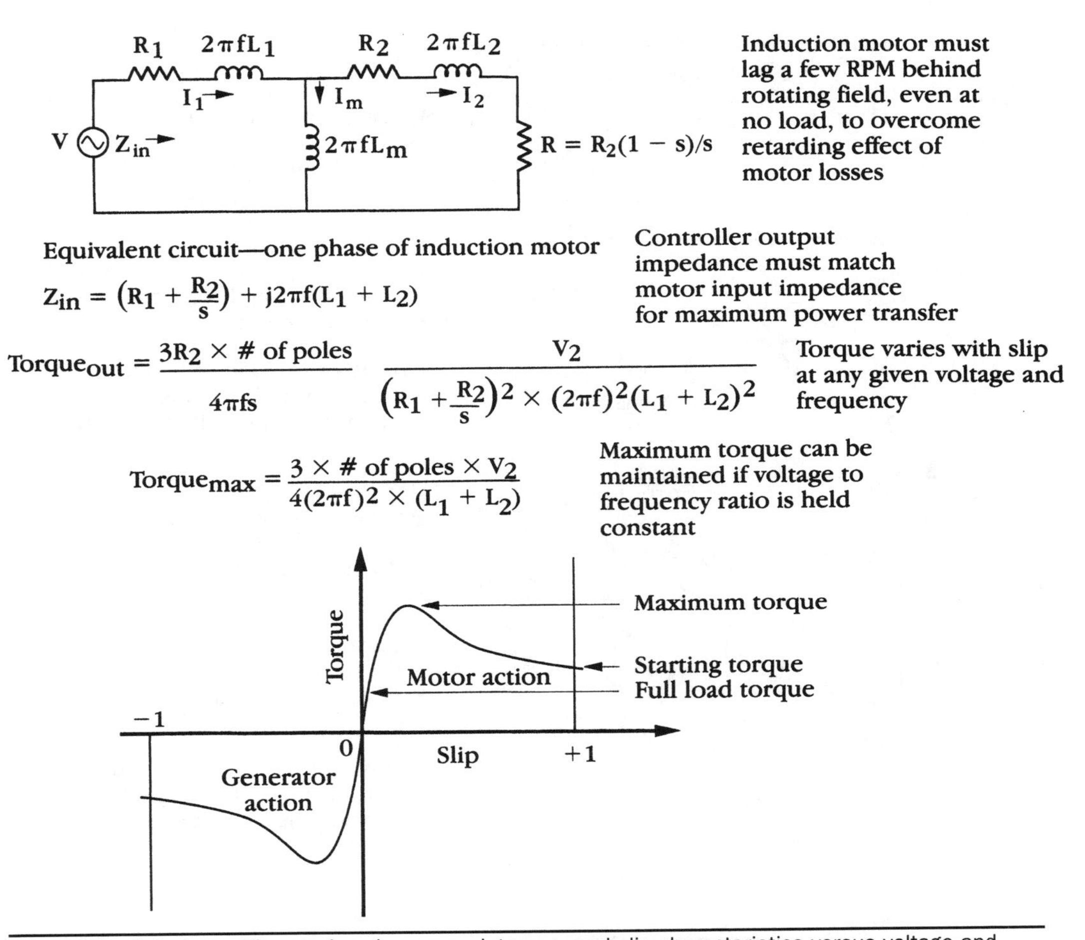

Figure 6-9 Polyphase AC motor's unique speed, torque, and slip characteristics versus voltage and frequency.

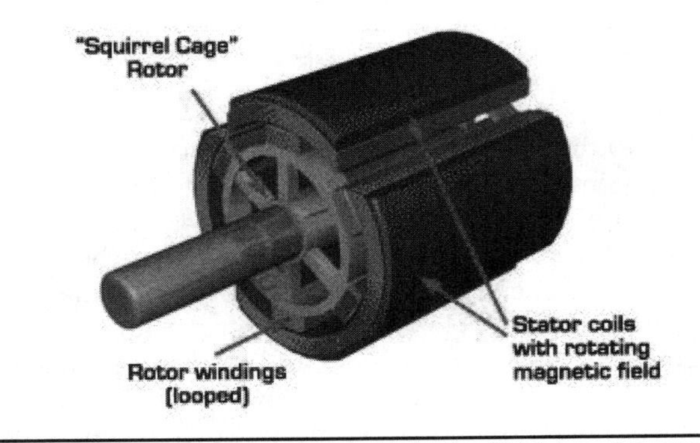

Figure 6-10 AC electric motor (Courtesy of Zero Emission Vehicles of Australia).

torque-slip curve), and generates power (instantaneously runs with negative slip in the generator region, supplying power back to the source).

Wound-Rotor Induction Motors

You might also encounter the wound-rotor induction motor in your EV power source quest. A wound rotor's windings are brought out through slip rings—conductive rings on the rotor's shaft—through brushes (analogous to DC motor construction) to an external resistance in series with each winding. The difference is that slip rings are continuous with no commutator slots. The advantage of the wound-rotor induction motor over the squirrel cage induction motor is that resistance control can be used to vary both the motor's speed and its torque characteristics. Increasing the resistance causes maximum torque to be developed at successively higher values. Along the way the wound rotor has better starting characteristics and more flexible speed control. What you give up is efficiency at an increased complexity and cost.

Today's Best EV Motor Solution

Different motor types in their numerous variations give you an almost unlimited bag of tricks in terms of solutions, but you have to figure out which one to use. If someone tells you there is only one motor solution for a given application, ask another person. Their not-so-surprising and different answer ought to convince you that there are no black-and-white motor solutions—only shades of gray. You can probably find three or four good motors for any need and you should look for and ask the questions that uncover these alternate solutions. On any given motor solutions list, you might choose the motor with the lowest price because that is your most important criteria. Another person might choose the motor with the lightest weight. Still another might choose the motor with the best startup torque. Regenerative braking might be most important to another person, and so on. Just remember shades of gray. Following along with this theme, the solution recommended here is not *the* solution, it is just *a* solution that happened to work best in this case. However, for economy conversions, DC series wound has been noted to be the best approach. Other motors are getting harder to find and not every controller on the market can drive the motor. For efficiency and for regenerative braking (which buys about 25 percent range in city driving), AC motors are the best way to go.

Series DC Motors Are the Best Approach Today

Series DC motors are available from many sources, they work well, controllers are readily available, adapters to different vehicles are easily made or purchased for them, and the price is right. A series DC motor might not be the ultimate or even the best current solution, but it's one that most EV converters will have no trouble implementing today.

Don't read anything into the tea leaves about the appearance of WarP motors here. They are only one out of a large number of motor manufacturers (see Chapter 12 for a list to get you started) and the motor recommended is only one out of a number of motors they manufacture.

Today's Winner: WarP Motors

The New WarP ImPulse 9" Motor, as seen in Figure 6-11, is basically similar to the WarP 9" diameter series wound motor, only shorter. It's the same length as the 8" WarP motor but with a higher power and efficiency rating.

FIGURE 6-11 WarP ImPulse 9" motor for front-wheel-drive cars.

This motor has a "shorty" tail shaft housing from a Chevrolet Turbo 400 transmission fitted to the drive end-bell. The drive end-shaft is not the typical 1.125" single-keyed type, but rather a hefty 1.370", 32-tooth involute spline that is identical to the tail shaft spline of a Turbo 400 transmission. In other words, this motor was designed to replace a transmission and couple directly to a drive shaft!

The Advance FB1-4001

Another option, the Advance FB1-4001, shown in Figure 6-12, was designed to propel EVs. One of its big advantages that should not be underestimated is that you can acquire it new from a reputable vendor. A used compound wound DC aircraft starter motor for $200 might sound great, but powering EVs was never intended to be its true mission in life, and how much use has it seen already? With the FBI-4001 you get a motor that you can return if it doesn't work. In a reputable vendor you have someone to turn to for answers to questions, technical data, and more. A surplus dealer is rarely able to offer this capability.

As I researched for this edition of the book, I noticed that the electric vehicle called the Tango from Commuter Cars Corporation used the FB1-4001. When I spoke with Rick Woodbury, creator of the Tango and president of the company, we talked for over an hour about the state of electric cars. One of the greatest things that I had heard from him was that the first edition of this book helped him to create the vehicle. To paraphrase, "If it wasn't for this book, the Tango would not have been built." I am glad this book made such a contribution to the electric vehicle industry.

One of the more popular riders of the Tango is George Clooney, as pictured with his Tango in Figure 6-13. Hopefully his involvement and the wonderful aspects of the Tango will allow Rick to *Build More Of His Own Electric Vehicles*. Keep it up, Rick!!! Way to go!

Figure 6-14 shows the performance curves used in deriving Chapter 5's data, this time shown for values of 72 through 120 volts. Table 6-1 gives you the data from its S-2 DIN and ISO thermal tests. Another advantage to buying a new motor from a reputable

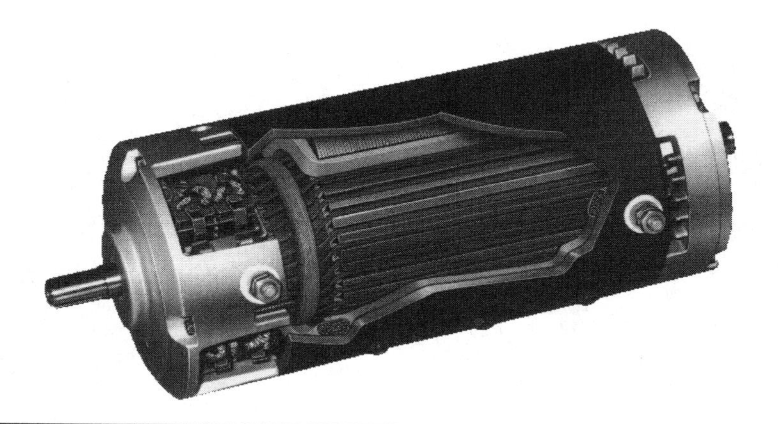

Figure 6-12 Advance series DC motor cutaway view.

dealer is that you have the curves and data you need to help you optimize your EV conversion.

There are better solutions. But regenerative braking was not important for the conversion we detail in Chapter 10, and there was already a matching controller available. We wanted good middle-of-the-road performance at a good price, as well as a product that any potential EV converter reading this book could use and get working the first time up to bat. This motor delivers all that and more.

Tomorrow's Best EV Motor Solution

While the series DC motor is unquestionably the best for today's first-time EV converters, the bias of this chapter toward AC induction motors was not accidental. Improvements in solid-state AC controller technology clearly put AC motors on the fast track for EV conversions of the future. AC motors are inherently more efficient, more rugged, and

Test Voltage	Time – On	Volts	Amps	RPM	HP	Peak HP	KW
75 volts – .03I	5 minutes	63.5	380	1900	27.0		20.3
75 volts – .03I	1 hour	68.0	240	2550	19.0		14.3
75 volts – .03I	continuous	69.0	210	2800	17.0		12.8
75 volts – .03I						42.0	
96 volts – .03I	5 minutes	88.0	360	3300	35.0		26.5
96 volts – .03I	1 hour	89.0	210	3600	23.0		17.3
96 volts – .03I	continuous	90.0	190	3900	20.0		15.0
96 volts – .03I						70	

Table 6-1 Data for Advance Model FBI-4001 Series DC Motor S-2 DIN and ISO Thermal Tests

Figure 6-13 The Tango uses the Advance FB1-4001; George Clooney is its most popular owner.

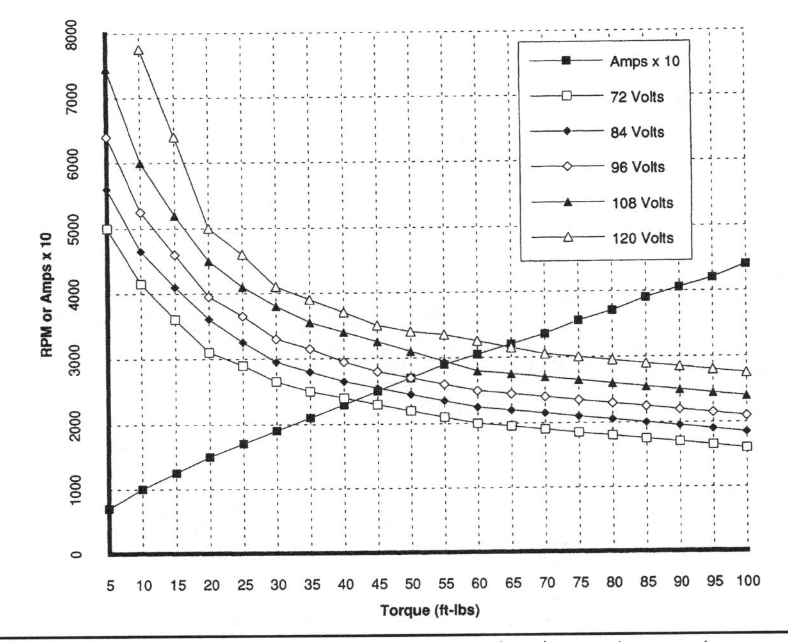

Figure 6-14 Advanced FB1-4001 series DC motor's speed and current versus torque curves.

less expensive than their DC counterparts; the reason why they are not in more widespread use today has to do with controllers, as you'll learn in Chapter 7.

These underlying facts made my surprise all the greater when I asked Darwin Gross (the "engineer's engineer" referenced in the Acknowledgments) the question, "What's the best electric vehicle solution for tomorrow?" His answer was brief, to the point, and thought-provoking. It forms the basis for this section. Right or wrong, the answer should get your creative juices flowing. Figure 6-15 tells the story at a glance.

The ideal EV drive for tomorrow has:

- Low weight
- Streamlined design
- Simple drivetrain (one or two speeds)
- DC to get started
- AC to run above 30 mph
- High frequency components (≥400 Hz)
- DC motor that gets 96 volts
- Three-phase AC motor that gets 400 volts
- Matching controller and motor impedance

Let's look more closely at some of the pieces.

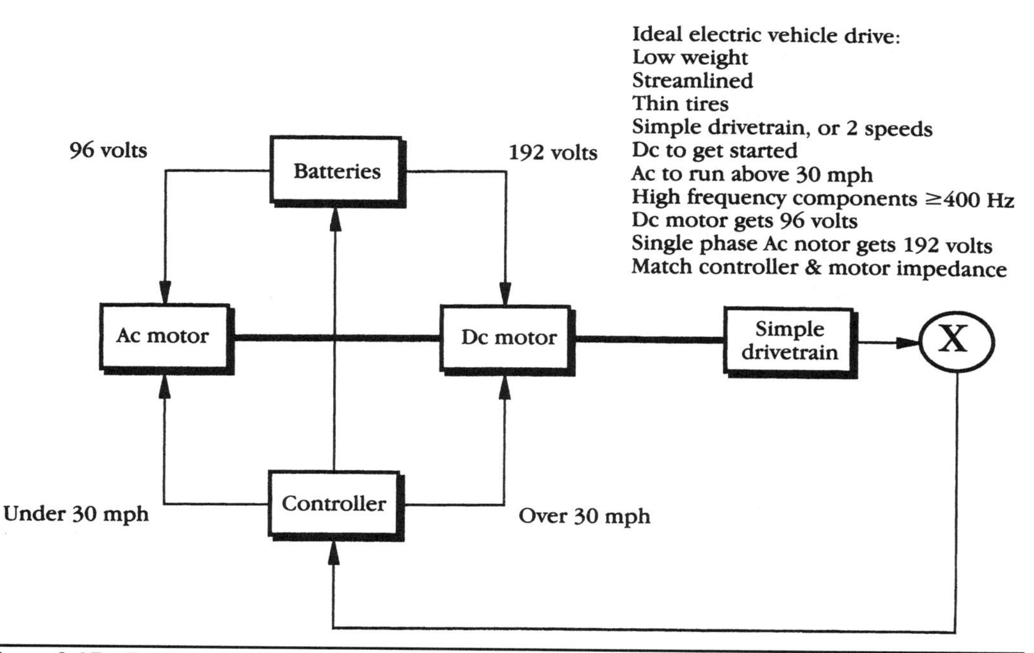

FIGURE 6-15 Tomorrow's best electric vehicle motor solution.

DC and AC Working Together

The series DC motor provides the best starting torque, and the AC induction motor is the most efficient at speed. By using them together with some mechanism to switch back and forth (for example, at around 30 mph on the level with a kickdown for hills), you get the best of both worlds. Only about 96 volts are needed to get the DC motor started in your average utility commuter EV. When the AC motor takes over, you can add the rest of the battery pack toward powering it—120, 144, 192 volts, or whatever.

Darwin emphasized, "Pick a good core material with some nickel in it to achieve 96 percent efficiency, and remember size and weight go down as the frequency does up."

Tuning

Electrical and mechanical systems deliver better performance when balanced and tuned up. Race car mechanics set up their cars for the average case, then tune and adjust them for maximum performance. Should you do anything less with your EV? Match the impedance of the motor to the controller for maximum power transfer, and go through everything else with a fine-tooth comb.

Keep It Simple

What's the simplest controller that can implement this approach? Do you even need to use solid-state designs? How few batteries can you get by with? Which are the smallest motors you can use? Can you do something innovative with their placement? How can you simplify the drivetrain? Can you do something better with the tires? You get the idea.

Conclusion

Currently, the most economical (and indeed common) option for electric vehicles is series DC technology. However, it is likely that all commutated motors will be phased out over the next decade or two, since a commutator's functionality can be replaced by clever electronics (which is getting cheaper every day), and with improved reliability and efficiency. At present, AC induction and permanent magnet brushless DC are the best technologies available, with efficiencies up to 98 percent, silent operation, and almost never requiring any servicing. They each have various advantages and disadvantages over one another. It will be interesting to see which one becomes the new standard in the years to come.

CHAPTER 7

The Controller

If automobiles had improved as much as electronics in the past few decades, they would go a million miles per hour, cost only pennies, and last for decades.

The controller is another pillar of every electric vehicle. If one area could take credit for renewed interest in EVs, the controller would be it. You can buy a controller, plug it in, and be up and running in no time—something earlier EV enthusiasts could only dream about. In the future, there will be only be further reductions in size and improvements in efficiency of the motor control electronics. While the motor may only benefit from small improvements due to technology changes, future motors may become distributed and located in the wheels themselves.

Your controller decision follows logically from your initial EV conversion choice: tire-smoker, high-range flyer, or around-town commuter. But, unlike the plethora of choices with electric motors, your controller choices are rather simple, and are dictated by the electric motor you use, your desire to make or buy, and the size of your wallet or purse.

In this chapter you'll learn what the different types of controllers are, how they work, and their advantages and disadvantages. Then you'll encounter the best type of controller to choose for your EV conversion today (the type used in Chapter 10's conversion), and the electric motor controller you're likely to be seeing a lot more of in the future.

Controller Overview

You can relax now. The useful but largely theoretical equations of Chapter 6 are behind you. Controllers can generate even more equations, but we'll save those for the engineers who are building them. The objective here is to give you a brief controller background and introduce you to a working controller for your EV with minimum fuss.

A controller basically is the brain or computer of an electric car. This computer "controls" or governs the performance of the electric motor. The controller integrates the motor speeds and expected battery range/speed through its energy density. Coordinating between the controller and the motor can help a car accelerate from 0 to 60 in 6 to 7 seconds (or less), which can determine the range of the car and the top speed. The controller controls both manual or automatic drive systems for starting and stopping, going forward or backward, governing the speed, regulating or limiting the torque, and protecting against overloads and faults.

Off-the-shelf sources mentioned in Chapter 12 are really the only way to include a controller into your electric vehicle. While a switch-based handmade controller is OK, safety is the key. Whereas, there are no interlocks or revlimiting features on a switch-based handmade controller. For those who do not know, revlimiting a technology inside the controller that regulates the highest speed the vehicle will be allowed to go. This basically acts as a rev limiter. Also, resistors waste energy. A controller from a reputable company that can provide safety and reliability is the number one priority.

The controller correctly matched to the motor will give it the right voltage. The weight of the motor magnets and size of its brushes determine the power and torque. You can buy a reliable controller from a number of sources today at a good price.

Controller Overview

A motor speed controller uses a microprocessor to drive a preamplifier and then power stage amplifier which controls the flow of power from the battery to the motor, with various feedback sensors to monitor the system's operation. Red arrows indicate pulse width modulated outputs, green indicates feedback sensors.

Switch Controller

Switch controllers are the oldest type. They have been used in EVs for over 100 years, and there are still lots of them in use today. Basically, they are a set of big switches that connect the batteries and motor(s) in various series/parallel combinations to get discrete speeds. For example, four 12 volt batteries could be connected for 12v, 24v, or 48v to provide slow, medium, and fast motor speeds. Switch controllers are cheap, simple, and efficient. Their main drawbacks are the stepwise speed control, and minimal feature set. You don't generally buy such a controller; instead, you buy standard parts and wire it yourself according to a basic wiring diagram (like an electrician wires a new room in a house).

Solid-State Controller

The DC brushed motor with solid-state controller is the most common type. These controllers have been used for decades in all types of electric vehicles, and are widely available off-the-shelf. Basically, they use solid-state switches (SCR, transistor, MOSFET, IGBT, etc.) to replace the mechanical switches in a switch controller. MOSFET stands for Metal Oxide Semiconductor Field Effect Transistor. These are more common among lower voltage controllers since low voltage MOSFETs tend to have very low on resistance (and hence power loss). MOSFET controllers are very efficient at low power levels, since power loss in a MOSFET is proportional to the square of the current. Their resistance also increases as they heat up, so when used in parallel they tend to automatically balance the load.

IGBT stands for Insulated Gate Bipolar Transistor, and it somewhat like a hybrid between bipolar transistors and field effect transistors. IGBTs have a constant voltage drop which makes them more efficient than MOSFETs at high power levels, though often less efficient for low power applications. One disadvantage with IGBTs is that, like all bipolar transistors, they are prone to thermal runaway and imbalances when used in parallel. As such IGBT controllers require good cooling systems (liquid cooling is common) and/or matched transistors to avoid imbalance. (Source: Zero Electric Vehicles Australia, www.zeva.com.au)

Electronic Controllers

Electronic switches can be switched on/off much faster, and don't wear out. The controller switches them on/off so fast (thousands of times per second) that the motor gets the average current rather than the peak or zero. This is called Pulse Width Modulation (PWM). PWM controllers are modestly priced, readily available in a wide variety of sizes, and provide smooth stepless motor control.

AC Controller

The AC motor with solid-state inverter is the most sophisticated type. They are used in high-end EVs where meeting performance objectives is more important than cost. An AC induction motor or AC PM motor (often called a "brushless DC" motor) is driven by an inverter that converts battery DC voltage into variable voltage variable frequency 3-phase AC. New EV sports cars and the recent auto company produced EVs all used AC drive systems. Like modern ICEs, these AC controllers are very complex and expensive, but offer the most advanced features (like cruise control and regenerative braking) and provide the best overall range.

Controller Choice

High reliability, high performance, and minimum cost are the key trade-offs that factor into your EV controller choice." You can pick one, or maybe two, but you can't have all three. Examples:

- High reliability, high performance—means high cost

  (Cafe Electric Zilla controller, no know failures, 300v 2000a, $3500)
- High performance, minimum cost—means low reliability

  (Logicsytems rebuilt Curtis controller, high failure rate, 156v 1000a, $1500)
- High reliability, low cost—means low performance

  (used GE EV-1 SCR controller, 30 years old and still working, 72v 400a, $1000)

If you've driven an EV, you'll find that most electronic controllers do *not* imitate the accelerator pedal response of a normal ICE vehicle! It is undesirable to do so!

The throttle response of most ICEs is unpredictable and very "lumpy." Some cars leap forward at the slightest touch; others gently creep ahead. As the pedal is pushed farther, the amount of speed increase for a given amount of pedal movement varies considerably. On some cars, there is hardly any difference between half throttle and full throttle; on others it's the difference between 30 mph and 120 mph. There is always a noticeable delay. It doesn't coast, and letting up the pedal produces variable amounts of engine braking. People just get used to these huge variations.

The throttle response of EV controllers is completely different. It is entirely predictable, repeatable, and instantaneous. Most people have to relearn it, but find it *better* than an ICE. The motor controller normally controls motor current and voltage, from 0 to maximum, as a direct function of throttle position, 0 to maximum (i.e. half throttle means half voltage and half current). Electric motors don't need to idle, so releasing the pedal means the motor is off (no creep). Pressing the pedal to a certain point sets a certain motor voltage, which means a certain speed. Pressing hard gives you a constant rate of acceleration (because the controller is in current limit; constant current = constant torque). Release the pedal, and it simply coasts (unless you have regenerative braking, in which case you can configure what it does).

DC Motor Controller—The Lesson of the Jones Switch

The easiest way to vary the DC voltage delivered to a DC motor is to divide its steady, nonvarying DC component into smaller pieces (see Figure 7-1). The average of the number or size of these pieces will be a resultant DC voltage that the motor "thinks" it is receiving. At the top left of Figure 7-1, the value of the voltage delivered to the motor is

$$V_{motor} = ((t_{on})/(t_{on} + t_{off})) \times V_{battery} = (t_{on}/T) \times V_{battery} = K_{dc}V_b \quad (1)$$

This equation shows that the voltage delivered to the motor is proportional to the amount of time the pulse is on versus the total length of the period. As $t_{off} >> t_{on}$ in the left-hand graph, the average voltage (V) the motor receives is only equal to a small

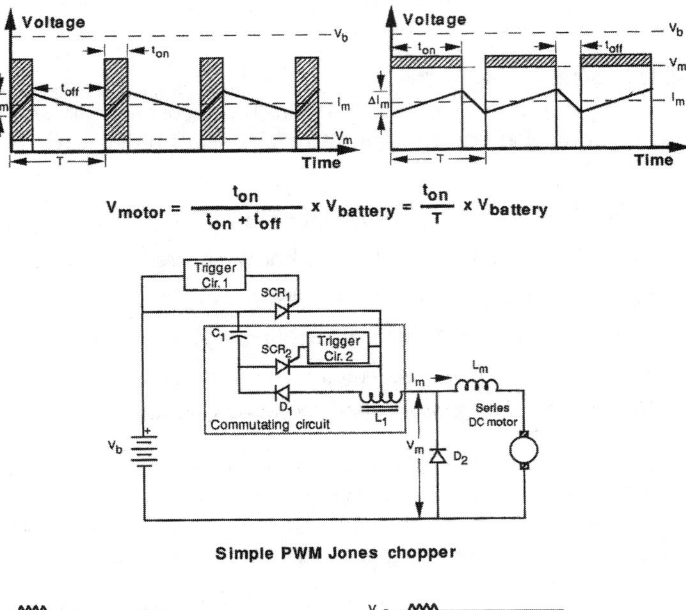

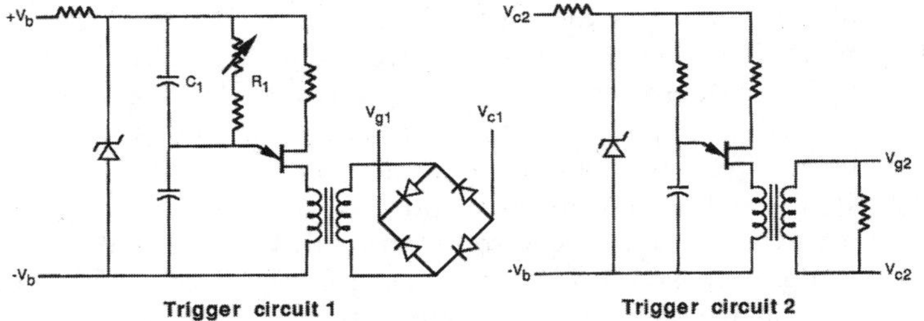

FIGURE 7-1 Summary of Jones switch PWM chopper controller characteristics.

value. But in the right half of the graph, $t_{on} >> t_{off}$ and V_m is a much larger value—nearly equal to V_b—for the battery voltage. If you vary t_{on} while holding T (the total period) constant, you have the principle of *pulse width modulation* or PWM. (If you vary T while holding t_{on} constant, you have *frequency modulation*.) For a fixed ratio, a duty cycle of K_{dc} might be defined (far right or the equation above). PWM, or *chopper control*, as it is customarily called, allows you easily to control a DC motor by electronically chopping the voltage to it.

The earliest solid-state DC motor chopper controllers used SCR devices. These were technically primitive compared to today's solid-state controllers, but typically offered greater efficiency, faster response, lower maintenance, smaller size, and lower cost than the motor-generator sets or gas tube approaches they replaced.

The Jones chopper circuit shown in the middle of Figure 7-1, adapted from an early GE SCR manual, is one of the simplest variations. Earlier, similar PWM circuits were also built in transformerless style using transistors. The Jones chopper is easy to look at and analyze, and its basic lessons apply to all PWM controllers.

It might be viewed as:

- An SCR (SCR_1) controlled by a commutating circuit (SCR_2 and trigger circuit 2) that controls the t_{on}/t_{off} time and a trigger circuit (trigger circuit 1) that controls the period T or frequency.
- The diode D_2 connected across the series DC motor's terminals—usually called a *free-wheeling diode*—performs two important functions in this and other circuits where the pulsed output is inductive.
- It "smooths" motor current I_m and enables it to continue to flow when SCR_1 is not conducting (top of Figure 7-1), and prevents the inductive-current-generated high-voltage spikes, which could potentially damage the transistors, from appearing across the motor when SCR_1 is turned off.
- Diode D_1 prevents the L_1–C_1 combination from oscillating. Other than the SCRs and diodes, the key circuit elements are the center tapped autotransformer (both windings have the same number of turns, inductance, etc.), whose inductance is simply labeled L_1 in Figure 7-1, the capacitor labeled C_1, and the combined series motor and external inductance labeled L_m.

The frequency of the chopper is controlled by the R_1–C_1 combination (where R_1 is the sum of the fixed and variable resistors in that branch of the circuit) in trigger circuit 1 shown at the bottom left of Figure 7-1. In this case, a unijunction transistor forms the heart of a relaxation oscillator whose period is controlled by the RC time constant such that

$$f = 1/(R_1C_1)$$

where R is resistance in ohms and C is capacitance in farads.

For a 2-kHz chopping frequency, you might choose $C_1 = 0.1\ \mu F$,

$$R_1 = 1/(2 \times 10^3 \times 0.1 \times 10^{-6}) = 5 \times 10^3 = 5 \text{ kilohms}$$

The switching frequency in a modern controller is derived from a crystal, not an RC oscillator. Stability under varying load and environmental conditions becomes more critical as power levels increase. This rudimentary example is simply for demonstration purposes.

From a design standpoint, the important elements in the circuit are

- The battery voltage V_b
- The maximum motor current I_m to be commutated
- The turn-off time of SCR_1
- The voltage rating of SCR_1
- The values for C_1, L_1
- The value for L_m to minimize the ripple current in the motor armature ΔI_m.

Motorola still has this product, which is now made by ON Semiconductor (www.onsemi.com/pub/Collateral/MC33033-F.PDF).

These are the basics. You have a highly modular, very superior solution to the Jones chopper, and you aren't even breathing hard. As newer components are introduced, you can take advantage of them just by plugging them into the drive or power stages. You can easily change the frequency or even the motor speed control circuitry if it suits your purposes better.

As for the options, in Figure 7-2 there are two additional circuit boxes:

- Current limit control
- RPM feedback control

Current limit control monitors motor armature current 1m and feeds a signal back to the shutdown command input on the PWM chip if it goes beyond a preset level.

The Jones chopper circuit was a popular early electric vehicle controller, even though it was an incomplete circuit. Like equivalent circuits of its vintage, it was inexpensive and easy to make (with relatively few components).

However, the disadvantages almost outweigh buying off-the-shelf controllers. They are relatively unsophisticated. Jones Chopper is even potentially dangerous to your series

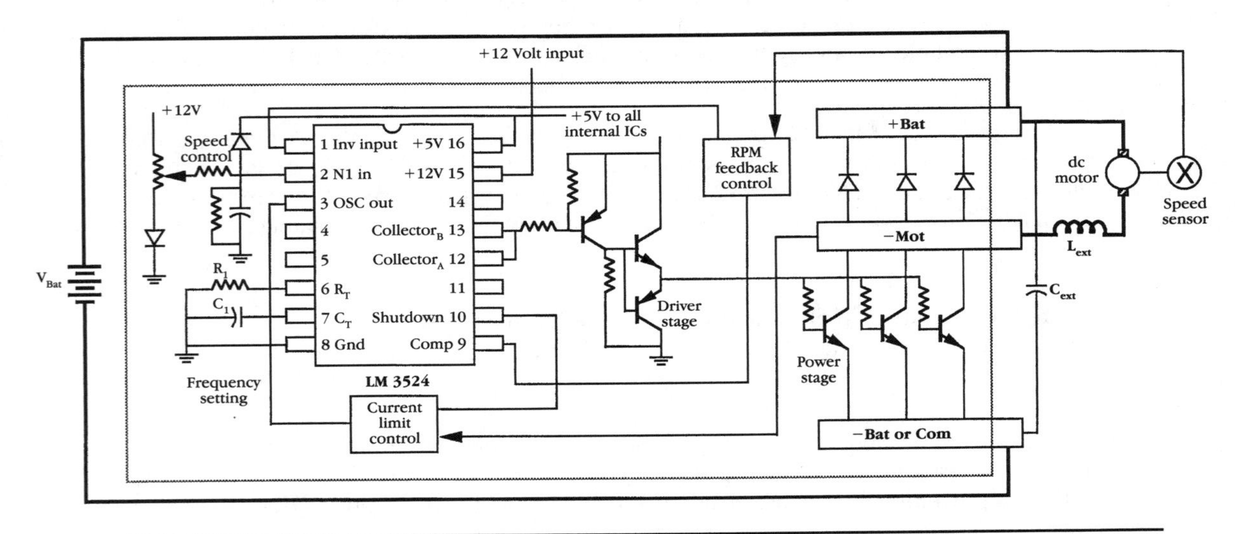

FIGURE 7-2 Real controller design using LM3524.

DC motor because it has no overcurrent sensing or limiting features; in other words, no recourse or shutdown mechanism if one or both of your SCRs has commutation problems, or other discrete components in the circuit fail outright or drift out of tolerance with temperature and age. Now let's move on to more modern solutions, and address the implied question of what can be done to improve the performance of the PWM circuit.

In other words, can you build your own controller? Absolutely. But an off-the-shelf controller gets you up and running quickly, and puts your EV conversion on the road with the least fuss. That's why you will also find a number of professional conversion shops using them as well.

An Off-the-Shelf Curtis PWM DC Motor Controller

You can do something today that EV converters of a decade ago could only dream about—pick up the telephone and order yourself a brand-new DC motor controller from any one of a number of sources. You can have it in your hands a few days later, mount it, hook up your EV's electrical wiring and throttle control to it, and be up and running with virtually a 99 percent chance of everything working the first time.

Like series DC motors, today's DC controllers are readily available from many sources, they work great, most of them are of the PWM variety, they are easily installed in different vehicles, and the price is right. A modern, off-the-shelf PWM DC motor controller is not the ultimate, but it's pretty close to the best current solution. More important, it's one that most EV converters will have no trouble in implementing today. After you do your first conversion, become the acknowledged genius in your neighborhood; once you know what you really like and don't like, you can get fancy and exotic.

The DC PWM controller recommended here is from Curtis PMC, a division of Curtis Instruments, Inc. of Mt. Kisco, New York. As with the motors, don't read anything into its appearance here. Curtis is only one of a large number of controller manufacturers from the list in Chapter 12, and the recommended controller model is only one out of a number they manufacture.

The Curtis PMC model 1221B-7401 DC motor controller, shown in Figure 7-3, is already very familiar to you. Its features include:

- PWM-type controller
- MOSFET-based technology
- Runs at a constant switching frequency of 15 kHz
- Requires use of an external 5-kilohm throttle potentiometer
- Automatic motor current limiting
- Thermal cutback at 75 through 95 degrees C
- High pedal lockout (prevents accidental startup at full throttle)
- Intermittent-duty plug braking
- Overvoltage and undervoltage protection
- User-accessible adjustments for motor current limit, plug braking current limit, and acceleration
- Comes in a waterproof heat sink case

Figure 7-3 Curtis PMC 1221B DC motor controller.

The controller is also well matched in characteristics to the Advance model FBI4001 series DC motor, particularly in the impedance area (you read about the importance of this for peak power transfer). If the Curtis PMC DC controller characteristics sound familiar, it's because they use the PWM IC technology you've been reading about throughout this chapter, and bring all these benefits to you with additional features in a rugged, preassembled, guaranteed-to-work package at a fine price. The same reputable vendor comments from the Chapter 6 also apply here.

Installation and hookup is a breeze. If you look closely at the controller terminals, you'll notice the markings M–, B–, B+ and A2 appear (listed clockwise from the lower left when facing the terminals). You already know that the first three correspond to the –Mot, –Bat, and +Bat markings on the terminal bars in Figure 7-2. The A2 marking means *Armature 2*, the opposite end of the armature from the *Armature 1* winding that is normally connected to the +Bat, or in this case the B+ terminal. Anything else you might want to know is covered in the Curtis PMC manual that accompanies the controller.

AC Controllers

AC overwhelmingly has benefits that make it a winner, in spite of complications involved. In general, AC motor controllers require more protection devices to isolate against noise, yet DC motors make way more noise than AC!

Chapter 6 showed that the speed-torque relationship of a three-phase AC induction motor is governed by the amplitude and frequency of the voltage applied to its stator windings (the upper left part of Figure 7-4 depicts this relationship). The best way to change the speed of an AC induction motor is to change the frequency of its stator voltage. As you can see in Figure 7-4, a change in frequency results in a direct change in speed, and if you change the frequency in proportion to the voltage (both at ¼, ½, ¾, etc.), you get the speed-torque curves shown.

Knowing the voltage and frequency ratio that you want to maintain allows you to calculate the voltage, current, and output torque relationship for any values of input voltage and frequency using vector math and lookup tables. In simpler terms, if you feed the speed and torque values you want to some sort of "smart box," it can provide the voltage and frequency necessary to generate the proper motor control signals.

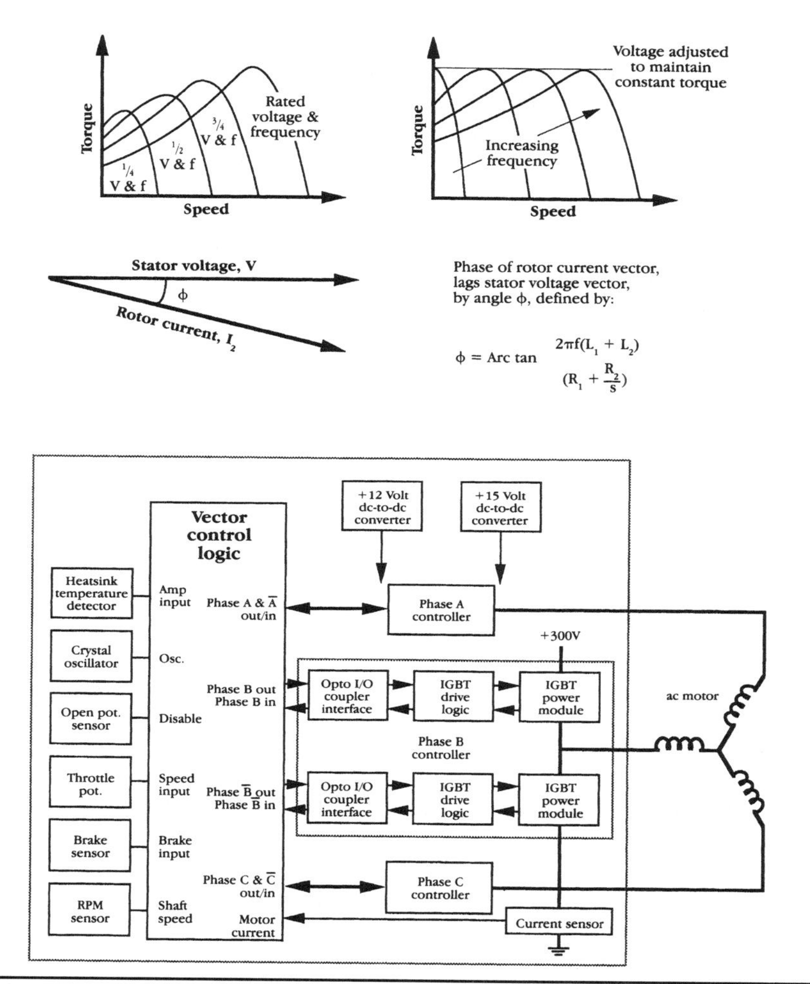

FIGURE 7-4 AC controller design objectives and block diagram.

Unfortunately, since most of our names are not Tesla or Steinmetz, all this additional complexity means a lot more sleepless nights in the garage for even the most enthusiastic build-it-yourselfers, and is the real reason why three-phase AC induction motors—despite their enormous advantages—will probably not reach the do-it-yourself EV converter. Meanwhile, let's look at the best solution for your EV conversion project of today.

Today's Best Controller Solution

If you've previously read Chapter 6 on motors, you already know this book recommends a series DC motor as the best motor solution for today's EV converters. This greatly simplifies our choice of motor controllers—we only have to choose from those in the DC motor universe. But there are still numerous DC controller vendors and models to choose from, and you have to figure out which one to use. The same words of advice apply in the controller area as in the motor area: If someone tells you there is only one controller solution for a given application, ask another person. Like motors, you can probably find three or four good controllers for any application because there are only shades-of-gray controller solutions. And, as with the recommended motor, the controller recommended here is not the solution, it is just a solution that happened to work best in this case.

While the series DC motor and PWM controller are unquestionably best for today's first-time EV converters, the bias of this book is toward AC controllers and motors. However, one of the best EV technologies today is a DC controller. On one side, AC induction motors are inherently more efficient, more rugged, and less expensive than their DC counterparts. This translates to more driving range from a given set of batteries, and less probable failures and the possibility of graceful degradation when one does happen, but not less expense! The package isn't cheap and the motor alone isn't either. These benefits come at a price. That's why nearly every newly designed commercial EV today utilizes one or more AC induction motors or its closely related cousin, the brushless DC motor.

No new electric car uses DC drive. However, some of the greater conversions out there use DC controllers. DC is cheap and very suitable for a budget conversion. Again, for a conversion that is less expensive, a DC motor works.

What's in the labs today will be available to you in the not-too-distant future and, beyond that, continued improvements in solid-state AC controller technology could put AC motors in every EV conversion of the future. Let's look at developments in two areas—systems and components—that virtually guarantee this outcome.

Now, let's look at the future—today!

Zilla Controller (One of the Best DC Controller for Conversions)

The Zilla controller (Zilla), shown in Figure 7-5, is by far one of the most popular and powerful motor controllers available for electric vehicles. The Zilla has exceptionally high power density. According to the conversion specialists and people in the EV community today, the Zilla is ranked as one of the best controllers out there on the market. This controller can make your EV have the most amazing pickup and torque like no other on the market (or even on your block).

We like that Zilla cared about safety in their literature because it shows that a reasonably priced, efficient, and off-the-shelf controller can still be safe. From carefully monitoring that the controller comes up to voltage, communicates properly, and checks the integrity of the output stage before engaging the main contractor, to the dual microprocessors that cross-check and have independent means of shutting the system off, there is no other DC EV controller that approaches this level of security.

It is no surprise that all of the world's quickest electric vehicles use Zilla controllers, but the safety features allow them to also excel in street applications. The Z1K in particular has become very popular for street conversions due to its superior feature set and low price. Table 7-1 shows the specifications of this popular controller.

Zilla Models

The Zilla comes in two current ratings and up to three voltage ratings:

Figure 7-5 Zilla Controller Model Z2K-HV (Courtesy of Café Electric, LLC).

- The **Z1K-LV** is the lowest cost unit. While it regulates up to 1,000 amps of motor current to run on lead-acid battery packs of 72 to 156 volts, this unit is more than twice as powerful as the common type of controllers for most "normal" car conversions. Because Café Electric wants people to convert cars to have affordable options, they use "the sales of higher voltage controllers to partially subsidize production of the Z1K-LV units." This keeps the amount of these units delayed for a reason.
- The **Z1K-HV** can regulate up to 1,000 amps of motor current and run on lead-acid battery packs of 72 to 300 volts nominal. This unit in a typical car with a shifting transmission and a powerful battery pack of at least 200 volts will result

Maximum nominal input voltage range for Lead-Acid batteries: 72 to 348 volts
Absolute maximum fully loaded input voltage range: 36 to 400 volts*
Maximum motor current at 50°C heat sink temperature: 2000 Amps for Z2K, 1000 Amps for Z1K
Maximum Battery Current at 200V: 1900 Amps for Z2K, 950 Amps for Z1K
Maximum Battery Current at 300V: 1770 Amps for Z2K, 885 Amps for Z1K
Maximum Battery Current at 400V: 1600 Amps for Z2K, 800 Amps for Z1K
Continuous motor current @ 50°C coolant temp and 100% Duty Cycle: over 600 Amps for Z2K, 300 Amps for Z1K
Peak Power: 640,000 Watts for Z2K, 320,000 Watts for Z1K
PWM frequency 15.7 kHz
Power devices IGBT
Voltage Drop: <1.9 volts at maximum current.
* At this time we are suggesting not exceeding 375 volts on the EHV models; we hope to bring that back up to 400 volts with further testing.

Table 7-1 Zilla Specifications

in a very quick EV. These units receive priority in production and so will be built quicker than the Z1K-LV units. If you are in a rush to get a controller, you might consider one of these even if you are running a lower voltage car.

- The **Z2K** is the unit for those looking for the most power available. It regulates up to 2,000 amps of motor current while running off a lead-acid battery pack of 72 to 300 volts nominal (348V for the EHV model). This model is used in all the world's quickest EVs. *There is nothing wrong with running a Z2K in a street car, and many people do it, but be sure to turn it down first or you will just end up breaking other parts of the car!*
- There are also the **Z1K** and **Z2K HV** models, which can be ordered as an EHV model. These give a nominal input voltage rating of 348 volts for those truly wishing to push the limits of sanity in their EV.

These Zilla controllers are so powerful that the batteries have trouble supplying the high currents of these controllers, and motors can get damaged by them as well. That is why it helps to have appropriate engineers to assist you with the formulas needed to calculate proper voltage limits. While the book is a great resource, be *sure* to set the current limits to reasonable values for your setup before testing.

With the new highly capable batteries available today such as a 32-series automotive class lithium ion cell from A123 Systems are more energy dense and able to deliver speed and range. Combine A123 and the Zilla controller and you will have a full performance ride of your life. (Source: www.A123Systems.com).

ZAPI

ZAPI (www.zapiweb.com/main.htm or www.zapiinc.com) has a family of series controllers that are designed to perform with standard DC traction motors. Microprocessor-based, these controllers utilize the latest in solid-state MOSFET technology including high frequency operation, microprocessor logic, digital adjustment, diagnostics, and fault code storage. This controller is very similar to the Curtis controller but with many more features, programming options, and regenerative braking. I personally found this controller to be very reliable. For their controllers a hand-held microprocessor-based tool is used for the adjustment, testing, and diagnosis of ZAPI control systems. The tester feature provides a visual display of critical operating parameters such as battery voltage, motor voltage, motor current, controller temperature, accelerator voltage, and hours of operation.

Metric Mind Engineering

There is also a company called Metric Mind Engineering, which has high-end AC liquid-cooled synchronous and induction three-phase motors for EVs. Some of these motors have a max power of 150 kW, run at over 400V, and have an rpm of 10,000. This same company has motor controllers that can put out a maximum power of 212 kW.

Electric Vehicles' Controllers Help to Dispel All Myths about Electric Cars Today

Electric vehicle conversion specialists today are making the greatest electric vehicles on the market. Each one is a new challenge, a new dimension, but the reality of it is that from the smaller conversion companies like GrassRoots EV or Left Coast Electric to the bigger companies like AC Propulsion and Tesla, they show us that today we can have what we want in an electric car. It moves the entire market forward toward faster, more efficient, and more powerful cars. Here are two companies that say it all.

AC Propulsion Inc. to the Rescue—Today

While others have only dreamed or talked about it, AC Propulsion Inc. has done it—designed an integrated AC induction motor and controller that has been installed into numerous prototype EVs. In fact, the October 1992 issue of *Road and Track* with the picture of AC Propulsion's Honda CRX smoking its tires has become a collector's item among EV aficionados. AC Propulsion's Cal Tech alumnus co-founders Alan Cocconi (of GM's prototype Impact AC propulsion system fame) and Wally Rippel (of Chapter 3's Great Electric Vehicle Race of 1968 fame) just had a better idea and did something about it.

The Burbank, California Alternate Transportation Exposition of September 1992 gave me the privilege of seeing an AC Propulsion Honda CRX, shown in Figure 7-6, close up. While AC Propulsion's AC-100 EV controller is complex (and fills the engine compartment), its drivetrain is simple (only 1st gear is installed), and the results are astonishing. Driving it is an absolute breeze and a big surprise. After a small preflight

Figure 7-6 Alan Cocconi in front of AC Propulsion's Honda CRX EV conversion.

checklist from Alan—"You do have a driver's license, don't you?"—he slipped into the passenger's seat while I took over the wheel. After turning the key, one of the three buttons on the left of the dash selects forward, neutral, or reverse. Once on the highway, the induction motor behaves like the smoothest automatic transmission imaginable, with one big difference. All the motor's torque is instantly available at nearly any speed. After Alan counseled, "Don't be afraid to step on it," I did, and was abruptly pushed back into my seat while silently forming the word "Wow" on my lips. When I turned to look at Alan, he was grinning from ear to ear. Next, Alan directed my attention to the regeneration lever on the right of the dash. All the way in one direction is full regeneration. Take your foot off the accelerator pedal and the vehicle slows down immediately, without even touching the brake. Push the lever all the way in the other direction and you have no regeneration. Take your foot off the accelerator pedal and the vehicle coasts and coasts.

After I returned to earth, Alan gave me the walk-around tour, and talked about the batteries (AC Propulsion's Honda CRX uses 28 conventional 12-volt, deep-discharge, lead-acid batteries that produce 336 volts), controller, motor, charging philosophy, and temporary instrumentation (to monitor performance).

Back at their exhibit-hall booth, AC Propulsion employee Tiffany Mitchell showed off the other benefit of the design (see Figure 7-7)—the incredibly small size of the motor! This 4-pole AC induction motor puts out 110 ft-lbs. of torque at anywhere from 0 to 5,000 rpm, and an astonishing 120-hp maximum between 6,500 rpm and 10,000 rpm (maximum rpm is 12,000), yet it weighs only 100 lbs. Recall that Chapter 6's recommended series DC motor put out 70 hp maximum and weighed 143 lbs. Directly behind and above the motor on the booth table was the AC controller. It includes a 100-kVA PWM-style inverter; battery-charging circuitry; control logic; interfaces for control pedals, dash instruments,

FIGURE 7-7 AC Propulsion's Tiffany Mitchell shows off an AC induction motor in front of an AC controller/inverter of the 1990s.

FIGURE 7-8 Picture of the AC-150 (Courtesy of AC Propulsion).

and lighting; an auxiliary 12-volt DC power supply; and numerous interlocks for operator safety. If you pop the lid on the controller, you can see it uses off-the-shelf components (MOSFET power devices) and standard fabrication techniques.

AC Propulsion's own literature says it best about the newer AC-150 shown in Figure 7-8: "The AC-150 drivetrain has consistently received praise the world over for its innovative design and jaw-dropping performance. First available in 1994, the AC-150 drive system retains the first generation's 150 kW (200 hp) rating, but has fewer parts, is 30 percent smaller, 8 pounds lighter and packages more functions inside the electronics enclosure than the original AC-150. By far one of its most attractive features is the integral 20-kW bidirectional grid power interface. The integrated grid interface was originally developed to serve as a high-power battery charger for battery electric vehicles. With the bidirectional capability, many new applications are opened up for electric drive vehicles of all types, including distributed generation, selling grid ancillary services, automated battery diagnostic discharge testing, and using vehicles to provide uninterruptible backup power to homes or businesses."

The AC Propulsion design also contains its own state-of-the-art battery charger (eliminating the need for an additional external charger), whose 20-kW capability at unity power factor allows you to fully recharge in only one hour from a 240-volt, 40-amp AC outlet, and overnight from a conventional 120-volt AC source. Below in Figure 7-9 is a close-up view of the AC Propulsion 120kW power electronics drivetrain inside of their newest vehicle the eBox. Figure 7-10 shows how the connections are made to the 120kW amplifier in the eBox to the motor fan, the chassis, the 260 VAC power, the motor signals, and the motor power. They have made the controller setup with easy connections as shown in Figure 7-9, which is centrally located in the middle of the area where an ICE engine would be located.

If you have the money you can buy one of the AC Propulsion systems for your own EV conversion project today ($25,000 plus for their system)—however, it costs several times Chapter 4's conversion budget by itself. But, given the propensity of solid-state devices to double in price performance every few years, the trail blazed by AC Propulsion has an obvious destination. AC Propulsion's Burbank showing for Bob Brant was a consciousness-expanding experience of what the future will be like for all EV converters.

Figure 7-9 AC Propulsion 120kW power electronics drivetrain in the eBox.

Figure 7-10 Close-up of some of the connections to the 120kW "amplifier" on the eBox.

Today, some of the best high-performance electric cars use the AC Propulsion systems—from the Wrightspeed x1 electric sports car, to the Venturi Fetish high-performance car out of Italy, to the Tesla.

Tesla

What can we say, but that there is Tesla Motors. This new electric car on the market has an amazing AC drive of a new design with a new controller, new motor, and new battery sub-system.

The Tesla Roadster delivers full availability of performance every moment you are in the car, even while at a stoplight. Its peak torque begins at 0 rpm and stays powerful at 13,000 rpm.

This makes the Tesla Roadster six times as efficient as the best sports cars while producing one-tenth of the pollution. Figure 7-11 shows you that performance versus an internal combustion engine and the numbers speak for themselves. The Tesla is a great "build your own" EV sports car for the masses today.

The connection with the motor and batteries in an intelligent and efficient manner. This makes the Tesla Roadster six times as efficient as the best sports cars while producing one-tenth of the pollution. Figure 7-11 shows you that performance versus an internal combustion engine and the numbers speak for themselves.

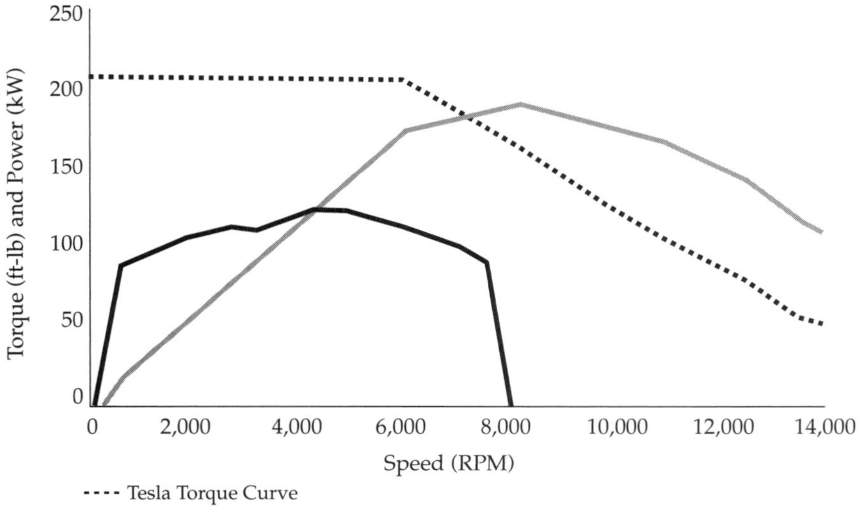

The black line represents the torque capabilities of a gasoline engine, which has little torque at low rpm and can only deliver reasonable horsepower within a narrow rpm range. In contrast, the dotted line demonstrates how the Tesla Roadster's electric motor produces high torque at 0 rpm, delivers constant acceleration up to 6,000 rpm, and continues to provide high power up to 13,000 rpm. The gray line shows the shaft power from the Tesla Roadster's electric motor as it builds steadily with increasing speed to a peak of 189 kW at around 8,000 rpm.

FIGURE 7-11 Performance and torque versus an internal combustion engine (Courtesy of Tesla Motors).

Conclusion

In simple terms, computers think in binary logic: 1's and 0's, on and off, yes and no. Rather than complex feedback systems that home in on the results you want (engines), you can implement a simple logic approach that moves you there directly (controller).

Also, as a result of improving technology, you can save on components while not compromising on safety and dollars. Cell phone and camera makers, with whom miniaturization has been raised to an art form, were ecstatic with the technology and have been improving on every new generation. What that means to EV converters is smaller and less expensive electronics. So, there is a great future for EVs and EV conversions today. Going forward with each new generation of technology improves our chances for all electric drive.

There are more controllers than ever before available; AC and DC, high quality industrial and cheap Chinese throwaways; new, used, and surplus. There are vast quantities of surplus parts and controllers are readily available. You can get all the part you need at pennies on the dollar. The Internet, search engines, and email discussion lists have made information on EVs far easier to find.

The skills and tools needed to build your own EV controller have never been easier to get. There are manufacturer's application notes with complete working circuits, inexpensive motor controller demo boards that only need to be scaled up, power transistor modules that handle all the tricky high power wiring for you, etc.

The perfect EV controller design has not been created yet. There is no open source controller project equivalent to Linux, for example—but there are people working on it! The main thing stopping most people is simply their own apathy. Whether you believe you can do it, or cannot do it—you're right! (Source: Lee Hart, 3/12/08)

CHAPTER 8
Batteries

"More than 85 percent of an average car's gasoline energy is thrown away as heat."
—Dr. Paul MacCready, *Discover*, March 1992

Today's batteries, motors, and controllers are all superior to their counterparts decades ago. Contrary to those who say you'll need a different type of battery before EVs are suitable at all, today's conventional lead-acid batteries of the deep-discharge variety are perfectly adequate for your EV conversion. (Note: While lithium-ion batteries and nickel batteries are being used more readily, this chapter will focus mostly on lead-acid.)

Improved lead-acid batteries are routinely available from numerous suppliers at a good price. Assuming a proper system design, if you install batteries correctly and maintain them conscientiously, you don't have to worry about replacing them for tens of thousands of miles. Future batteries will be lighter and more powerful, but can hardly be more convenient than they are today.

In this chapter you'll learn about how batteries work and the language used to discuss them. You'll be introduced to the different battery types, and their advantages and disadvantages. Then we'll look at the best type of battery for your EV conversion today, the lead-acid type used in Chapter 10's conversion, and look at probable future battery developments.

Battery Overview

Your EV's chassis involved mechanical aspects, and its motors and controllers dealt with electrical ones. Its batteries will now take you into the chemical area. While there are all sorts of battery developments going on in the labs, the objective here is to give you a brief battery background, and introduce the lead-acid batteries you'll be working with on your EV conversion. Many good books about batteries are available both at introductory and the more advanced levels for those who want more data.

Because your EV battery pack—a collection of 16 to 24 6-volt (or 12-volt equivalent) individual lead-acid batteries—represents the single largest replacement cost item, and quite possibly is also your largest initial expense item, it's worth spending some time learning about batteries so you can choose and use them wisely.

Batteries are the life's breath of your EV, and every EV converter should be familiar with them on three levels. To graduate from battery class you need to:

- Understand what goes on inside a battery

- Become familiar with a battery's external characteristics
- Learn the pros and cons of working with real-world batteries

Knowledge in these three areas is a sound business investment that can save you time and money. Using the tried-and-true, 100-year-old lead-acid battery as an example, let's look at each of the three areas in turn, starting with what goes on inside a battery.

Inside Your Battery

A battery is a chemical factory that transforms chemical energy into electrical energy. You don't need to know all the detailed inner workings of your battery, but you should have a basic understanding of its elements and processes. We'll start with a little history and overview, and then get into the pieces and parts.

Your first acquaintance with chemical batteries probably occurred in high school biology class when your instructor reproduced Luigi Galvani's 1786 experiment by placing a voltage across a frog's leg and making it twitch. Alessandro Volta "went to school" on the Galvani phenomenon. He reasoned that if a voltage across two dissimilar metals produced a reaction in the frog's leg, two dissimilar metals in a conductive solution would produce a voltage, and the first battery—his Voltaic pile—was born in 1798.

Battery improvements have steadily occurred ever since, but the basic principles have remained unchanged. Battery action takes place in the cell, the basic battery building block, that transforms chemical energy into electrical energy. A cell contains the two active materials or electrodes and the solution or electrolyte that provides the conductive environment between them. There are two kinds of batteries: in a primary battery, the chemical action eats away one of the electrodes (usually the negative), and the cell must be discarded or the electrode replaced; in a secondary battery, the chemical process is reversible, and the active materials can be restored to their original condition by recharging the cell. A battery can consist of only one cell, as in the primary battery that powers your flashlight, or several cells in a common container, like the secondary battery that powers your automobile starter.

Active Materials

In chemical jargon, the active materials are defined as electrochemical couples. This means that one of the active materials, the positive pole or anode, is electron deficient; the other active material, the negative pole or cathode, is electron rich. The active materials are usually solid (lead-acid) but can be liquid (sodium-sulfur) or gaseous (zinc-air, aluminum-air). Table 8-1 gives a snapshot comparison of a few of these elements.

When a load is connected across the battery, the voltage of the battery produces an external current flow from positive to negative corresponding to its internal electron flow from negative to positive. The observed voltage in a galvanic cell is the sum of what is happening at the anode and cathode. To make an ideal battery, you'd choose the active material that gave the greatest oxidation potential at the anode coupled with the material that gave the greatest reduction potential at the cathode that were both supportable by a suitable electrolyte material. This means you'd like to pair the best reducing material—lithium (+3.045 volts with respect to hydrogen as the reference electrode)—with something that just can't wait to receive its electrons, or the best

Periodic Table Group	1A Light Metals	2A Light Metals	8 Heavy Metals	8 Heavy Metals	8 Heavy Metals	1B Low-Melting Metals	2B Low-Melting Metals	4A Low-Melting Metals	6A Non metals	7A Non-metals
Element– Symbol	Lithium Li	Beryllium Be								Fluorine F
Number– Voltage	#3 +3.045	#4 +1.85								#9 –2.87
Element– Symbol	Sodium Na	Magnesium Mg							Sulfur S	Chlorine Cl
Number– Voltage	#11 +2.714	#12 +2.37							#16 +0.51	#17 –1.36
Element– Symbol	Potassium K	Calcium Ca	Iron Fe	Cobalt Co	Nickel Ni	Copper Cu	Zinc Zn		Selenium Se	Bromine Br
Number– Voltage	#19 +2.925	#20 +2.87	#26 +0.44	#27 +0.277	#28 +0.246	#29 –0.337	#30 +0.763		#34 +0.78	#35 –1.065
Element– Symbol	Rubidium Rb	Strontium Sr		Rhodium Rh	Palladium Pd	Silver Ag	Cadmium Cd	Tin Sn	Tellurium Te	Iodine I
Number– Voltage	#37 +2.925	#38 +2.89		#45 –0.6	#46 –0.987	#47 –0.7995	#48 +0.403	#50 +0.136	#52 +0.92	#53 –0.536
Element– Voltage	Cesium Cs	Barium Ba			Platinum Pt	Gold Au	Mercury Hg	Lead Pb		
Number– Voltage	#55 +2.923	#56 +2.90			#78 –1.2	#79 –1.68	#80 –0.854	#82 ≠0.126		

TABLE 8-1 Element Oxidation Voltages Comparison

oxidizing material—fluorine (–2.87 volts with respect to hydrogen)—with something that just can't wait to give electrons to it.

In practice, many other factors enter the picture, such as availability of material, ease in making them work together, ability to manufacture the final product in volume, and cost. As a result of the trade-offs, only a few electrochemical couple possibilities make it into the realm of commercially produced batteries that you will meet later in the chapter.

Electrolytes

The electrolyte provides a path for electron migration between electrodes and, in some cells, also participates in the chemical reaction. The electrolyte is usually a liquid (an acid, salt, or alkali added to water), but can be in jelly or paste form. In terms of chemistry, a battery is electrodes and electrolyte operating in a cell or container in accordance with certain chemical reactions. Figure 8-1 shows the chemistry of a very

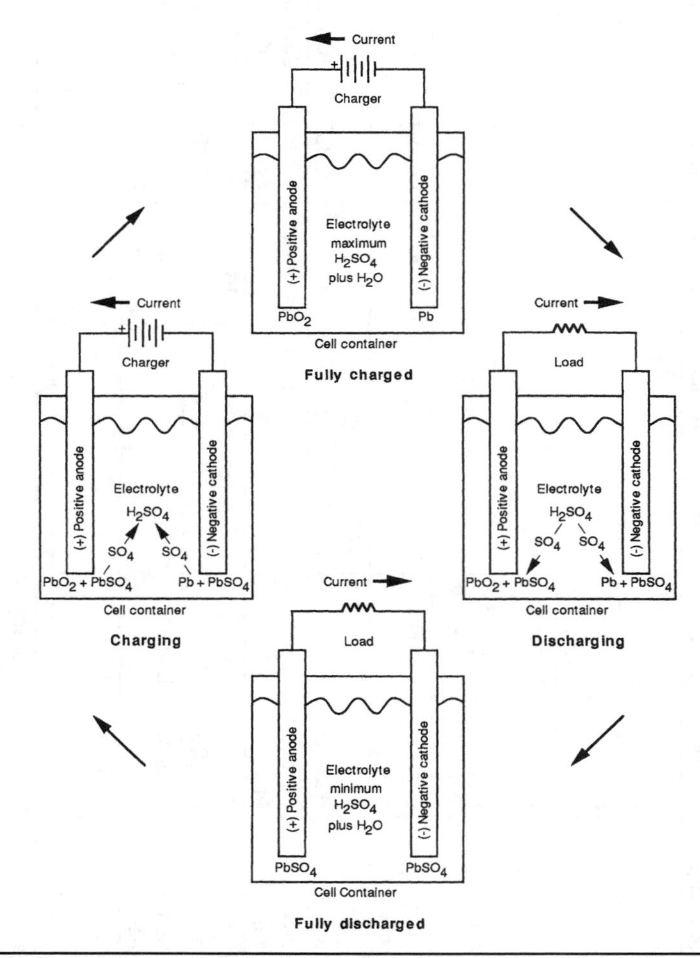

FIGURE 8-1 Chemistry of a simple lead-acid cell.

simple lead-acid battery cell that will be examined in the next few sections as it undergoes the four stages: fully charged, discharging, fully discharged, and charging. It consists of an electrode made of sponge lead (Pb), another electrode made of lead peroxide (PbO_2), and an electrolyte made of a mixture of sulfuric acid (H_2SO_4) diluted with water (H_2O).

Overall Chemical Reaction

Combining active material elements into compounds that further combine with the action of the electrolyte significantly alters their native properties. The true operation of any battery is best described by the chemical equation that defines its operation. In the case of the lead-acid battery, this equation is given as

$$Pb + PbO_2 + 2H_2SO_4 \leftrightarrow 2PbSO_4 + 2H_2O$$

The left side of the equation represents the cell in the charged condition, and the right side represents the discharged cell. In a charged lead-acid battery, its positive anode plate is nearly all lead peroxide (PbO_2), its negative cathode plate is nearly all sponge lead (Pb), and its electrolyte is mostly sulfuric acid (H_2SO_4) (see the top of Figure 8-1). In a discharged condition, both plates are mostly lead sulfate ($PbSO_4$), and the acid electrolyte solution used in forming the lead sulfate becomes mostly water (H_2O) (see the bottom of Figure 8-1).

Discharging Chemical Reaction

The general equation gives a more accurate view when separately analyzed at each electrode. The discharging process is described at the anode as

$$PbO_2 + 4H^- + SO_{4--} + 2e^- \rightarrow PbSO_4 + 2H_2O$$

The discharging process is described at the cathode as:

$$Pb + SO_{4--} - 2e^- \rightarrow 4\ PbSO_4$$

When discharging, the cathode acquires the sulfate (SO_4) radical from the electrolyte solution and releases two electrons in the process. These electrons are acquired by the electron-deficient anode. The electron flow from negative cathode to positive anode inside the battery is the source of the battery's power and external current flow from positive anode to negative cathode through the load. In the process of discharging (right of Figure 8-1), both electrodes become coated with lead sulfate ($PbSO_4$)—a good insulator that does not conduct current—and the sulfate (SO_4) radicals are consumed from the electrolyte. At the same time the physical area of the sponge-like plates available for further reaction decreases as it becomes coated with lead sulfate; this increases the internal resistance of the cell, and results in a decrease of its output voltage. At some point before all the sulfate (SO_4) radicals are consumed from the electrolyte, there is no more area available for chemical reaction and the battery is said to be fully discharged.

Charging Chemical Reaction

The charging process is described at the anode as

$$PbSO_4 + 2H_20 - 2e^- \rightarrow PbO_2 + 4H^- + SO_{4--}$$

The charging process is described at the cathode as:

$$PbSO_4 + 2e^- \rightarrow Pb + SO_4^{--}$$

The charging process (left of Figure 8-1) reverses the electronic flow through the battery and causes the chemical bond between the lead (Pb) and the sulfate (SO_4) radicals to be broken, releasing the sulfate radicals back into solution. When all the sulfate radicals are again in solution with the electrolyte, the battery is said to be fully charged.

Electrolyte Specific Gravity

The specific gravity of any liquid is the ratio of the weight of a certain volume of that liquid divided by the weight of an equal volume of water. Or the specific gravity of a material can be expressed as its density divided by the density of water, because the density of any material is its mass-to-volume ratio. Water has a specific gravity of 1.000. Concentrated sulfuric acid has a specific gravity of 1.830—1.83 times as dense as water. In a fully charged battery at 80 degrees F, water and sulfuric acid mix in roughly a four-to-one volume ratio (25 percent sulfuric acid) to produce a 1.275 specific gravity, and the sulfuric acid represents about 36 percent of the electrolyte by weight.

While specific gravity is not significant in other battery types, it is important in lead-acid batteries, because the amount of sulfuric acid combining with the plates at any one time is directly proportional to the discharge rate (current × time, usually measured in ampere hours), and is therefore a direct indicator of the state-of-charge.

State-of-Charge

Battery voltage, internal resistance, and amount of sulfuric acid combined with the plates at any one time are all indicators of how much energy is in the battery at any given time. Frequently this is given as a percentage of its fully charged value; for example, 75 percent means that 75 percent of the battery's energy is still available and 25 percent has been used.

Traditionally, the specific gravity of the electrolyte was used as a measurement. Today, because voltage can be used to determine a battery's state-of-charge and a hydrometer, the device used to measure specific gravity, can introduce inaccuracy and contaminate a battery's cells, state-of-charge is determined electronically.

Gassing

As charging nears completion, another phenomenon takes place: hydrogen gas (H_2) is given off at the negative cathode plate and oxygen gas (O_2) is given off at the positive anode plate. This is because any charging current beyond that required to liberate the small amount of sulfate radicals from the plates ionizes the water in the electrolyte and begins the process of electrolysis (separating the water into hydrogen and oxygen gas). While most of the hydrogen and oxygen gas recombines to form water vapor (the main reason that periodic replenishing of the water is needed in this battery type), the presence of flammable and potentially explosive hydrogen gas strongly indicates that charging be conducted in a well-ventilated area, and that you avoid lighting a cigarette.

Equalizing

Over time, the cells of a lead-acid battery begin to show differences in their state-of-charge. Differences can be caused by temperature, materials, construction, electrolyte, even by electrolyte stratification (the tendency of the heavier sulfuric acid to sink to the lower part of the cell) causing premature aging of the plates in that part. The only cure for these differences is to use a controlled overcharge, equalizing the characteristics of the cells by raising the charging voltage even higher after the battery is fully charged, and maintaining it at this level for several hours until the different cells again test identical. Obviously, this produces substantial gassing, so the precautions of a well-ventilated area and no smoking definitely apply.

Electrolyte Replacement

Adding sulfuric acid to a discharged lead-acid cell or battery does not recharge it—it only increases the specific gravity without converting the lead sulfate of the plates back into active material (another reason why measuring voltage is preferred to electrolyte replacement). Only passing a charging current through the cell restores it to its fully charged condition.

Sulfation

The finite life of the battery is caused by the fact that all of the sulfate (SO_4) radicals cannot be removed from the plates upon recharge. The longer the sulfate radicals stay bonded to the plates, the harder it is to dislodge them. To postpone the inevitable as long as possible, the battery should be kept in a charged state, and equalizing charging should be done regularly.

Outside Your Battery

When you hook up to the closed container of a battery, it exhibits certain external physical and electrical properties. You should be relatively familiar with these because they are useful to you. We'll review the terms you've already become familiar with from Chapters 6 and 7, and then move into the battery-specific areas.

Basic Electrical Definitions

You recall the example of the gallon plastic drinking water jug from Chapter 6. Let's relate it to the key electrical definitions:

Voltage

The battery's force (that is, its potential or voltage) corresponds to the water's force (its pressure or potential to do work): the height of the water in the water jug example. When you hook up a light bulb to a battery, the bulb lights up. When you hook two batteries in a series to double the voltage, the bulb lights even brighter.

There is another important aspect to the voltage of a battery. In the water jug analogy, the pressure of the water coming out of the jug goes down as you take more and more water out of the jug. In the same way, battery voltage goes down as you use the battery—as you use up its capacity. This important battery characteristic will be covered in more detail later in the section.

Current

The current (the rate of electron flow) corresponds to the rate of flow of the water coming out the bottom of the jug. When you doubled the voltage, you sent twice as much current through the wire and the light bulb became brighter.

Resistance

The resistance corresponds to the size of the hole controlling the rate of flow of the water coming out the bottom of the jug. A battery's voltage is directly related to current flow by resistance via the Ohm's Law equation you met in Chapter 6:

$$V = IR$$

where V is voltage in volts, I is current in amps, and R is resistance in ohms. Actually, there are two resistances: the external resistance of the load (the light bulb in this case) and the internal resistance of the battery. The battery's internal resistance is important in battery efficiency (heating losses), power transfer, and state-of-charge determinations.

Power

Electrical power is defined as the product of voltage and current:

$$P = VI$$

where V is voltage in volts, I is current in amps, and P is the power in watts. To use a 100-watt light bulb instead of a 50-watt light bulb requires twice the amount of power from the battery—twice the current at the same battery voltage. If the Ohm's Law equation is substituted into the previous equation,

$$P = I^2R$$

this equation defines the power losses in the resistances in the circuit—either external load or internal battery.

Efficiency

Battery efficiency is

$$\text{Efficiency} = \text{Power Out}/\text{Power In}$$

The principal battery losses are due to heat. These come from resistance and chemical sources: internal resistance of the battery determines its heating or PR losses when charging and discharging; chemical reaction between the lead and the sulfuric acid produces heat (called an *exothermic reaction*) during charging; and chemical reaction absorbs heat (called an *endothermic reaction*) during recharging.

While PR losses are present whether charging or discharging—because they are proportional to the square of current flow—battery heat rise is higher during charging (because PR heating losses add to the internal heat-generating chemical reaction) and lower during discharging (because IR heating losses are balanced by the internal heat-absorbing chemical reaction). Given the PR relationship, charging or discharging at a lower current rate obviously contributes to keeping battery losses lower.

Battery Capacity and Rating

Capacity and rating are the two principal battery-specifying factors. Capacity is the measurement of how much energy the battery can contain, analogous to the amount of water in the jug. Capacity depends on many factors, the most important of which are

- Area or physical size of plates in contact with the electrolyte
- Weight and amount of material in plates

- Number of plates, and type of separators between plates
- Quantity and specific gravity of electrolyte
- Age of battery
- Cell condition—sulfation, sediment in bottom, etc.
- Temperature
- Low voltage limit
- Discharge rate

Notice the first four items have to do with the battery's plates and electrolyte—its construction; the next two items concern its history; and the last three depend on how you are using it at the moment. We'll get into all the details, but keep in mind that the most truthful thing you can say about battery capacity is: *it depends*.

Battery capacity is specified in ampere-hours. A battery with a capacity of 100 ampere-hours could in theory deliver either 1 amp for 100 hours or 100 amps for 1 hour. This doesn't help you any more than would drawing a straight line on a map if someone asked you for a destination. You need the second coordinate, the second factor-rating.

A battery's rating is the second specifying factor. It refers to the rate at which it can be charged or discharged. It is analogous to how fast the sink will fill up with the water from the jug. In equation form:

$$\text{Battery Rating} = \text{Capacity}/\text{Cycle Time}$$

In this equation, the rating is given in amperes for a capacity in ampere-hours and a cycle time in hours. In practical terms, a battery with a capacity of 100 ampere-hours that can deliver 1 amp for 100 hours (known as a C/100 rate) would not necessarily be able to deliver the much higher 100 amps for 1 hour (known as a C/1 rate). You can only get the water out of the jug so fast.

Requesting 10 amps from a fully charged 100 AH capacity battery reflects a C/10 rate; this same request reflects a much lower C/40 rate from a 400 AH battery. In other words, smaller batteries have to deliver energy faster in relation to their size, or larger batteries have lower discharge rates in relation to their capacity.

Capacity of commercial batteries is standardized by the Battery Council International (BCI) into several usable figures. Two figures, a 20-hour capacity and a reserve capacity, are usually given for every battery depending on its application.

- **20-Hour Capacity**—This is a battery's rated 20-hour discharge rate—its C/20 rate. Every battery is rated to deliver 100 percent of its rated capacity at the C/20 rate, if discharged in 20 hours or more. If a battery is discharged at a faster rate, it will have a lower ampere-hour capacity.
- **Minutes at 25 amps Reserve Capacity**—This is the number of minutes a fully charged battery can produce a 25-amp current. This is the automotive starting battery rating that tells you how long your starter battery will power your automotive accessories if your fan belt breaks and disconnects the alternator; in other words, how many minutes you have to get to the nearest gas station.
- **Minutes at 75 amps Reserve Capacity**—This is the number of minutes a fully charged battery can produce a 75-amp current. This is the golf cart battery rating, because 72 minutes translates to about the amount of time it takes to

play two rounds of golf. So this figure tells you how long your batteries will power your golf cart: two rounds, three rounds, etc.

- **Three-Hour Reserve Capacity**—This is the BCI standard currently coming into vogue covering EV users. It is defined as 74 percent of the 20-hour rate. Because three hours translates to the average amount of time an EV might be in daily use, commuting, shopping, etc.:

$$\text{3-Hr Reserve Capacity} = 0.74 \times \text{20-Hr Reserve Capacity}$$

The Gentle Art of Battery Recharging

The objective with batteries is to maintain a balance. How fast batteries are filled and emptied are critical factors determining both their immediate efficiency and ultimate longevity. Where the batteries are filled and emptied, relative to their state of charge, are equally critical factors.

Because urban driving patterns for EVs are highly intermittent, battery discharge rates will vary all over the map. While energy is drawn out of your battery pack a lot harder than C/20 on startup and acceleration, you're only doing this momentarily, and the urban driving cycle usually implies that an EV's battery pack is given a certain amount of "rest" between discharge requests. The bottom line is

- Avoid placing continuous, heavy, C/1-type loads on your batteries anywhere in their state-of-charge cycle. A battery pack that can deliver 100 percent of its capacity when discharged in X time might only deliver 50 percent of its capacity when discharged in X/3 time. Remember the example of the water flowing out of the jug—the faster you take it out, the less pressure there is to push out the remaining amount.
- Avoid over-discharging your batteries when they're below 20 percent state-of-charge. High-rate discharging below the 20 percent state-of-charge can greatly reduce battery life or even destroy them.
- Unlike with discharging, you can control the destiny of your batteries during the charging process. In fact, it's vital that you do, because both overcharging and undercharging shorten battery life. Continually overcharged or too rapidly charged batteries can be destroyed; constantly undercharged batteries become sulfated and inefficient. Chapter 9 covers modern battery rechargers that can help you. The top of Figure 8-2 shows the ideal battery charging curve.
- Confine heavy charging within the 20 percent to 90 percent of the state-of-charge range, because a lead-acid battery's ability to store energy is reduced when almost full or nearly empty. Below 20 percent and above 90 percent, C/20 is the most efficient rate (divide the capacity of your battery in ampere-hours by 20) to charge your batteries. In the 20–90 percent range, C/10 delivers the fastest rate at which it's efficient to charge a lead-acid battery; it wastes more heat than at the C/20 rate, but saves time. Below 90 percent, control charging by limiting the current so as not to charge nearly empty batteries too rapidly. Above 90 percent, limit voltage so as not to overcharge the batteries (or possibly damage other attached electronic devices).

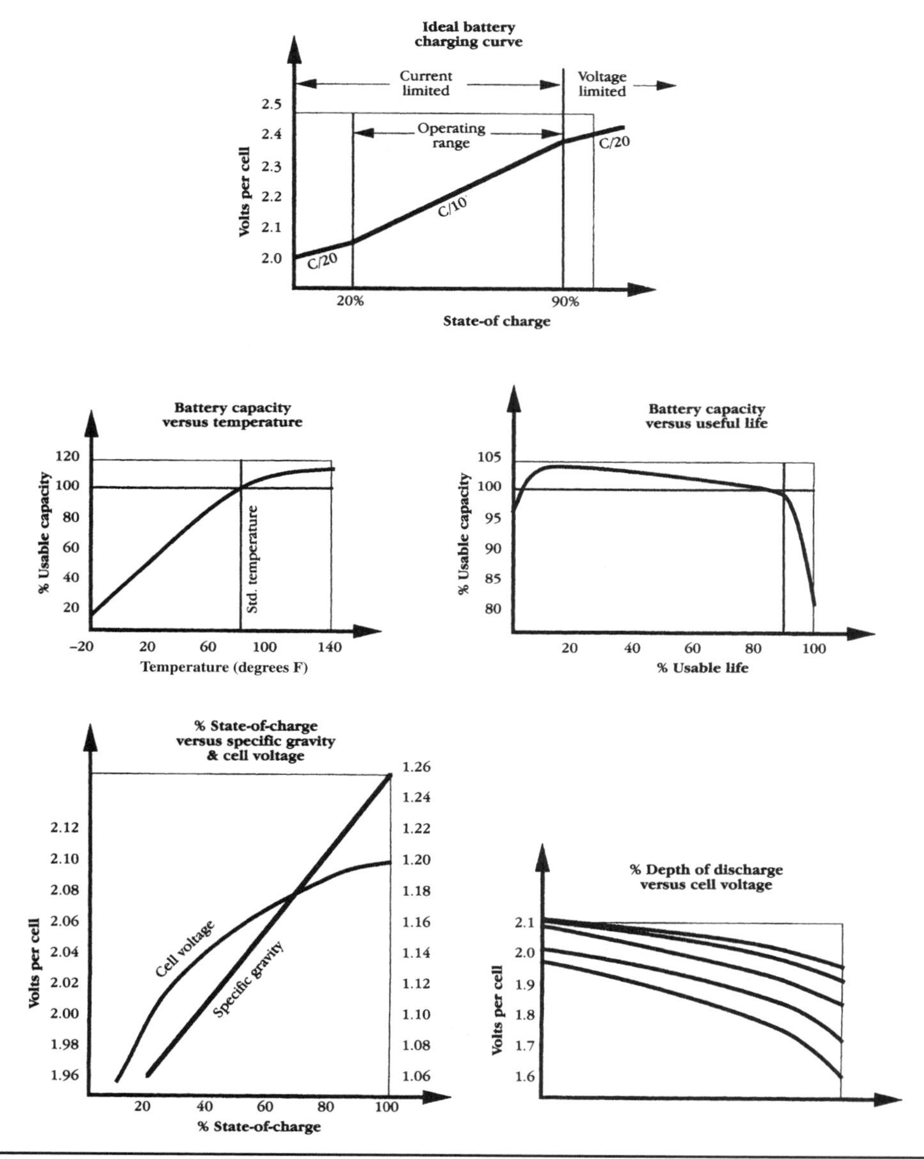

FIGURE 8-2 Lead-acid battery charging and discharging characteristics.

Temperature Determines Performance for Lead-Acid Batteries

Because the energy stored in a lead-acid battery is dependent on a chemical reaction, everything about it is affected by temperature: capacity, voltage, current, etc. Because the lead-acid chemical reaction is most efficient at 78 degrees F, most battery manufacturers rate their batteries at this "standard" temperature. The middle-left diagram in Figure 8-2 shows that the output of a battery is strongly affected by the temperature. Notice that only about 70 percent of a battery's capacity is available at 32 degrees F, while about 110 percent of battery capacity is available at 110 degrees F. The obvious conclusion here is that EV converters in colder climates need to opt for the next larger model battery model in the line, while their Sun Belt counterparts can either enjoy their extra power, or downsize a notch if they live in the desert.

Notice also that batteries don't freeze at 32 degrees F because of the concentration of sulfuric acid (H_2SO_4) in the electrolyte. This concentration increases (the specific gravity is higher) with increasing temperature and vice versa, so it's important to keep low-temperature operation batteries near fully charged at all times; letting the electrolyte freeze in a lead-acid battery can result in permanent battery cell damage.

Age Determines Performance for Lead-Acid Batteries

Battery capacity is also highly dependent on age. The middle right diagram in Figure 8-2 shows that the battery's capacity starts at about 95 percent when brand-new, rises to about 105 percent after it has been used for about 20 percent of its lifetime, stays fairly level, then drops off rapidly after 90 percent of its lifetime. One observation is that a brand new EV battery pack will not give you as good a result as one that's been used awhile. Another observation is that once you begin to see battery performance go down significantly, it's time to think about buying another set.

Battery Charge—Use It or Lose It

Because every battery has an internal resistance, it will discharge by itself if it sits around doing nothing. Temperature and battery age are the two main determinants of how rapidly this takes place—increasing temperature and age hasten the process. A 5 percent capacity loss per week is the average for lead-acid batteries (or 50 percent in 10 weeks) whether you use them or not, so periodically recharge the batteries in your stored EV. This is also the reason why buying the batteries is the last step in your EV conversion process.

An examination of Figure 8-2's temperature curve suggests its corollary—that batteries kept at cold temperatures don't discharge themselves at all. True, but you also can't get the energy out of the battery for your own use at cold temperatures! If you are planning to store your batteries for a few months, 40 degrees F is the best temperature. Fully charge your batteries before storage and warm them up before using them.

State-of-Charge Measurement—Volts or Specific Gravity

The lower left graph of Figure 8-2 shows why specific gravity has been used as a battery state-of-charge indicator for so long: it's an easy-to-use-and-understand straight line. Unfortunately, it doesn't show you that temperature directly affects specific gravity (specific gravity measures higher at lower temperatures). In addition, the device used to measure it, the bulb hydrometer, is prone to calibration, compensation, and readout errors because you're typically measuring in the range from 1.100 to 1.300 to three decimal places. And if you use a hydrometer on a regular basis, it's virtually guaranteed that you'll contaminate one or more battery cells.

A digital voltmeter readout accurate to 3F digits (at least to 0.1 volt) is today's preferred method for measuring state-of-charge. Thanks to modern electronics, you can observe voltage levels, current levels, and/or have the voltmeter readout drive the battery charger electronics directly. You can even monitor single-cell voltages if your battery type has external cell straps, all without the trouble of opening your batteries, dealing with sulfuric acid and hydrometers, etc. Since voltage also varies with temperature (lower temperatures produce lower voltages), you can either make a little chart to help you with your individual readings or rig a circuit to do it for you or your charger automatically.

Whether you use specific gravity or the voltage method, to get the most accurate state-of-charge reading, let the batteries "rest" for several hours (2 hours minimum, 6 hours is better, 24 hours is probably optimum, if you can afford it) before taking measurements. Monitor one or a few batteries rather than the whole pack; check ambient temperature and odometer reading at the same time; and keep a logbook. Do it at a convenient time, do it consistently, and make a simple graph. Your diligence will reward you with a beautiful record of your EV's battery health.

Discharge Not in Haste

The maxim "haste makes waste" is absolutely true when discharging batteries. The lower-right graph of Figure 8-2 shows that the faster you discharge your batteries, the lower the voltage (the less capacity) you have. If you take more and more from less and less, eventually you wind up with nothing at all—a polite way of saying over-discharging kills battery life.

A corollary of this action (depth-of-discharge) affects the number of charge/discharge cycles your batteries can deliver. The number of cycles you can expect from your batteries is approximately given by the equation:

$$\text{Battery Life Cycles} = K_d/\text{Depth-of-Discharge In}$$

This equation says that the number of battery life cycles is inversely proportional to the depth-of-discharge ratio. If you consistently discharge your batteries to 90 percent, you're going to get less cycles out of them than running them down to the 50 percent depth-of-discharge area. In numbers, K_d might be about 12,000 for starting batteries, 24,000 for deep-cycle batteries, and 30,000 or more for industrial batteries. These values reflect the fact that heavier-duty batteries deliver more cycles or support heavier depth-of-discharge rates better than starting batteries.

What this equation doesn't say is K_d will vary with each and every individual user, because every user's application is different. If a manufacturer's literature mentions they obtained 750 cycles out of one of their batteries, it's no guarantee that you will. On the other hand, you might even do better.

Because of gassing and loss of plate material as you go above the 90 percent charging point (10 percent depth-of-discharge), liability for battery damage as you go below the 20 percent charging point (80 percent depth-of-discharge), and the fact that every lead-acid battery has a finite lifetime, the best operating guidance translates to operating your deep-cycle batteries at the middle of this range—roughly the 40 to 60 percent depth-of-discharge range—for optimum balance between cycle-life, depth of discharge, and the actual physical (calendar) battery life. Heavier-duty industrial batteries can target the 60 to 80 percent maximum depth-of-discharge range for most efficient operation.

Lead-Acid Batteries

The practical aspects of lead-acid batteries affect all EV converters. You need to be intimately familiar with:

- Characteristics you should be aware of when buying
- Steps you should take during installation
- Maintenance you should perform during ownership

The intent of this section is not to make you a battery professional, but to provide you with practical knowledge so that you're prepared to buy, install, and maintain your batteries.

Battery Types

As you learned at the outset, there are two major classes of batteries: primary or nonrechargeable, and secondary or rechargeable. Unless your EV's task is to operate on the moon (like the Lunar Rover you read about in Chapter 3) or some other specific mission, you are unlikely to require the services of a nonrechargeable battery.

Among rechargeables, there are lead-acid batteries and there are all the rest. In a nutshell, there are no alternatives to the lead-acid battery for the casual EV converter today, because the disadvantages of the other two choices far outweigh the benefits.

Nickel-Cadmium Batteries

NiCad batteries are the type you'd use in your portable computer, shaver, or appliance, and are unquestionably better than lead-acid batteries in their ability to deliver twice as much energy pound for pound; they also have about 50 percent longer cycles. But the nickel-cadmium electrochemical couple delivers a far lower voltage per cell (1.25 volts), meaning you need more cells to get the same voltage. It is far more expensive (four times as much and up). There are fewer sources for the heavy-duty EV-application batteries (cadmium itself is harder to obtain and has generated environmental concerns). Finally, most of the nickel-cadmium technology development is taking place overseas (England, France, Germany, Japan).

Nickel-Iron Batteries

The "Edison battery" used in early 1900s EVs is even a poorer choice. It offers a higher cycle-life (about twice as many), delivers slightly more energy pound for pound (about a third more), and is very rugged mechanically. But the nickel-iron electrochemical couple delivers only slightly more voltage per cell than a NiCad (about 1.3 volts) and has a high internal resistance and self-discharge rate (10 percent per week). Its performance degrades significantly with temperature (both above and below 78 degrees F). It's far more expensive (four times as much and up), there are few sources for them (they're only made in Europe and Japan), and there is little technology development taking place.

All the battery development going on in the labs (which we'll look at briefly later in the chapter) is great, but you can't buy one. Your choice boils down to the good old lead-acid battery. But all lead-acid batteries are not created equal. Confining our discussion to the larger sizes suitable for the heavy-duty EV application, you have three types to choose from.

Starting Batteries

These are the kind used to start the engine in every internal combustion engine vehicle in the world today. The average starting battery spends only a few seconds of time turning over your vehicle's electric starter motor and the rest of its time being recharged by the alternator under the lightest of loads (unless you are driving at night in the rain with all the electrical accessories on). While they are great for this "high-power output for a short period of time" application, they are not suitable for use in your EV (other than for powering its accessories) because this battery type has thin plates that are only lightly loaded with active material. Used in an EV, it would give you only the shortest deep-cycle discharge life—you'd be lucky to get 100 cycles out of it. Even on a brief trip, if you stomped down too hard (or for too long) on the accelerator pedal, you'd be lucky to make it back to your own driveway.

Deep-Cycle Batteries

These are what you need. The low end of the capacity range might go into a golf cart–type or low-speed electric vehicle. The upper end of the capacity range goes into your EV. They can also be found in manufacturers' catalogs under the Marine heading. Any of these are a step up from starting batteries; they have much thicker plates and are specifically designed for a deep-discharge cycle-life in the 400 to 800 cycles (and up) ballpark.

Industrial Batteries

These monsters go into forklift pallet and stationary wind- or solar-generation applications. While they give great depth-of-discharge results on paper, have 1,000 cycles and up cycle-life, and make great counterweights for forklifts, their weight and size generally make them unsuitable for EV applications. Your mission is to go after the deep-cycle batteries that might be found under the golf cart, marine, or electric vehicle catalog headings.

Battery Construction

From a manufacturing viewpoint, a lead-acid battery is one of the most efficient things going. While lead is definitely something you don't want in anything you drink or consume (you don't even want it in the paint on the wall inside your home), the EPA loves lead-acid batteries because more than 97 percent of these batteries are recycled and 100 percent of every battery is recyclable.

Battery construction makes this possible. Used lead batteries are gathered at collection points, and then sent to smelting specialists where they are disassembled. The lead is melted, refined, and delivered to battery manufacturers and other users; the plastic is ground up and sent to reprocessors who make it into new plastic products; and the acid is collected and either reused or treated.

How a battery is constructed affects which battery you buy. Figure 8-3 shows the details.

Plates

Battery plates are formed on a wire-like grid of lead alloy (antimony is sometimes used to stiffen the lead); a mud-like lead oxide, sulfuric acid, and water paste is applied to them and allowed to harden. An expander is added to the negative cathode plate that prevents it from contracting in use. The plates are then "cooked" in a dilute sulfuric acid solution by sending a forming charge through them that changes the positive anode plate to a highly porous, chocolate-brown lead dioxide material, and changes the negative cathode plate to a gray sponge lead. The positive and negative plates are assembled into a "sandwich" with separators—thin sheets of electrically insulating material that is still porous to the electrolyte—and held in place inside the battery by

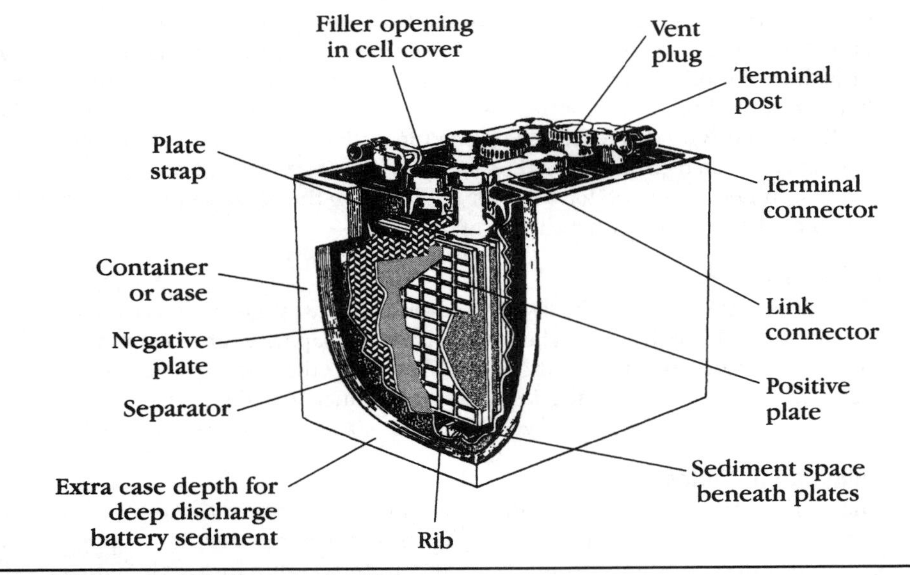

FIGURE 8-3 Lead-acid battery construction.

the plate straps. How thick and heavy these plates are, combined with the efficiency of their design (how much area is exposed), and how efficient the separator is, all collectively determine the capacity of your battery. While the initial way to tell the capacity of a battery is to pick it up, the only real way to tell is by using it.

Case or Container

The case is a plastic or hard rubber one-piece rectangular container with three or six cells molded into it. Each cell has molded-in ribs running across the width of its bottom or down the long dimension of the battery (see Figure 8-3). The plates are mounted at right angles to the ribs, whose multifold purpose is to stiffen the case, support the plates in a non-electrically-conductive manner, and act as collection channels for the active material shed from the plates. A battery is usable until the active material it sheds makes a pile that eventually reaches the plates and shorts them out. All other things being equal, a deeper-case battery will outlive a shallow-case battery because it allows a higher pile of active material to accumulate. Larger industrial batteries are cost-effectively rebuilt just by opening them, dumping the used active material, cleaning out any cell residue, and replacing the plates, separators, and electrolyte.

Cell Connectors or Links

These connectors can be inside the battery (through or over the cell partitions) or outside the battery via the link connector (see Figure 8-3). External connectors found in older and larger battery types allow you access to individual cell voltage measurements; improved battery reliability has made individual cell measurements less necessary with modern batteries.

Filler Opening and Vent Plugs

These openings allow you to refill your battery with distilled water or electrolyte solution. Vent plugs are baffled to allow gas to escape but not an accidental electrolyte splash. Individual vent plugs are threaded to screw into the vent well but might also be ganged into multiple opening press-fit caps. (Note: Most batteries used today for conversions are maintenance free, which means you do not need to even refill the batteries with water. Even low-speed electric vehicles like the GEM use maintenance-free batteries for an easier consumer experience. It is more prevalent in golf carts.)

Terminal Post

These might be taper top, side terminal, or L-type on starting batteries, but they are usually of the stud type on deep-discharge batteries (a heavy duty post with a bolt and a washer). This is because high currents soften the lead terminal post, and taper top terminals might shrink away from their connectors and cause increased resistance and intermittent problems.

Battery Distribution and Cost

Cost has been scrupulously avoided until this point, because a battery's "suggested retail price" is like your automobile sticker price—it bears little relationship to the real-world price.

Battery manufacturers have multi-tiered international distribution networks. This means they have wholesale and retail distributors across the United States and around the world. They sell to these distributors in wholesale bulk—750 batteries at a time and up, etc. Wholesale distributors or battery specialists might in turn sell to retailers like golf cart dealers, etc. There are shipping costs involved at each step, along with whatever prevailing local rules or conditions exist.

The point is, price is negotiable, and you should expect a discount of 25 to 30 percent from suggested retail list price, especially for your 16- to 24-battery order. As with any other large ticket item, it pays to shop around. While you can open your telephone directory and find battery retailers in virtually any medium-size and up city in the United States, it might pay to drive the extra 100 miles or so to deal with the large volume battery dealer in the biggest city near you. On the other hand, if your local dealer's largest deals are six batteries at a time and you come in with your bid for 20 batteries, you might have some real negotiating leverage.

Get at least three bids on the same battery model from the same manufacturer, and compare them. If you want to get more creative at the expense of spending more time, compare several manufacturers this way. Be sure to add in all shipping, setup, deposits (we'll cover this next), and tax charges so you have an apples-to-apples comparison in your bids. For example, shipping charges from a nonlocal source might cancel out their advantage over a higher-priced local source.

Battery Core Deposit

Certain states and localities do more than just suggest you recycle your batteries—they make it mandatory. Dealers are required to charge you a "core deposit" on your batteries that gets refunded to you when you bring them in for recycling—a larger version of the soda bottle deposit. Obviously, this adds to the price of your battery purchase. Another reason why one price cannot be quoted.

The flip side of this is your no-longer-usable batteries have an intrinsic scrap value. Like your used car when you buy a new one, your used batteries have a small but well-defined trade-in value. The lead commodity price quote from your local newspaper times the weight of your batteries gives you an approximate figure to deal with.

Battery Installation Guidance

If you think of what you will be doing with your batteries, it will help you during the installation process. In general, there are three areas of concern:

- **Safety**—This always has to be number one. You do not want a loose battery in the back seat becoming a missile in the event of a sudden stop, or Uncle Fred's cigar torching your EV as he casually examines it while it's recharging, or periodic battery maintenance overfills burning acid holes in your interior upholstery. You do want to mount your batteries in a location (or locations) where they can be securely tied down and well ventilated; where they are not harmful to anyone or anything and vice versa; and where they can be reached easily for servicing. This is the advantage of choosing a pickup chassis as an EV conversion platform—it already takes care of all these aspects.
- **Tightness**—All electrical and mechanical battery connections should be tight when you finish installation, and you should periodically recheck them. Loose electrical connections lead to reduced efficiency, invite corrosion, and create other needless intermittent electrical problems. Clean all connections before tightening them down. Loose mechanical connections mean premature aging of your battery by vibrating it to death. (Tight, by the way, means snug tight, or snug tight with a lock washer—something that doesn't break off later when you try to remove it in corroded form.)
- **Measurement and Watering**—Ideally, you want to install a system to minimize your service labor at the time you install your batteries. Even if you don't do this at first, plan your installation to accommodate it as a future retrofit. Electrical measurement is relatively easy—you just need convenient voltage and current pickoff points and an analog meter, digital meter, or automatic charging circuit to help you do the rest. Automatic watering is a little trickier but at a minimum involves being able to conveniently get at your battery tops—which means your batteries are not strewn all over the inside of your EV conversion in random fashion.

TLC and Maintenance for Your Batteries

The tender loving care and maintenance of your batteries will reward you many times over. All it really takes is a plan, a schedule, and the discipline to do it. The plan starts with a notebook or logbook. The data you need are voltage from a battery (or two or three—not the whole pack), the odometer reading, the date, and a comments section listing the water you added and anything unusual you noticed. The schedule is weekly or biweekly. Other than the log sheet, four areas are involved:

- **Safety**—This appears again because it is important. Always recharge your batteries in a well-ventilated area; don't smoke, light a match, or make an electrical spark around them; and strongly encourage all your friends and visitors to do the same. Working around batteries means you need to invest in protective eyeware (safety glasses), protective rubber gloves, and old clothes whose increasingly holey (yup, acid makes holes in them) appearance doesn't bother you.
- **Distilled Water**—(Note: Not as relevant, but some conversions do use batteries in which you need to maintain the water levels.) Steam-distilled water is the

only kind of water you ever want to put in your batteries. Just cover the plates; don't overfill the cells till the electrolyte overflows the battery top and causes a mess. Ideally, you want a tube from your water jug, or a small cup or glass with a pouring spout, or a clean funnel—something that makes it easy for you to pour without spilling. If any electrolyte is spilled, clean it up immediately and neutralize the spill area and its surroundings with baking soda per the next section's instructions. Remember that the battery electrolyte is a strong acid that eats metal, upholstery, clothing, shoes, and people without discrimination.

- **Corrosion and Tightness**—Make a visual inspection of your batteries, the battery connections, and the battery compartment. Look at, touch, and pull on things. The battery tops (and anything else in the battery compartment) should be kept clean of dust, dirt, corrosion, and splashed battery acid. Nip any one of these in the bud immediately. An old toothbrush and a box of baking soda work wonders. Use a solution comprising two tablespoons of baking soda added to a small glass of water (one pound per gallon is the ratio), applied to the battery tops and terminals. Never use it in the battery cells—be sure to keep them tightly capped during cleaning. Diligence with baking soda and toothbrush will neutralize any acid and keep the batteries clean. Touch and pull to check that none of your electrical connections have worked loose. Tighten any loose connections immediately.
- **Measurement**—Use a digital voltmeter or hydrometer to give you a readout of the battery state-of-charge. Remember to monitor your batteries in a "rested" condition, and try to shoot for the same rest period in all your measurements, or make a note of any discrepancies.

Today's Best Battery Solution

You already know this book recommends lead-acid batteries as the best solution for today's EV converters. You also know what type of lead-acid battery to buy and a lot about its characteristics. Your choice is made even easier because there are only a certain number of battery vendors in your immediate geographic area to choose among. Unlike buying motors, controllers, and other parts, you're not likely to be ordering your batteries by mail. Your choice basically comes down to who offers the best price on the batteries you want, and what capacity, rating, voltage, size, and weight you need.

In a slight departure from the previous chapters, we're going to recommend one manufacturer, then look at several alternative offerings from their line to give you the flavor of the real choices you will encounter. The batteries recommended are from the Trojan Battery Company of Santa Fe Springs, California. As with the motors and controllers, don't read anything important into their appearance here. They are only one of a large number of battery manufacturers. A list of battery manufacturers appears in Chapter 12, but in this case, which battery distributors are operating in your geographic area is the more important factor.

Before getting into the actual batteries, let's add a few more definitions to your already expanded battery vocabulary:

- **Power Density (Orgravimetric Power Density)**—Also known as *speck power*, this is the amount of power available from a battery at any time (under optimal conditions), measured in watts per pound of battery weight. It translates directly to the acceleration and top speed performance your EV can get out of its batteries.

- **Energy Density (Orgravimetric Energy Density)**—Also known as specific energy, this is the amount of power available from a battery for a certain length of time (under optimal conditions), measured in watt-hours per pound of battery weight. It translates directly to the range performance your EV can get out of its batteries.
- **Volumetric Power Density**—This is a factor more of interest to the technical battery community working across different battery chemistry types. It is power density measured in watts per gallon or watts per cubic foot volume rather than weight.
- **Volumetric Energy Density**—Ditto here. This is energy density measured in watt hours per gallon or watt-hours per cubic foot—again volume rather than weight.

You will find these useful both for this section's comparisons as well as those made in the "Future Batteries" section. Now let's look at the winning batteries.

Five Trojan Battery Solutions

The Trojan Battery Company has been innovating golf cart battery solutions since the 1950s; their appearance here should not surprise you. Electrical vehicle batteries today are substantially superior to those of only a decade ago. You can pick from 6-volt or 12-volt solutions, and the distribution network has evolved to give you more service at better prices.

We're going to look at three 6-volt and two 12-volt alternatives from Trojan. The Trojan T-125 model—one of the 6-volt alternatives—is shown in Figure 8-4. Notice its

FIGURE 8-4 Trojan T-125 6-volt deep-cycle battery.

FIGURE 8-5 Trojan 5SHP 12-volt deep-cycle battery.

rugged construction, and the stud-type terminal posts with bolts and nuts. This case and construction is common to all family members in this 6-volt line. Figure 8-5 shows you a 12-volt unit, the 5SHP model case mockup previewed by Trojan at the September 1992 Burbank Alternate Transportation Expo. You might (or might not) have the EV label on the batteries you buy from your distributor.

Table 8-2 gives you the details of this lineup of five recommended EV battery choices from Trojan. Other than suggested list price—an area that we'll save for special discussion—this is all from published data that you can get from your local dealer.

It lays out everything you need, but doesn't quite give it to you in the form you need it—yet. Figure 8-6, also drawn from published data, shows the actual capacity versus time performance charts; notice the similar performance of the 6-volt and 12-volt data groups. You can use this data to determine the results of applying actual loads to any of the batteries you choose.

Figure 8-7 is from actual Trojan data on the T-105 model battery calculated 6/29/92. It is the real-life example of an equation shown earlier:

$$\text{Battery Life Cycles} = K_d/\text{Depth of Discharge In}$$

In this figure K_d is around 28,000, so it shows that Trojan technology is pushing its deep-cycle batteries into the industrial battery area. In other words, the T-105 model and its other family members are heavy-duty deep-discharge batteries.

To figure out how many batteries you need, first determine the voltage at which you are going to operate your EV conversion. This voltage is established from your chassis, motor, and controller trade-offs, and heavily influenced by ultimate use, longest range, or fastest acceleration.

Our objective here is to pick the best battery for Chapter 10's actual pickup conversion, so the operating voltage of 120 volts has been selected. Assuming you want all your batteries of the same type, and also that you're not going to use any tricky series-parallel wiring combinations, this means you'll either require 20 of the 6-volt batteries or 10 of the 12-volt batteries, all wired in series to obtain the 120 volts. When you wire your batteries in series, the total capacity available—the total ampere-hours—is the same as that available from any one battery. The total watt-hours is simply the total voltage times the total ampere-hours. The total weight, cubic feet, and cost is

Trojan Battery Model	Nominal Voltage	20 AH Capacity	Minutes @ 25 Amps	Minutes @ 75 Amps	3 AH Capacity	Weight in Pounds	Energy Density watt-hours/lb	Length	Width	Height	Suggested List Price
T-105	6 volts	217	419	107	161	61	15.5	10.375	7.125	11.1875	107.76
T-125	6 volts	235	477	125	174	66	15.6	10.375	7.125	11.1875	115.67
T-145	6 volts	244	530	145	181	71	15.0	10.375	7.125	11.5	168.85
27TMH	12 volts	117	200	50	87	60	15.2	12.75	6.75	9.75	106.86
5SHP	12 volts	165	272	78	122	86	14.2	13.5625	6.75	11.5	220.50

TABLE 8-2 Comparison of Recommended Trojan Electric Vehicle Batteries

Trojan Battery Model	Nominal Voltage	Quality in Vehicle	Vehicle Voltage	Battery Total AH Capacity	Battery Total Watt-Hours	Battery Total Weight	Battery Total Cubic Feet	Battery Total Cost @ 70%
T-105	6 volts	20	120	217	26040	1220	9.57	1508.64
T-125	6 volts	20	120	235	28200	1320	9.57	1619.38
T-145	6 volts	20	120	244	29280	1420	9.84	2369.90
27TMH	12 volts	10	120	117	14040	600	4.86	748.02
5SHP	12 volts	10	120	165	19800	860	6.09	1543.50

TABLE 8-3 Comparison of Trojan Electric Vehicle Battery Trade-Offs

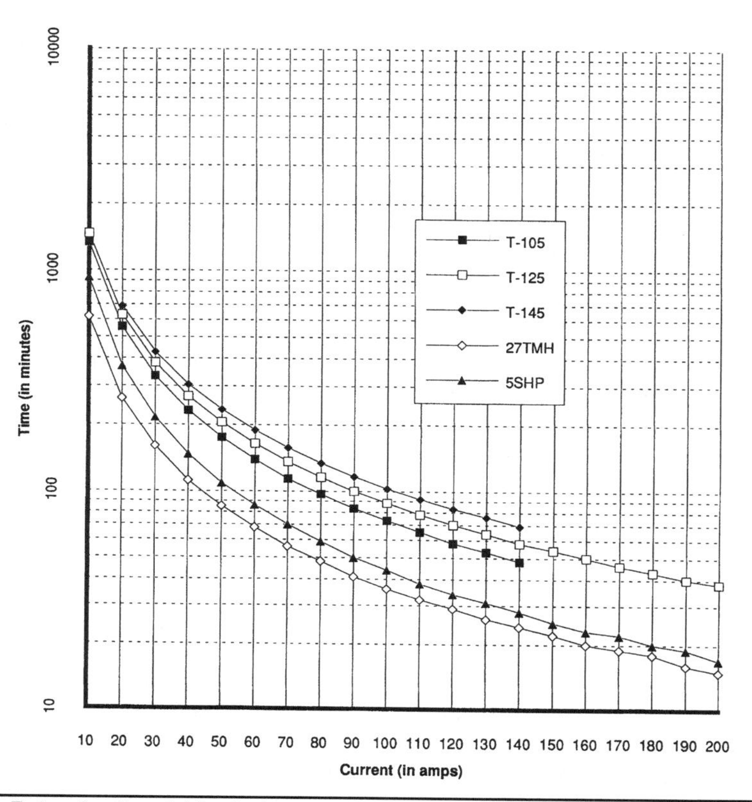

FIGURE 8-6 Trojan 6-volt and 12-volt deep-cycle battery family time versus current curves.

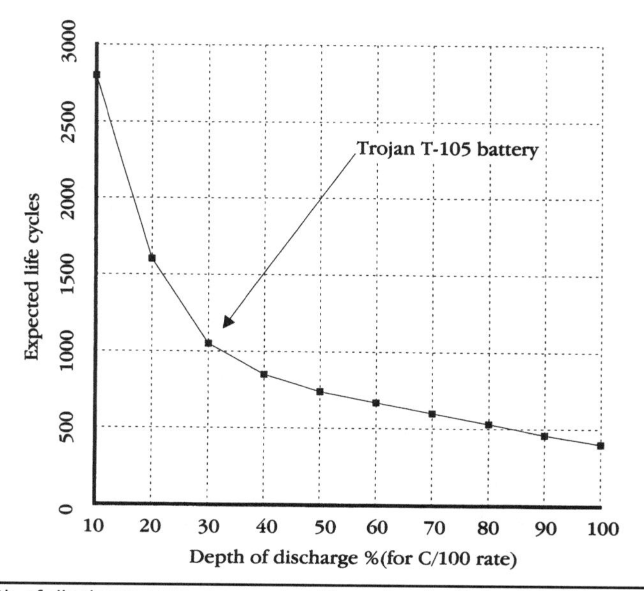

FIGURE 8-7 Depth of discharge versus expected life cycles for Trojan T-105 battery.

simply the sum of the individual 10- or 20-member battery set. Table 8-3's summary data is laid out in much more usable format.

Your choices are actually quite simple and logical at this point. Notice the 6-volt options give you from 26.0 to 29.3 kWh of energy-total on-board battery kilowatt hours. The two 12-volt choices give you either 14.0 or 19.8 kWh. Looking at the weight column, the 6-volt choices weigh from 1,220 to 1,420 lbs., while the 12-volt choices weigh either 600 or 860 lbs. In total cubic feet, although this figure would obviously translate to a larger actual mounting space required, the 6-volt batteries require from 9.6 to 9.8 cubic feet, while the 12-volt batteries need either 4.9 or 6.1 cubic feet.

Cost is another matter. Some figure had to be used here, so the manufacturer's suggested list price as of August 2008 was plugged in, discounted 30 percent (70 percent of list), and multiplied by either the 10 or 20 number appropriate for the battery string. As mentioned earlier, your costs will vary from these figures, so look at the cost here only for comparative purposes.

With the cost disclaimer out of the way, we can proceed. If you want the best range, choose the highest on-board energy. If you want the most acceleration, choose the lowest weight. If cost is a factor, choose the lowest cost. As all of these usually are factors, dividing energy and cost by weight is usually a good way to see what gives you the most for the least. Table 9-4 does this and gives you two clear winners: the T-125 for the 6-volt group, and the 27TMH for the 12-volt side. Notice the T-125 gives you about 1 kWh less energy, saves you 100 lbs., and costs you $750 less than its bigger T-145 brother—it's the best deal in the 6-volt group and the best choice for Chapter 10's pickup conversion. On the 12-volt side, the 27TMH gives you 40 percent less energy at 14.0 kWh, but saves you 260 lbs. and costs less than half that of its bigger 5SHP brother. It would be the ideal solution for a lighter-weight vehicle. Or if you just wanted to tool around your neighborhood in Chapter 10's pickup truck and win all the local drag races, then ten 27TMH batteries weighing in at 600 lbs. save you a cool 720 lbs. over the 20 6-volt T-125 battery solution.

Tomorrow's Best Battery Solution—Today

If you pick up a battery book (or read a battery chapter in any book) from any decade of the 1900s, it's interesting to note that every one states the best battery of the future is "just around the corner." The reality is a little different. The reality is that things move very slowly in the battery world—events are usually measured in decades, not years. So while this section will talk about the latest and greatest in batteries, don't expect any of these at your battery dealer soon—if at all.

Trojan Battery Model	Total kwh/lbs	Energy Decision Criteria	Total $/lbs	Cost Decision Criteria
T-105	21.34		1.237	
T-125	21.36	Highest 6-volt energy/lb	1.227	Lowest 6-volt cost/lb
T-145	20.62		1.669	
27TMH	23.40	Highest 12-volt energy/lb	1.247	Lowest 12-volt cost/lb
5SHP	23.02		1.795	

TABLE 8-4 Trojan Battery Final Trade-Offs

While lead and sulfuric acid would not be my initial choice for any construction project—one of the heaviest elements mated with one of the nastiest compounds—reliability, performance, and cost all weigh heavily in the lead-acid battery's favor. In fact, lead-acid's suitability for so many applications has greatly diminished even the need to search for alternatives.

But the recent save-the-environment, reduce-oil-dependency, and let's-try-electric-vehicles changed consciousness altered the pattern. Government, industry, and university laboratories all over the planet reflect this change. Pouring money on a problem never guarantees a solution, but it does guarantee that a lot more will be happening, and that some of what happens will be usable and good. Let's look at future battery trends starting with the consortium that's pushing the outside of the battery envelope—the USABC.

In 1991, Ford, General Motor, and Chrysler (joined by the Department of Energy and the Electric Power Research Institute) had a better idea—the United States Advanced Battery Consortium (USABC). In short, the Electric and Hybrid Vehicle Research, Development and Demonstration Act of 1976 recognized the need for battery development; the Department of Energy defined and funded it; and the USABC focused the efforts.

The near-term result was that a plethora of projects was honed down to just three high-energy battery research areas that could deliver significant vehicle range and power advantages: lithium polymer, lithium metal sulfide, and nickel metal hydride. Some of these batteries are considered the best of the best battery technologies on the market.

Future Batteries: The Big Picture

Table 8-5, adapted from an SAE paper, shows the entire story at a glance. Notice that 11 different battery technologies are on the list, and they are not equal. In very general terms, higher energy density and power density are desirable and look easy to do—on paper. Getting both at the same time, along with high cycle-life and low cost in a battery that operates efficiently over a range of temperatures, can be manufactured and supported by infrastructure, and causes no harm to people or the environment, has proven to be a bit more elusive. Notice that none of the batteries developed thus far—even the tried and proved lead-acid battery—even approaches its theoretical specific energy value. We still have a long way to go.

Lead-Acid

The big money involved in the lead acid battery business made some major improvements to the lead-acid batteries of the early 2000s superior to their 1990s counterparts. Along the way to higher specific energy and specific power, lead-acid batteries evolved to add sealed as a regular option. While it is higher in cost, it is less efficient (versus convenience of not watering). The flow-through (the conventional type you're accustomed to) has improved by greater plate thickness, improved separators, and higher specific gravity electrolyte solution). The other options available are tubular (electrode improvement) and gelled (electrolyte improvement, which I have used in some of the electric vehicles I have worked with).

Nickel-Cadmium

Close on the heels of lead-acid today, this battery type promises to be even better in the future. Its advantages over lead-acid today (less pronounced decrease in capacity under

Battery Type	Nominal Cell Voltage (volts)	Operating Temp Range (degrees C)	Life Cycles	Theoretical Spec Energy (wh/lb)	3 Hr Rate Spec Energy (wh/lb)	Energy Density (kwh/cu ft)	Specific Power 30 sec pulse (watts/lb)	Power Density 30 sec pulse (kw/cu ft)
Lead Acid	2.1	35 – 70	600	79.5	15.9	2.55	72.7	8.50
Nickel Cadmium	1.25	30 – 50	2000	99.1	25.0	3.40	86.4	9.35
Nickel Metal Hydride	1.4	20 – 60	600	84.1	29.5	4.96	68.2	11.33
Nickel Zinc	1.6	40 – 65	250	155.0	27.3	2.83	59.1	2.83
Nickel Iron	1.25	40 – 80	800	121.4	22.7	3.40	52.3	6.51
Sodium Sulfur	2.08	300 – 400	350	345.5	38.6	3.26	54.5	5.10
Sodium Nickel Chloride	2.59	250 – 350	1000	360.0	59.1	4.81	76.3	6.37
Zinc Bromine	1.8	0 – 45	500	194.5	31.8	1.98	38.6	3.26
Zinc Air	1.62	25 – 65	70	595.5	59.1	1.84	22.7	1.84
Lithium Iron Disulfide	1.66	400 – 450	500	295.5	75.0	6.80	170.5	1.56
Lithium Polymer	3.5	0 – 100	300	248.2	72.7	7.36	90.9	5.95

TABLE 8-5 Comparison of Future Electric Vehicle Battery Trade-Offs

high-discharge currents, higher cycle-life, slower self-discharge rate, better long-term storability, and improved low-temperature performance) will continue to provide markets to fund development aimed at improving its disadvantages over lead-acid: higher cost and environment-related cadmium issues.

Nickel Metal Hydride

The environmentally benign alter ego to the nickel-cadmium battery is also flat-out superior to it in specific energy and specific power comparisons, and should become the preferred alkaline battery of the future (Figure 8-8). The USABC certainly thought so and is invest its research and development dollars toward this technology.

In addition, when talking about nickel metal hydride batteries, Ron Freund's study in the next section tells everyone how mid-range battery packs stand the test of time. Personally, the RAV4 EV was the most reliable EV to get me from White Plains, New York to lower Manhattan and back to White Plains, New York. That should tell it all, but let's hear about Ron's report of his experience with the RAV4.

Lithium Polymer

This battery's specific energy and specific power numbers evoke nothing but envy from its competitors and mouth-watering anticipation from its advocates. The whole lithium group has shown great promise in the labs, and smaller lithium polymer batteries have greatly impressed users in the computer industry. But lithium still has to deliver on its promise when packaged in the giant, economy, suitable-for-powering-EVs size (see Figure 8-9). Stay tuned for future developments here—the USABC certainly will.

FIGURE 8-8 Nickel metal hydride battery

Figure 8-9 Lithium batteries ready to go into a Porsche for great speed and great range (Courtesy of EVPorsche.net).

Sodium-Sulfur

This technology has strong proponents in England, Germany, Japan, Canada, and the United States. This is good because its opponents have a continuous field day merely by labeling it and using scare tactics. Yes, sodium is combustible in air. Yes, sulfur is also on the head of matches. Yes, maintaining a "thermos bottle" at 350 degrees C to contain its molten sodium and sulfur electrodes and beta alumina electrolyte is inefficient. Yes, it can explode and/or do nasty things if punctured in an accident. So can internal combustion engine vehicles. On the other hand, it continues to be one of the most promising advanced battery systems for EV propulsion; its specific energy and specific power numbers are greatly superior to those for lead-acid batteries; and pilot plant battery production has already started. While many problems remain to be solved, perhaps low-cost production the largest among them, sodium sulfur is a hot technology in more ways than one.

Sodium Metal Chloride

If sodium sulfur is great, then all its advantages at a lower operating temperature with still better specific energy and specific power numbers has to be greater yet. And sodium metal chloride battery technology is. Throw in cells that can be assembled in the discharged state, have higher open-circuit voltage and better freeze/thaw and failure-mode characteristics than sodium sulfur's, and you have a real winner.

Lithium Iron Disulfide

The promise of lithium-iron disulfide batteries on the high temperature side is equally mouth-watering. This battery's specific energy numbers are the best of all, and its specific power numbers simply leave all others in the dust.

Lithium Ion

This has to be the most popular long range battery on the market today (Figure 8-10). The advances is this technology make it the next best solution for the present and the future. The lithium ion moves between an anode to the cathode during discharge and the cathode to the anode during recharging. It is extremely popular in consumer electronics and now power tools. It is light, has a slow energy degradation and no memory issues.

A123 Systems out of Watertown, Massachusetts has developed an amazing lithium ion battery being used in plug-in hybrid technologies from Hymotion™ using a Nanophosphate™ chemistry that was coordinated with the US Department of Energy. It is being used in hybrid electric transit buses (Orion buses), airplanes for when the airplane is docked at an airport for power and will soon be in hybrid electric cars so that they can even be more efficient than they are today. There was recently an electric motorcycle developed called the KillaCycle that uses the A123 batteries. It did a race that was shown on the cable show Planet Green and went 168 mph in 7.824 seconds. Bill Dube (Owner) and his team must be proud of their accomplishments! Keep it up!

Batteries and the RAV4 EV Battery Experience

Ron admits that clearly the RAV4-EV is best fit for those driving less than 100 miles per day.

Quick History

In 1997, the RAV4-EV was made by Toyota Motor Corporation mostly for the California Air Resources Board (CARB) Zero Emissions Mandate (ZEV). The cars were originally leased to commercial entities (such as the New York Power Authority) and other electric utilities. In 2002 they were sold to people for about $40,000 each. Ron added that CARB

FIGURE 8-10 Litium ion battery.

surveyed vehicles drove over 5 million ZEV miles. This translated to approximately 2,900 fewer tons of carbon dioxide released (assuming the CA average of 17 mpg and 19.54 lbs. per gallon). The batteries made the difference in drivers displacing additional gasoline trips and reducing their carbon footprint.

I love the RAV4 EV driving experience because of the batteries. I use to drive it all over NYC and Westchester County. All the time in New York City and White Plains, I remember people asking me when they could get a RAV4 EV and does the vehicle have a good range. All I could say was, hopefully soon and I can go from White Plains down to a meeting in New York City and back on a single charge. That is what batteries can do today!!

Batteries

The RAV4 batteries were nickel metal hydride (NiMH), which was an improvement (higher capacity) over the ECD model used in the GM EV-1 Gen II. The traction pack performed admirably, when air cooled, over 100 freeway miles at 65 miles per hour in good condition. Even with over 100,000 miles, this pack still could deliver this same kind of freeway performance. Traveling in mountainous areas meant incorporating energy expenditures of about 10 percent for every 1,000 feet of elevation climbed. (Regeneration helped reduce losses.)

Squelching Concern About Battery Degradation

As we all know, a car when left for long periods of time can deteriorate, with fuel fouling and more. In an electric car, prolonged periods of non-use might see a deterioration of maximal range if lead-acid technology were used. (Note: This is the main reason the most popular batteries on the market will not be the battery of the future.) Those batteries need regular exercise to maintain capability.

The RAV4 batteries never had this type of degradation. According to Freund, "If left in the cold winter of western Pennsylvania or at the LAX airport while owners were away, full capability was immediately available upon resumption of service. This is clearly much slower than stored hydrogen (for comparison to FCEV's)."

In general, Toyota proved with an electric car that batteries have to be an issue regarding range.

"Furthermore," Freund concluded, "given the indisputable environmental and national security benefits of shifting even a small fraction of our many petroleum-driven miles to the electric grid, and given the outstanding performance of the RAV4 EV with NiMH batteries—it is a tragedy that this choice does not exist in the market today. CARB and Toyota should celebrate this product success and take steps to allow Californians to choose whether to drive on petroleum or electricity. Plug-in cars provide that choice, and as battery and energy storage technology continues to inevitably improve, the inclusion of electricity in the transportation of tomorrow is inescapable."

So the next time someone tells you that an electric car will not work because of its batteries, you can tell them the truth (which can only set us free—from oil).

CHAPTER 9

The Charger and Electrical System

An efficient charger is an indispensable part of any electric vehicle.

The charger is an *attached* and *inseparable* part of every electric vehicle battery system. Discharging and recharging your batteries are opposite sides of the same coin; you cannot have one without the other. As you learned in Chapter 8, how you recharge your batteries determines both their immediate efficiency and ultimate longevity. As with motors, controllers, and batteries, technology has also made today's chargers superior to their counterparts a decade ago.

Because your motor, controller, batteries, and charger are also inseparable from the electrical system that interconnects them, it too is covered in this chapter, along with the key components needed for its high-voltage, high-current power side and its low-voltage, low-current instrumentation side.

In this chapter you'll learn about how chargers work and the different types, meet the best type of charger to choose for your EV conversion today (used in Chapter 10's conversion), and look at likely future charging developments. You'll also look at your EV's electrical system in detail and learn about its components so when you meet them again during Chapter 10's conversion process, they will be familiar to you.

Charger Overview

Chapter 8 dealt with discharging and recharging; now we'll take a closer look at the recharging side. It's a wise business decision to invest a few hundred dollars in a battery charger that gets the most out of a battery pack that can cost a thousand dollars or more and might be replaced several times during your EV ownership period. An efficient charger is an indispensable part of any EV.

The objective here is to give you a brief background and get you into the recommended battery charger for your EV conversion with minimum fuss. You have three battery charger choices today: build your own, buy an offboard charger, or buy an onboard charger. We'll look at each area in turn and give our recommendations. Let's start with a look at what goes on during the lead-acid battery discharging and charging cycle to understand what has to be done by the battery charger.

Battery Discharging and Charging Cycle

As you already know from Chapter 8, batteries behave differently during discharging and charging—two entirely different chemical processes are taking place. Batteries also behave differently at different stages of the charging cycle. Let's start with a look at an actual battery, then look at the discharging and charging cycle specifics.

What You Can Learn from a Battery Cycle-Life Test

Figure 9-1 shows cycle-life test results for the Trojan 27TMH deep-cycle lead-acid battery we looked at in Chapter 8. Two parameters are being monitored versus number of cycles: the minutes at 25 amps capacity, and the end of charge current.

The capacity parameter measurement is an actual version of the graph you saw in Figure 8-2. There are more wiggles in the real battery's data curve, but the resemblance between the two graphs is striking, and there should be no revelations for you here. This battery didn't actually "fail" at the end of 358 cycles; that's just a name assigned (by a battery test engineer) to the point at which this battery dropped below 50 percent of its rated capacity.

The *end of charge current (EOCC)* might be new to you. Notice it's quite low early in the battery's cycle-life (around one amp) but rises steadily until at some point around mid-life it shoots up to its limit value (around 20 amps in this graph).

What does this mean to you?

- It means that a battery's charging current fluctuates widely over its lifetime.
- It means that you can quickly kill a new battery that only requires a small amount of current to kick in its charging cycle by placing an unregulated voltage source without any current control across it.

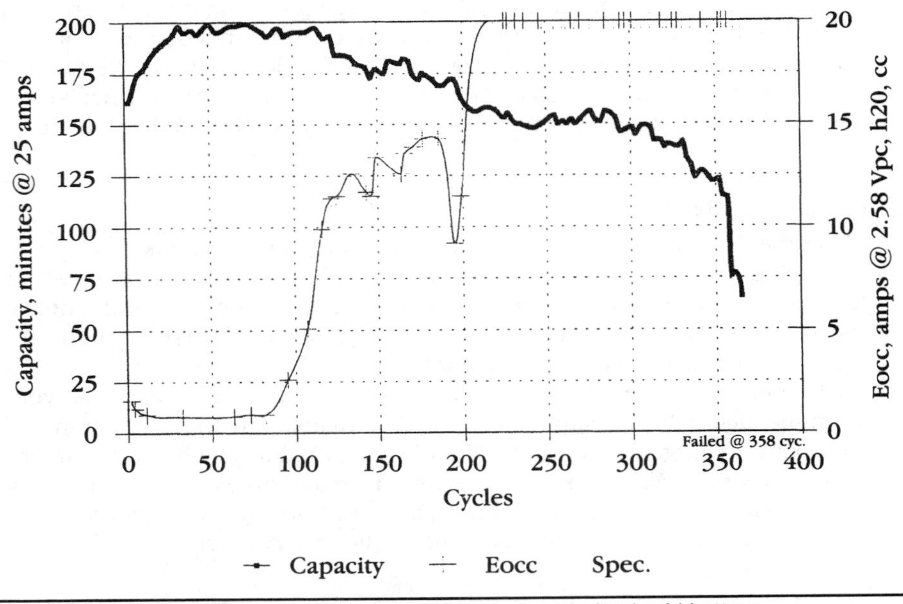

FIGURE 9-1 Cycle-life test results for Trojan 27TMH deep-cycle lead-acid battery.

- It means you have to crank up the voltage and current when charging a more mature battery. Both of these mean that you cannot plug a charger into your battery, set it, and forget it because a battery's charging needs also change from cycle to cycle and with temperature and depth of discharge.

Sealed lead-acid batteries, with a small amount of calcium added to eliminate the need for rewatering, don't exhibit this characteristic; their EOCC is relatively flat so you can be a little more tolerant with them. (They are also more expensive—you pay your maintenance costs up front.)

Battery Discharging Cycle

Let's observe the discharge cycle first, to contrast what is happening to the parameters with what goes on during charging. Capacity, cell voltage, and specific gravity all decrease with time as you discharge a battery. Figure 9-2 shows how these key parameters change (a standard temperature of 78 degrees F is presumed):

- **Ampere-Hours**—The measure of the battery's capacity and percent state-of-charge (the area under the line in this case) are shown decreasing linearly versus time from its full charge to its full discharge value.
- **Cell Voltage**—Cell voltage predictably declines from its nominal 2.1-volt fully charged value to its fully discharged value of 1.75 volts.
- **Specific Gravity**—Specific gravity decreases linearly (directly with the battery's discharging ampere-hour rate) from its full charge to its full discharge value.

Battery Charging Cycle

Battery charging is the reverse of discharging. Figure 9-2 again shows you how the key parameters change:

- **Ampere-Hours**—This is the opposite of the discharging case, except that you have to put back slightly more than you took out (typically 105 to 115 percent more) because of losses, heating, etc. The area under the line increases linearly versus time from its fully discharged value to its fully charged value.
- **Specific Gravity**—Specific gravity increases wildly over time as a battery is charging, so making specific gravity measurements during the charging cycle is not a good idea. At the early part of the charging cycle, specific gravity increases slowly because the charging chemical reaction process is just starting. Specific gravity increases rapidly as the sulfuric acid concentration builds, and gassing near the end of the cycle contributes to its rise.
- **Cell Voltage**—Voltage also increases wildly over time as a battery is charging, so making voltage measurements during the charging cycle is not a good idea either. Notice cell voltage jumps up immediately to its natural 2.1-volt value; slowly increases until 80 percent state-of-charge (approximately 2.35 volts); increases rapidly until 90 percent state-of-charge (approximately 2.5 volts); and then builds slowly to its full charging value of 2.58 volts.

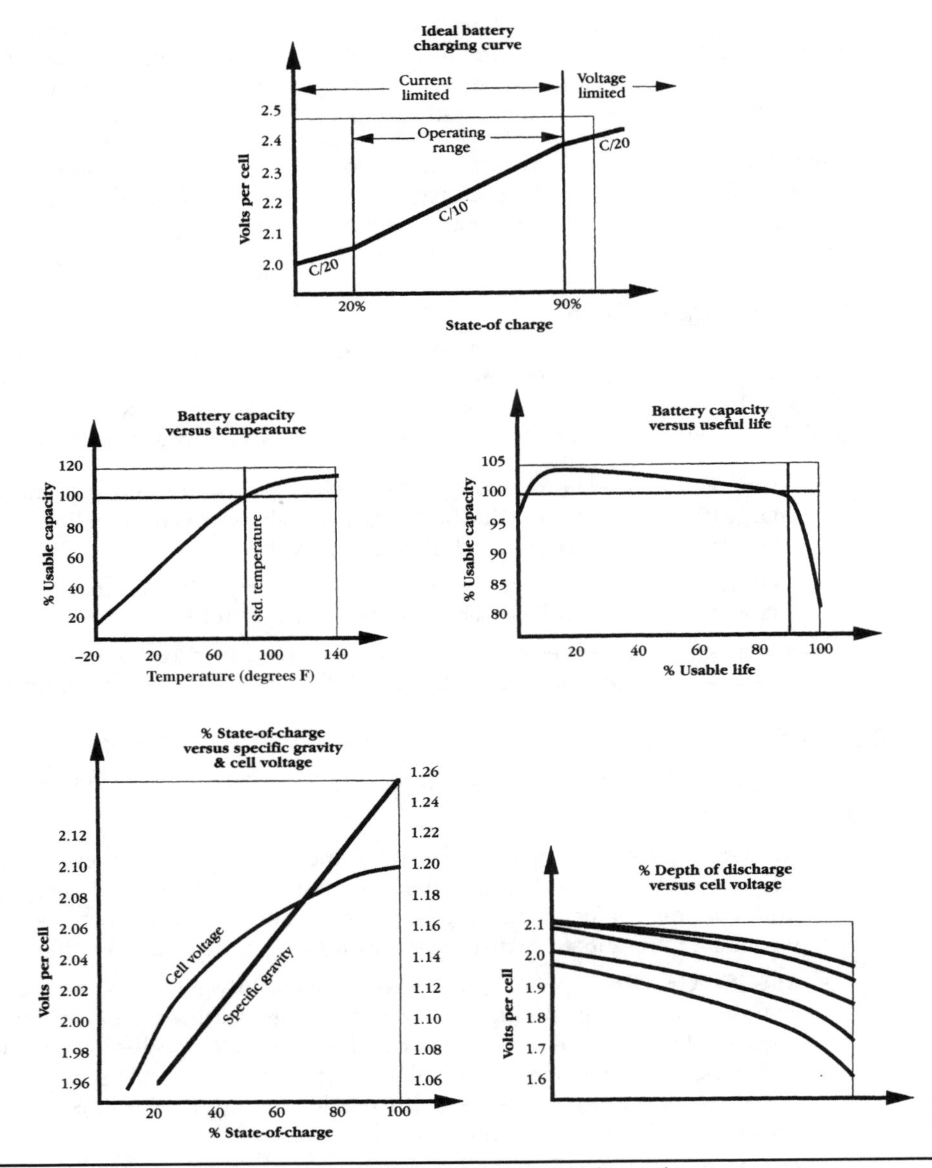

FIGURE 9-2 Graphic summary of battery discharging and charging cycles.

The Ideal Battery Charger

Battery charging is the reverse of discharging, but the rate at which you do it is critical in determining its lifetime. The basic rule is: Charge it as soon as it's empty, and fill it all the way up. The charging rate rule is: Charge it slower at the beginning and end of the charging cycle (below 20 percent and above 90 percent).

When a lead-acid battery is either almost empty or almost full, its ability to store energy is reduced due to changes in the cell's internal resistance. Attempting to charge

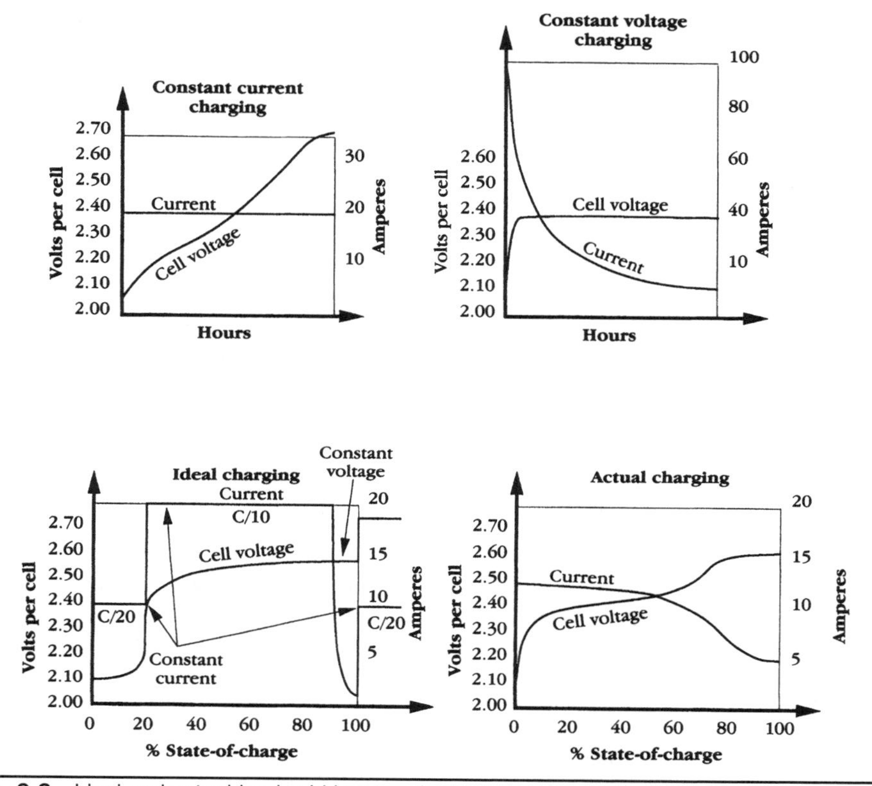

FIGURE 9-3 Ideal and actual lead-acid battery charging characteristics.

it too rapidly during these periods causes gassing and increased heating within the battery, greatly reducing its life. Ideally, you limit battery current during the first 90 percent of the charging cycle and limit battery voltage during the last 10 percent of the charging cycle. Either method by itself doesn't do the job. (Figure 9-3 shows why.)

The graph at the upper right in Figure 9-3 shows constant voltage charging in the ideal case. The constant voltage is usually set at the level where gassing causes a decrease in current flow through the battery with time as the battery charges. Unfortunately, with no restrictions on current, this method allows far too much current to flow into an empty battery. Feeding 100 amps or more of charging current to a fully discharged battery can damage it or severely reduce its life.

Let's look at the ideal approach during all four state-of-charge phases—0 to 20 percent, 20 to 90 percent, 90 to 100 percent, and above 100 percent. Figure 9-3 shows the results.

Charging Between 0 and 20 Percent

The first 20 percent of a fully discharged battery's charging cycle is a critical phase and you want to treat it gently. You learned in Chapter 8 that all batteries have a standardized 20-hour capacity rating. Every battery is rated to deliver 100 percent of its rated capacity at the C/20 rate. During the first 20 percent of the charging cycle, you ideally want to charge a battery at no more than this constant-current C/20 rate. To determine the first 20 percent charging current,

Charging Current = Battery Capacity/Time – C/20

For a 200 ampere-hour capacity battery, charging current would be

Charging Current – 200/20 = 10 amps

In other words, you would limit this battery's initial charging current to 10 amps. You can blast your battery with 200 amps and charge it in 1 hour, but you will prematurely age and kill it—it will not deliver its full, useful life to you. The graph at the lower left in Figure 9-3 shows the result of current-limited C/20 rate charging during the first 20 percent part of the charging cycle. The voltage rises gently and your battery is very happy.

Charging Between 20 to 90 Percent

In the middle of the charging cycle, you can charge at up to the C/10 rate. This is the fastest rate that efficiently charges a lead-acid battery. This rate is not as efficient as the C/20 rate—more energy is wasted in heat if you recall the I_2R losses—but it gets the charging job done faster. At even less efficiency (and more risk to your batteries), you can bump it up to higher C/5 or C/3 rates during this period of recharging if time is essential to you, and if you closely monitor the battery's temperature so its operating limits are not exceeded and you don't wind up "cooking" it. Charging current would be 20 amps at the C/10 rate for the 200 ampere-hour battery. Figure 9-3 shows that voltage, after a step increase when current settings were changed, rises slowly to its 90 percent state-of-charge value of approximately 2.50 volts.

Charging Between 90 and 100 Percent

At this point, you want to drop back to the C/20 rate or, ideally, switch to a constant-voltage method. If you switch to constant voltage set at the deep-cycle battery's full charging value of 2.58 volts, Figure 9-3 shows the result—current provided to the battery drops rapidly during this last 10 percent of charging and your battery is very happy while receiving its full charge.

Charging Above 100 Percent (Equalizing Charging)

You learned about equalizing charging in Chapter 8. It is needed to restore all cells to an equal state-of-charge (to "equalize" the characteristics of the cells); to keep the battery operating at peak efficiency; to restore some capacity of aging batteries; to restore float-charged or shallow-charged batteries to regular service; and to eliminate effects of sulfation in idle or discharged batteries. Equalizing charging is controlled overcharging at a constant-current C/20 rate with the charging voltage limit raised to 2.75 volts. It is done after the battery is fully charged, and maintained at this level for 6 to 10 hours.

Equalizing charging should not be done at rates greater than C/20. Equalizing charging should be done every 5 to 10 cycles or monthly (whichever comes first); only in well-ventilated areas (with no sparks or smoking) as it produces substantial gassing; and only while close attention is being paid to electrolyte level, as water consumption is substantial during rapid gassing periods. Figure 9-3 shows the step increase in the voltage to 2.75 volts and the increase in current back to the C/20 level.

Now let's look at the time involved in using the ideal approach to charge our hypothetical 200 ampere-hour capacity battery:

10 amps (C/20) for 5 hours = 50 AH

20 amps (C/10) for 7 hours = 140 AH

10 amps (C/20) for 1 hour = 10 AH

Totals: 13 hours = 200 AH

This approach requires 13 hours to charge a 200 ampere-hour capacity battery. Provided you don't exceed battery temperature, you could charge at the C/5 rate during the middle of the cycle (40 amps for 3.5 hours) and reduce total time to 9.5 hours.

The Real-World Battery Charger

The starting battery in your internal combustion engine vehicle is recharged by its engine-driven alternator, whose output is controlled by a voltage regulator. The starting battery is discharged less than 1 percent in its typical automotive role. The entire rated output power of the alternator is placed across it, and the voltage regulator makes sure voltage doesn't climb above 13.8 volts—simple. Why not use the same setup to recharge your deep-cycle EV batteries?

The answer is voltage, current, and "boom." This approach is not used because the voltage of your EV's deep-cycle battery pack string is probably 96 volts or more, but the alternator is typically set up for driving a 12-volt starting battery. Assuming you could adjust your alternator or buy the correct voltage model, and had a suitable electric motor around to turn it, the full alternator output applied to a completely discharged deep-cycle battery pack would deliver too much current and charge it far too quickly. You'd damage and/or destroy your batteries in short order—that's the "boom" part. The voltage regulator would not only be useless in stopping this, but would prevent raising the voltage to 2.58 volts per cell for the final 10 percent of the cycle, and would not allow the voltage to be further raised to 2.75 volts for the equalizing charge process.

The solution is AC power, a transformer, a rectifier, a regulator of either the "electronically smart" or "manually adjustable" variety, and a timer: an accurate description of today's EV battery charger.

The x pattern in the graph at the lower right in Figure 9-3 shows what most actual battery chargers deliver. Using a variation or combination of constant-current, constant-voltage, tapering, and end-of-charge voltage versus time methods, all battery chargers arrive at a method of current reduction during the charging cycle as the cell voltage rises. Fortunately, you can buy something off the shelf to take care of your needs. But you have to investigate before buying to make sure a given battery charger does what you want it to do.

Battery chargers are "sized" using the formula:

$$\text{Charging Current} = (\text{Battery Capacity} \times 115\ \text{Percent})/(\text{Time}) + \text{DC Load}$$

In this equation—very similar to the equation earlier in the chapter—the charging current determines the size charger you need, the 115 percent is an efficiency factor to take losses into account, and the DC load is whatever else is attached to the battery (this is zero, assuming you disconnect your batteries from your EV's electrical system while recharging). You're already familiar with battery capacity and time. You can plug chargers up to 20 amps into your standard household 120-volt AC outlet. Higher current capacity chargers require a dedicated 240-volt AC circuit—the kind that drives your household electric range or clothes dryer. Let's take a look at your actual options: build your own, and several different types of off-the-shelf chargers you can buy.

Also, three-phase power is found in industry and the power grid, not in the home. Homes are usually single-phase. Very high-voltage power lines have six wires plus a lightning ground at the top. Opposing wires are the same phase. AC electric cars usually use three-phase motors where the voltage and frequency must be changed as driving conditions vary.

Whether to have an on-board or an off-board charger is another consideration. On-board gives you driving convenience (charge whenever you like); is light in weight; and is low in power consumption. Some commercial EV controllers, like the AC Propulsion model discussed in Chapter 7, even share components between the on-board inverter-controller and battery charger for minimum parts count and maximum efficiency (current flow decides the circuit function automatically—nifty!). Off-board gives you high power capability that translates to minimum charging time in a permanent charging station, which can incorporate many additional features.

Below are two of the most popular chargers on the market today. They are the standard for the industry and are really accepted by the marketplace.

The Manzita Micro PFC-20

Some say, "More expensive than the Zivan, but worth it." A Manzita Micro is pictured in Figure 9-4. Here are some of the specifications:

- The PFC-20 is designed to charge any battery pack from 12 volts to 360 volts nominal (14.4 to 450 peak). It is power factor–corrected and designed to either put out 20 amps (if the battery voltage is lower than the input voltage) or draw 20 amps from the line (if the line voltage is lower than the battery voltage). The buck enhancement option on the PFC-20 will enhance the output to 30 amps. There is a programmable timer to shut off the charger after a period of time set by the user.
- For installation instructions, go to www.manzanitamicro.com/install pfc20revC no photos.doc

FIGURE 9-4 The Manzita Micro PFC-20, an extremely versatile battery charger.

The Zivan NG3

The Zivan is a very popular charger. It is used by the two notable electric vehicles called Corbin Sparrow from Corbin Motors and the GEM, a low-speed electric vehicle from Global Electric Motorcars, a Chrysler Company. Personally, I have used this charger before with the GEM car and it works well (see Figure 9-5).

The Zivan is a notable charger in the electric vehicle community. What I like the most about the Zivan charger is that it is microprocessor controlled and protected against overload, and short circuiting. The charger can also determine if the batteries have problems and let you know so you can keep a properly maintained battery pack at all times. It also has thermosensor options that allow the electric output to adapt to any type of battery. This increases the batteries' life and reduces the water needed to add to the batteries. They also operate at 85 to 90 percent efficiency for maximum savings of electric power costs, which is really efficient. See Table 9-1.

Other Battery Charging Solutions

Against the numerous trade-offs available to you with today's commercially available battery chargers, along with the additional options of the build-your-own approach (using today's advanced electronics components), are the still newer developments coming down the road for tomorrow and beyond. These include rapid charging techniques, induction charging, replacement battery packs, and infrastructure development. Let's take a brief look at each.

Rapid Charging

A number of modern papers have discussed the alternatives of rapid charging. The Japanese are perhaps the most forward-thinking in this area. In short, if charging your EV's battery pack in eight hours is good, then accomplishing the same result in four hours is better. Fortunately, you can use pulsed DC current, alternating charge and discharge pulses, or just plain high-level DC to accomplish the results.

Unfortunately, you need to start with at least 240-volt AC (240V three-phase is better), and you probably overheat and shorten the life of any of today's lead-acid batteries in the process. However, if you design your lead-acid batteries to accommodate this feature (a larger number of thinner lead electrodes, special separators, and electrolyte), it becomes easy. Even nickel-cadmium batteries can be adapted to the

Figure 9-5 The Zivan NG3.

Code	Type	Nominal Battery Voltage	Output Current	Output Current wt-lb	Recommended Batteries Capabilities
F7 AV	NG3 12-100	12 V	100 A	120	500-1000 AH
F7 BQ	NG3 24-50	24 V	50 A	60	250-500 AH
F7BT	NG3 24-80	24 V	80 A	95	400-800 AH
F7 CR	NG3 36-60	36 V	60 A	70	300-600 AH
F7 EQ	NG3 48-50	48 V	50 A	60	250-500 AH
F7 GN	NG3 60-35	60 V	35 A	40	175-350 AH
F7 HM	NG3 72-30	72 V	30 A	35	150-300 AH
F7 IL	NG3 80-27	80 V	27 A	30	135-270 AH
F7 LL	NG3 84-25	84 V	25 A	30	125-250 AH
F7 MI	NG3 96-22	96 V	22 A	26	110-220 AH
F7 NH	NG3 108-20	108 V	20 A	25	100-200 AH
F7 PH	NG3 120-18	120 V	18 A	22	90-180 AH
F7 QG	NG3 132-16	132 V	16 A	19	80-160 AH
F7 RG	NG3 144-15	144 V	15 A	18	75-150 AH
F7 WG	NG3 156-14	156 V	14 A	17	70-140 AH
F7 SF	NG3 168-13	168 V	13 A	16	65-130 AH
F7 TF	NF3 180-12	180 V	12 A	14	60-120 AH
F7 UE	NG3 192-11	192 V	11 A	13	55-110 AH
F7 VE	NG3 216-10	216 V	10 A	12	50-100 AH
F7 XC	NG3 240-9	240 V	9 A	11	50-100 AH
F7 YB	NG3 288-8	288 V	8 A	10	50-100 AH
F7 YB	NG3 312-7	312 V	7.5 A	9	50-100 AH

TABLE 9-1 High-Frequency Battery Charger NG3 Single-Phase 230/115

process if your wallet is bigger. You'll hear more about this idea as time goes by. (Three-phase 240V will supply 50 percent more power than 480-volt single phase. Because adding one wire gives you three times the power, the world's power grid is three-phase. In a residential area, blocks are rotated between phases A, B, and C. Full-wave rectification of three-phase gives you DC with very little ripple. (Second, third, and fourth harmonics cancel.) In electric vehicles, three-phase motors can be reversed by swapping any two phases.)

Induction Charging

Many alternatives have been proposed for induction charging. General Motors' Hughes division is probably the most forward-thinking in this area. The photo in Figure 9-6, taken at the September 1992 Burbank Alternate Transportation Expo, shows the Hughes concept. You drop the rear license plate to expose the EV's inductive charging slot, drop the Hughes "paddle" into the slot, and turn on the juice. With other inductive floor-mounted or front-bumper-mounted approaches, you had to position your vehicle accurately. The Hughes approach makes it even easier than putting gasoline into a vehicle. There are no distances to worry about, no spills, and no risk of electrical shock.

Figure 9-6 Hughes inductive charging system paddle.

The process of putting electrical energy into your EV can be computerized with a credit card. EV charging kiosks—or whatever shape they assume—will someday be as familiar as telephone booths, and Hughes-style inductive charging will probably lead the way.

The Toyota RAV4 and General Motors EV1 used these chargers.

Replacement Battery Packs

Many attractive alternatives have also been proposed in this area. The EV racers at the Phoenix Solar & Electric 500 have used it for years. When they wheel into the pits, saddle-bag-style battery packs are dropped from their outboard mounting positions and fresh battery packs attached in their place. This same approach, with neighborhood "energy stations" replacing gasoline stations, also has a role in the future. Future EV designs could be standardized with underbody pallet-mounted battery packs. You wheel into the energy station, drop the old battery pack, raise the new one into place and latch it down, pay by credit card (probably a deposit for the pack plus the energy cost for the charge), and you're on your way within a few minutes time.

Beyond Tomorrow

Nothing stops you from using rapid charging, induction charging, and/or replacement battery pack techniques today. The first takes a special heavy-duty lead-acid battery. The other two can be done by an individual, but obviously require infrastructure development for widespread adoption. Articles have even been written about how you can charge your EV from a solar source today.

But serious infrastructure development is necessary for the real prize: roadway-powered electric vehicles. Numerous papers and articles have been written, all because of the mouth-watering appeal of the idea. A simple lead-acid-battery-powered EV has more than enough range to carry you to the nearest interstate highway. Once there, you punch a button on the dashboard, an inductive pickup on your EV draws energy from the roadway through a "metering-box" in your vehicle (so the utility company knows who and how much to charge), and you get recharged on your way to your destination without all the pollution, noise, soot, and odors.

If the roadway and your EV are both of the "smart" design, you can even get traveler information (weather, directions, etc.) while the roadway guides your vehicle in a hands-off mode, and you read the morning newspaper or scan the evening TV news. All it takes is infrastructure development—read that as M-O-N-E-Y.

Your EV's Electrical System

To say that the electrical system of an EV is its most important part is not an oxymoron—far from it. The idea is to leave as much of the internal combustion vehicle instrumentation wiring as you need intact, and to carefully add the high-voltage, high-current wiring required by your EV conversion.

Do a great job on your EV conversion's high-voltage, high-current wiring and it will reward you with years of trouble-free service. Do it in a sloppy manner and, aside from the obvious boom or poof (accompanied by smoke), you open yourself to a world of strange malfunctions. The hidden beauty of any home-built EV conversion is that there are only two places to lay blame if anything has gone wrong: look in the mirror and under the hood.

This section will cover the electrical system that interconnects the motor, controller, batteries, and charger along with its key high-voltage, high-current power, and low-voltage, low-current instrumentation components. Figure 9-7 shows you the system at a glance. We'll look at the components that go into high-current and low-current side separately, then discuss wiring it all together.

High-Voltage, High-Current Power System

The heavier lines of Figure 9-7 denote the high-current connections. When you put the motor, controller, battery, and charger together in your vehicle, you need contactor(s),

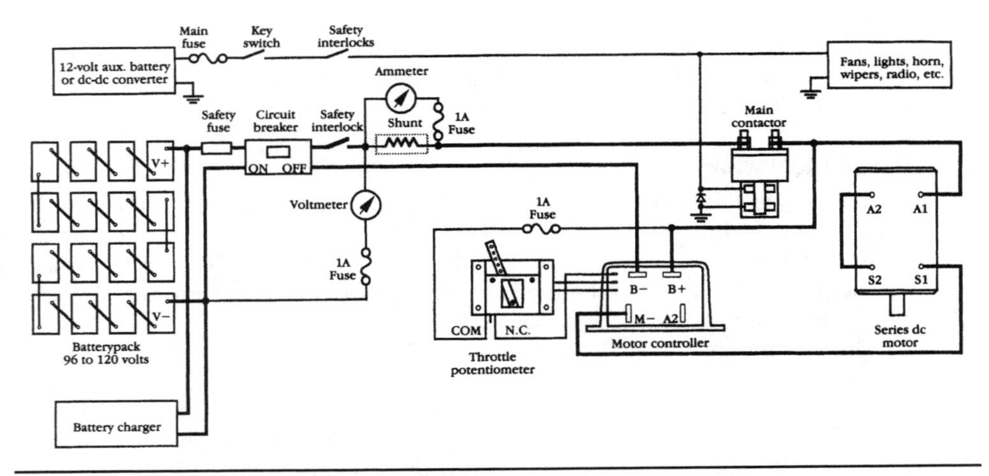

Figure 9-7 Electric vehicle basic wiring diagram.

FIGURE 9-8 Main contactor—single pole.

circuit breakers, and fuses to switch the heavy currents involved. Let's take a closer look at these high-current components.

Main Contactor

A contactor works just like a relay. Its heavy-duty contacts (typically rated at 150 to 250 amps continuous) allow you to control heavy currents with a low-level voltage. A single-pole, normally open main contactor, shown in Figure 9-8, is placed in the high-current circuit between the battery and the controller and motor. When you energize it—typically by turning the ignition key switch on—high-current power is made available to the controller and motor.

Reversing Contactor

This contactor is used in EVs when electrical rather than mechanical transmission control of forward-reverse direction is desired. The change-over contacts of this double-pole contactor, shown in Figure 9-9, are used to reverse the direction of current flow in the field winding of a series DC motor. When this contactor is used, a forward-reverse-center off switch is added to the low-voltage wiring system after the ignition key switch.

Main Circuit Breaker

A circuit breaker is like a switch and a resetting fuse. The purpose of this heavy-duty circuit breaker (typically rated at 300 to 500 amps) is to instantly interrupt main battery power in the event of a drive system malfunction, and to routinely interrupt battery power when servicing and recharging. For convenience, this circuit breaker is normally located near the battery pack. The switchplate and mounting hardware are useful—the big letters immediately inform casual users of your EV of the circuit breaker's function.

FIGURE 9-9 Reversing contactor—double pole, cross-connected.

Safety Fuse

The purpose of the safety fuse is to interrupt current flow in the event of an inadvertent short-circuit across the battery pack. In other words, you blow out one of these before you arc-weld your crescent wrench to the frame and lay waste to your battery pack in the process.

Safety Interlock

There is an additional switch that some EV converters incorporate into their high-current system, usually in the form of a big red knob or button on the dashboard—an emergency safety interlock or "kill switch." When everything else fails, punching this will pull the plug on your battery power.

Low-Voltage, Low-Current Instrumentation System

The instrumentation system includes a key switch, throttle control, and monitoring wiring. Key switch wiring, controlled by an ignition key, routes power from the accessory battery or DC-to-DC converter circuit to everything you need to control when your EV is operating: headlights, interior lights, horn, wipers, fans, radio, etc. Throttle control wiring is everything connected with the all-important throttle potentiometer function. Monitoring wiring is involved in remote sensing of current, voltage, temperature and energy consumed, and routing to dashboard-mounted meters and gauges. Let's take a closer look at these low-voltage components.

Throttle Potentiometer

This is normally a 5-kilohm potentiometer, but has a special purpose and important safety function. The Curtis model, designed to accompany and complement its controllers and to use the existing accelerator foot pedal linkage of your vehicle, is shown in Figure 9-10. The equivalent model, for replacement use or for ground-up vehicle designs not already having an accelerator pedal, is shown in Figure 9-11. With either of these, the Curtis model provides a high pedal disable option that inhibits the controller output if the pedal is depressed; that is, you cannot start your EV with your foot on the throttle—a very desirable safety feature. As the Curtis controller also contains

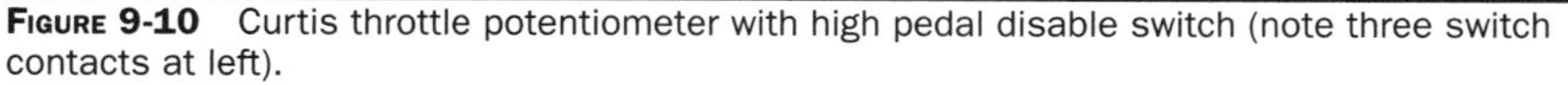

FIGURE 9-10 Curtis throttle potentiometer with high pedal disable switch (note three switch contacts at left).

FIGURE 9-11 Curtis replacement accelerator pedal throttle potentiometer.

a fault input mode that turns the controller off in the event an open potentiometer input is detected (for example, in the case of a broken wire)—a condition that would result in a runaway—you are covered in both instances by using the Curtis controller and throttle with high pedal disable option. Figure 9-11 shows that the throttle potentiometer wiring goes directly to the controller inputs, with the interchangeable potentiometer leads and the common and normally closed contacts wired as shown.

Auxiliary Relays

Figure 9-12 shows these highly useful auxiliary control double-pole, double-throw relays in both 20-amp-rated 12-volt DC (on right) and 120-volt AC coil varieties. A typical use for the AC coil type would be as a charger interlock. Wired in series with the on-board charger, AC voltage sensed on the charger's input terminals would immediately disable the battery pack output by interrupting the auxiliary battery key switch line, which, in turn, opens the main contactor. DC coil uses are limited only by your imagination: additional safety interlocks; voltage, current, or temperature interlocks; and controlling lights, fans, and instrumentation.

Terminal Strip

Whatever you're doing in the electrical department, a simple terminal strip like the one shown in Figure 9-13 makes your wiring easier and neater. Using one or more of these as convenient tie-off points not only reduces error possibilities in first-time conversion wiring, but also makes it simpler to track down your connections later if needed. Of course, it's only as valuable as your hand-drawn sketch of what function is on which terminal.

Figure 9-12 Auxiliary relays—120-volt AC (left) and 12-volt DC (right).

Figure 9-13 Terminal strip.

Shunts

Shunts are precisely calibrated resistors that enable current flow in a circuit to be determined by measuring the voltage drop across them. Two varieties are shown in Figure 9-14: the left measures currents from 0 to 50 amps; the right measures currents from 0 to 500 amps.

Ammeter

The most useful of all your EV on-board instruments is your ammeter. The dual-scale model shown in Figure 9-15 (top left) delivers 0- to 50-amp or 0- to 500-amp monitoring ranges at the flip of a switch. The higher range enables you to determine your motor's instantaneous current draw; it functions much like a vacuum gauge in an internal combustion engine vehicle—the less current, the higher the range, and so on. The lower range functions the same as the ammeter that might already be on the instrument cluster of your internal combustion conversion vehicle—it tells you the amount of current your 12-volt accessories are consuming.

Figure 9-14 Ammeter current shunts—50 amp (left) and 500 amp (right).

FIGURE 9-15 Instrumentation meters—ammeter (top left), voltmeter (top right), battery indicator (center), temperature (bottom left), and rotary switch (bottom right).

Voltmeter

The second most useful EV on-board instrument has to be your voltmeter. The dual expanded-scale model (Figure 9-15, top right) delivers 50- to 150-volt or 9- to 14-volt ranges at the flip of a switch. Expanded scale means only the required voltmeter range is used; the entire scale is expanded to fill just the range of voltages you use. The higher range enables you to determine your battery pack's instantaneous voltage (it functions like the fuel gauge in your internal combustion engine vehicle—the less voltage, the less range remaining. The lower range functions like the voltmeter that might already be on the instrument cluster of your internal combustion conversion vehicle—it tells you the status of your 12-volt system.

Battery Indicator

Users seem either to like the battery indicator a lot or find it redundant (with the voltmeter) so investigate before you invest. The Curtis Model 900 version (Figure 9-15 center) indicates battery pack state-of-charge as of your last full charge-up using a 10-

element LED readout. The battery indicator is wired directly across the battery as if it were a voltmeter. Proprietary circuitry inside the module then integrates the voltage state into a readout of remaining energy that's displayed on one of the 10 LED bars. While this is not as useful for those employing on-board charging, and certainly does not come close to being a battery energy management system, it is useful for those who charge only from a fixed site as a guide to when the next pit stop is required.

Temperature Meter

While your lead-acid-battery-powered EV won't run at all if it's very, very cold, you can use it again when the temperature rises. At the high-temperature end of the scale, you can cause permanent damage to your batteries, controller, or motor if their temperature limits are exceeded. A temperature gauge certainly falls into the nice-to-have rather than the mandatory category, but if you are so inclined, an easy way to keep tabs on temperature is by using a thermistor and a temperature gauge like the one shown in Figure 9-15 (bottom left). Or you can use multiple thermistors—one bonded to each object of interest (battery pack, controller, motor, etc.)—and monitor all by switching between them.

Rotary Switch

Rotary switches such as the four-pole, two-position switch shown in Figure 9-15 (bottom right) are ideal companions to your instrumentation meters for range, sensor, and function switching. While you can opt for only an ammeter and a voltmeter in your finished EV, you might want to check voltage and current at numerous points during the testing stage. A handful of rotary switches helps you out in either case.

Fans

Fans fall into the mandatory category for keeping temperature rise in check for engine compartment components, or for keeping the battery compartment ventilated during charging. Whether DC-powered full time from the key switch circuit, DC-powered intermittently via relay closure, or powered from an AC outlet, the spark-free brushless 12-volt DC motor fans and 120-volt AC motor fans are the type you want to choose.

Low-Voltage Protection Fuses

All your instrumentation and critical low-voltage components should be protected by 1-amp fuses (the automotive variety work fine) as shown in Figure 9-8. Whenever 25 cents can save you up to $200, it's a good investment.

Low-Voltage Interlocks

Many EV converters prefer to implement the kill switch referred to earlier in this section on the low-voltage side. Often it's easier because there are a number of interlocks already there—seat, battery, impact, etc. In addition, a low-voltage implementation takes just a simple switch, possibly a relay, and some hookup wire—a few ounces of weight at the most—while a high-current solution takes several pounds of wire plus bending and fitting, etc.

Figure 9-16 Sevcon DC-to-DC converters—72/12 model (left) and 128/12 model (right).

DC-to-DC Converter

Figure 9-16 shows you two DC-to-DC converter options. Most of today's EV converters will opt for the 128-volt to 12-volt model on the left (it operates from 78 volts to 126 volts input and delivers a nominal 13.5-volt output). The advantages of using one of these 25-amp units to power the key switch accessories, throttle potentiometer, and instrumentation boil down to one word: weight. Using a DC-to-DC converter in place of an auxiliary battery saves you 50 lbs. And most DC-to-DC converters give you a nice, stable 12-volt output, even with widely varying battery pack voltage swings. By this time, you already know it's inadvisable to draw power from anything less than all the batteries in the pack's string (or risk reduced battery life, etc.), so choose your DC-to-DC converter accordingly—no 12-volt to 12-volt models, please.

Auxiliary Battery Charger

Figure 9-17 shows you the other reason for using a DC-to-DC converter: you won't need this auxiliary battery charger. An auxiliary battery also has to be recharged, so you would need a separate charger supplying 12 volts (unless you opt for a dual-voltage charger like Lester's 12/108 model). If it's a deep-discharge accessory battery, using any automotive charger just won't do. On balance, the battery and charger wind up costing you the same as the DC-to-DC converter, but take continuing attention and deliver a less steady voltage with an added weight penalty.

You mount it on the wall of your garage, hooked up between your battery charger and its electrical service, as shown in Figure 9-18. Keep a logbook nearby. You can record

Figure 9-17 Auxiliary battery charger for on-board 12-volt instrumentation battery.

FIGURE 9-18 AC-style wattmeter goes between AC outlet and charger.

the wattmeter's reading, charge your EV's battery pack, do your tooling around, then come back and repeat the process. Over time, the wattmeter tells you your EV's energy use patterns, and can quickly tell you if something is amiss (dragging brake shoe, etc.) by deviations from the pattern. Plus you can use the results to show your wife/husband, friends, neighbors, and community just how much money you saved compared to an internal combustion engine vehicle.

Wiring It All Together

Five things are important here—wire and connector gauge, connections, routing, grounding, and checking. We'll cover them in sequence.

Wire and Connectors

This might be one of the last things you think about, but it's by no means the least important. While your wire size and connector type choices on the instrumentation side are not as important as the connections you make with them, all of these are important on the power side.

Working with AWG 2/0 cable gauge wire is not my favorite pastime—think of it as involuntary aerobic exercise—but its minimal resistance guarantees you a high-efficiency EV as opposed to the world's greatest moving toaster.

Minimal resistance means how the connectors are attached to the wire cable ends is equally important to the overall result. Crimp the connectors onto the cable ends using the proper crimping tool (ask your local electrical supply house or cable provider) or have someone do it for you. A dinky triangle contact crimp, which you can easily get away with when working in AWG 18 hookup wire, is fatal to your round ferrule AWG 2/0 connector. It will cause you a hot spot that sooner or later will melt (or be arc-welded) by the routine 200-amp EV currents. Meanwhile, you will get poor performance. If you are getting 20 miles per charge and your neighbor is getting 60 miles per charge with the identical setup, and you checked for the obvious mechanical-motor-controller-battery reasons, chances are it's in your wiring. Treat each crimp with loving attention and craftsmanship, as if each was your last earthly act, and you will be in heaven when it comes to your EV's performance.

Connections

On the power side, connections are important. These occur when your AWG 2/0 wire connectors attach to motor, controller, batteries, shunts, fuses, circuit breakers, switches, etc. Check to ensure surfaces are flat, clean, and smooth before attaching. Use two wrenches to avoid bending flat-tabbed controller and fuse lugs. Torque everything down tight, but not gorilla-tight. Check everything and re-tighten battery connections at least monthly.

Routing

Aim for minimum length routing on the power side. Leave a little slack for installation and removal, and a little more slack for heat expansion; then go for the line that's the shortest distance between the two points. On the instrumentation side, it's neatness and traceability that count: you want it neat to show off to friends and neighbors, you want it orderly so that you (or someone else) can figure out what you did.

Figure 9-19 shows the ideal layout of Don Moriarty's custom sports racers—everything neatly laid out on a giant heat sink backing plate (1/8-inch to 1/4-inch aluminum reinforced and cross-braced, etc.). This shelf can be hinged at the back to the firewall and pinned in the front (or vice versa). Gas shocks (of the rear trunk deck variety) can be added to make its 30 to 50 lbs. easy to lift up for access—a user-friendly touch. In Chapter 10, Jim Harris' "magic box" gives you another page for your idea book—a different approach that produces the same desirable results: minimum length combined with neatness and traceability.

Grounding

The secret of EV success is to be well grounded in all its aspects. "Well grounded" in electrical terms means three things:

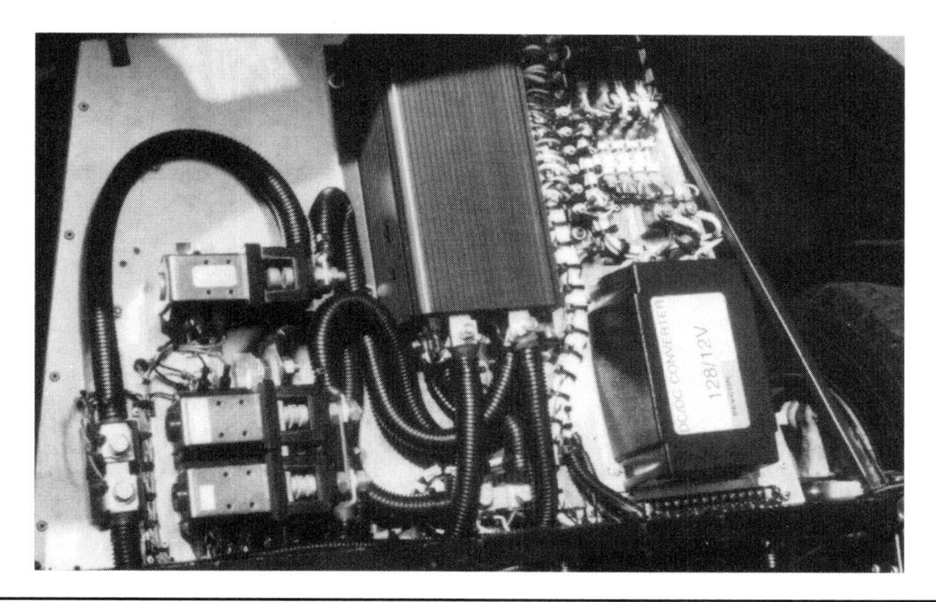

Figure 9-19 Dan Moriarty's nearly ideal layout of electric vehicle control and wiring of components.

- **Floating Propulsion System Ground**—No part of the propulsion system (batteries, controller, etc.) should be connected to any part of the vehicle frame. This minimizes the possibility of being shocked when you touch a battery terminal and the body or frame, and of a short circuit occurring if any part of the wiring becomes frayed and touches the frame or body.
- **Accessory 12-Volt System Grounded to Frame**—The 12-volt accessory system in most EV conversions is grounded to the frame, just like the electrical system of the internal combustion vehicle chassis it utilizes. The body and frame is not connected to the propulsion system, but it can and should be used as the ground point for the 12-volt accessory system, just as the original vehicle chassis manufacturer did.
- **Frame Grounded to AC Neutral When Charging**—The body and frame should be grounded to the AC neutral line (the green wire) when an on-board or off-board AC charger is attached to the vehicle. This prevents electrical shock when the batteries are being charged. To guarantee shock-free performance, transformerless chargers should always have a ground fault interrupter, and transformer-based chargers should be of the isolation type.

Checking

This is not a paragraph about banking. It's a paragraph about partnership. Whatever system you decide to use—continuity checking, verbal outcry, color coding, matching terminal pairs to a list, etc.—at least have one other human help you. It'll make the conversion go faster, plus chances are you'll find something that you alone might have overlooked. Speaking of chances, be sure your EV's drive wheels are elevated the first time you hook up your batteries to your newly wired-up creation so it doesn't accidentally "wander" through your garage door.

CHAPTER 10

Electric Vehicle Conversion

Pickup trucks and sports cars are the best-selling vehicles; they are also the best EV conversions.

Whether you've read through the whole book thus far or just picked up the book and skipped ahead to this chapter, you've finally made it to the chapter that tells you how to do it. The conversion process is what ties your electric vehicle together and makes it run. You've learned that choosing the right chassis to meet your driving needs and goals is a key first step. You've also learned about motor, controller, battery, charger, and system wiring recommendations. Now, it's time to put them all together. A carefully planned and executed conversion process can save you time and money during conversion, and produce an efficient vehicle that's a pleasure to drive and own after it's completed.

This chapter goes through the conversion process step by step with the assistance of a few conversion specialists. It also introduces the type of chassis to choose for your EV conversion today—the compact pickup truck. You'll discover that after the simple act of going through the conversion process, your efforts and results will perform even better—you are now an expert.

Conversion Overview

What do you do for a fresh point of view when constructing something mechanical, even if it's an electric vehicle? You go to a mechanic. We went to a master mechanic and machinist, Jim Harris, who has had more than 20 years' experience, owned his own auto repair shops, and was able to offer a completely fresh point of view on how to build an EV.

What do you do first? Start with a pickup truck. Pickup trucks are the most popular single-person transport vehicle for commuter use, and are useful for carrying many other loads as well. They make an excellent platform for EVs because they isolate the batteries from the passenger compartment very easily; the additional battery weight presents no problems for a vehicle structure that was designed to carry the weight anyway; and the pickup is far roomier in terms of engine compartment and pickup box space to do whatever you want to do with component design and layout. Pickup trucks, the best-selling vehicles, are also the best EV conversions.

FIGURE 10-1 Jim Harris' 1987 Ford Ranger EV conversion.

Keep in mind that, while we are talking about a step-by-step Ford Ranger pickup truck EV conversion here, the principles will apply equally well to any sort of EV conversion you do. And while the Ford Ranger pickup truck used in this chapter's conversion is also used by other conversion professionals, the Chevy S-10 and Dodge Dakota platforms also have their proponents.

Some of the greatest results were the first one Bob Brant wrote in the first edition of a project by Jim Harris on his first vehicle—the 1987 Ford Ranger shown in Figure 10-1.

FIGURE 10-2 Paul Little's electric Porsche (Courtesy of EVPorsche.com).

FIGURE 10-3 Joe Porcelli and Dave Kimmins of Operation Z's Nissan.

Figure 10-2 shows Paul Little's Porsche conversion, and Figure 10-3 shows the Nissan that Joe Porcelli and Dave Kimmins of Operation Z converted to electric.

The objective here is to get you into a working EV of your own with minimum fuss, converting from an internal combustion engine vehicle chassis. For those who are into building from-the-ground-up and kit-car projects, there are other books you can read, and the techniques discussed here can be adapted. The actual process for your EV conversion is straightforward:

- Before conversion (planning)—who, where, what, when
- Conversion (doing)—chassis, mechanical, electrical, and battery
- After conversion (checking)—testing and finishing

Figure 10-4 shows you the entire process at a glance. Let's get started.

Before Conversion

A little effort expended before conversion, in the who-where-what-when planning stage, can pay large dividends later because you've thought out what you need beforehand and don't have to go running around at the last minute. Let's look at the individual areas.

Arrange for Help

This is the "who" part. Help comes in two flavors: inside help and outside help.

Inside Help

Whether it's lifting out the engine, pulling the high-current wiring, installing the batteries, or just halving the assembly time and having someone to talk with while you work, an inside helper can work wonders. It's strongly suggested you schedule one for all the heavy tasks and for any/all others as it suits you.

Outside Help

This involves subcontracting out entire tasks to professionals who are more competent and faster at doing their specialty. Excellent candidates for outside help would be in internal combustion engine removal and fabrication of electric motor mounting components.

Arrange for Space

This is the "where" part. Your work area can be owned or rented or borrowed, but it has to be large enough, clean enough, heated or cooled enough, with light and appropriate electrical service available to take care of your needs during the length of time your conversion requires. Let's run down the checklist of what you should look for.

Ownership or Rental

You'll need the space for 80 to 100 hours of work time or more—one to three months of calendar time and up. If you're using your own two-car garage for the project, household rules of companionship dictate you don't tell your wife or husband, "I just need it for a little while, honey—you don't mind parking your Porsche in the street, do you?" Plan ahead. It's going to take at least a couple of months.

An alternative solution is to rent the space you need. This can be anything from a neighbor's unused garage to an oversized public storage locker to space at a local garage. At some point your conversion is going to look like a dinosaur with its bones strewn all about; at another it's going to be very messy and greasy; at another you might need to attach a winch to an overhead beam that can support the weight of the engine you are removing; and at most times you are going to have upwards of several thousand dollars' worth of components sitting around—consider all these needs.

Another alternative is to do it as a two-step. If you have the internal combustion engine parts removed by a professional, then tow your chassis body to your work area for the conversion, there is less need for a large work area and heavy ceiling joists, and it won't get as dirty.

Size

Realistically, an EV conversion will overflow the needs of any single-car garage or storage locker. While it can be done, you will be cramped, so why not do it right to begin with. Unless you opt for the two-step approach just mentioned, you need at least a two-car garage or equivalent amount of space.

Heating, Cooling, Lighting, and Power

Suit yourself here. You're going to be working in this spot for several months. Why not do it in comfort and convenience?

Arrange for Tools

This is the "what" part. What kind of tools are you going to need to do the job? If you have to get into areas and tools that you are unfamiliar with, maybe you are best served by subcontracting those tasks to an outside professional. A little forethought in this area lets you quickly sort out those tasks you want to farm out versus those you want to do yourself—all just by using the tools criteria.

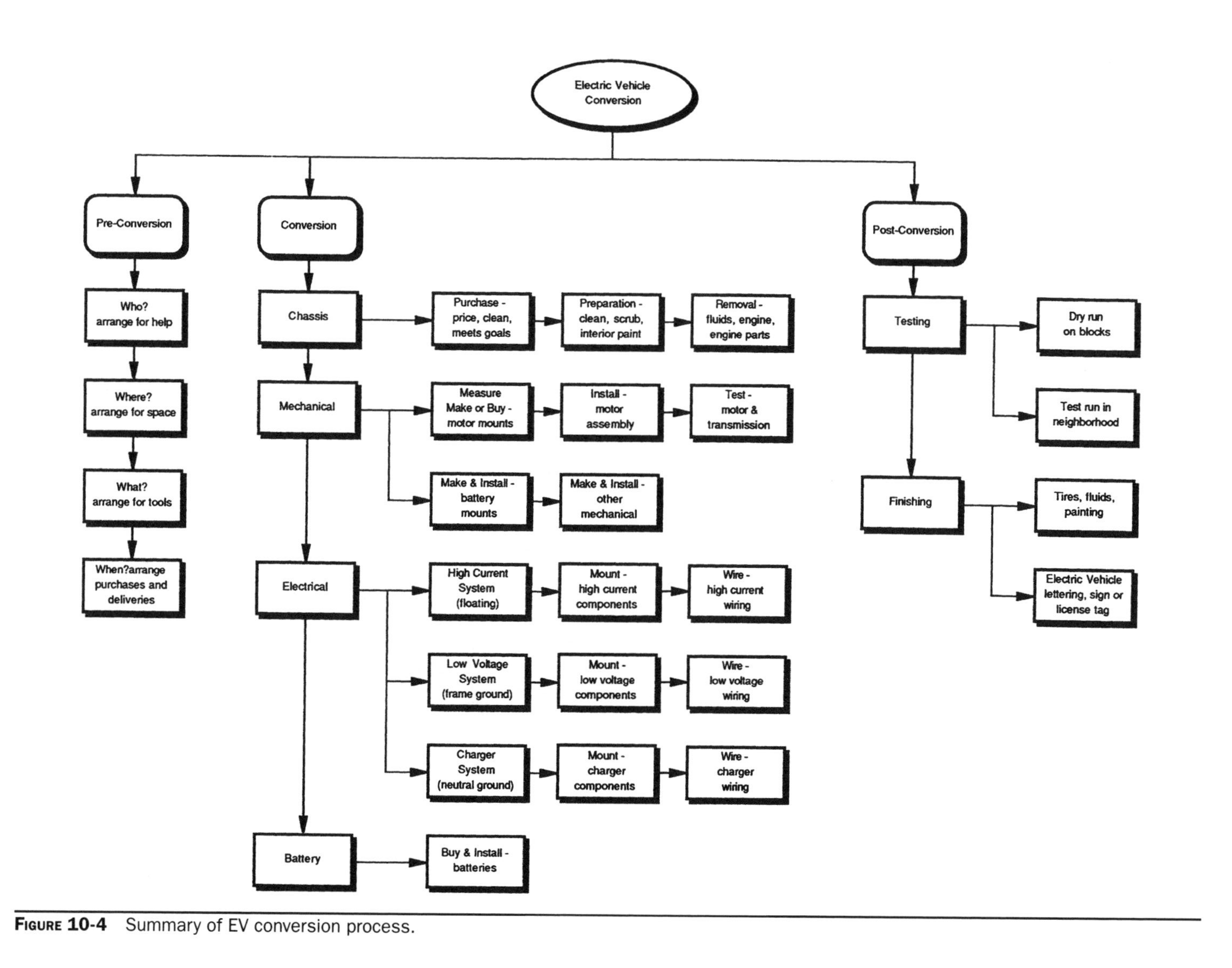

FIGURE 10-4 Summary of EV conversion process.

Arrange for Purchases and Deliveries

This is the "when" part. Ideally, you have a just-in-time arrangement—the exact part you need comes magically floating through the door exactly when you need it. Reality might fall somewhat short of this, but nothing stops you from setting it up as your goal by thinking about what you are going to need by when in advance.

Conversion

Conversion planning can pay even greater dividends. As you saw from Figure 10-4, there are four parts to the conversion, or doing stage, and each is further subdivided:

- **Chassis**—Purchase, preparation, removal of internal combustion engine parts
- **Mechanical**—Motor mount fabrication, motor installation, battery mounts, and other mechanical parts fabrication and installation
- **Electrical**—High-current, low-voltage, and charging system components and wiring
- **Battery**—Purchase and installation of batteries

A simple way of looking at the procedure is: buy and clean up the chassis, remove all the internal combustion engine parts, make or buy the parts to mount the motor and batteries, mount and wire the electrical parts, then buy and install the batteries. Let's look at the individual areas.

Chassis

The chassis part involves the purchase, preparation, and removal phases. In other words, you do everything necessary to get the chassis you're going to convert ready for conversion. Let's take a closer look at each step.

Purchase the Chassis

The first step in a step-by-step conversion is to purchase the vehicle you're going to convert to electric. As an added incentive, take another look at Figures 10-1, 10-2, and 10-3. These are the finished conversion photographs. Notice, in each case, the vehicle is absolutely stock on the outside (Jim Harris did add a smooth nose in place of the 1993 model's grille). Paul Little did a job on a Porsche like no other. The Operation Z crew built a great car from a Nissan that is also like no other. All of them exemplify what building your own electric vehicle is all about!

(Note: Bob Brant liked to mention that the only giveaway is the lettering. If you're going for acceleration, make sure to make the word "ELECTRIC" on the back of your tailgate large.)

The chassis purchase details were covered in Chapter 5; the bottom line is to get the most for the least. You want the stripped 4-cylinder (6-cylinder in some models and years), manual transmission, least weight version. Ideally, you don't need the engine so you can do a trade-out with the dealer or selling party on the spot. Just make sure everything else is as close to perfect running condition as possible—less work for you later. Most of all, make sure there's little or no rust; clean is nice, too. Mechanical parts you can replace, but a rusted body is nearly impossible to deal with; dirt and crud just exacerbate the condition and possibly hide additional problems.

FIGURE 10-5 When you start with a new vehicle...

On the other hand, you can start with a showroom-new vehicle like that shown in Figure 10-5 and own a conversion you'll be really proud of. In addition, removing everything from its crowded engine compartment—as shown in Figure 10-6—is easier because all the parts are clean and without accumulated road grime.

FIGURE 10-6 ...you start with a full but clean engine compartment.

Purchase Other Components

Once you pick your chassis, you can make other parts decisions based on your overall performance goals: high mileage, quick acceleration, or general-purpose commuter. Expected vehicle range and acceleration can be calculated from the vehicle's total weight, tire rolling resistance, aerodynamics, and the amount of energy and power available from the motor and batteries.

One of the easiest and quickest ways to do your conversion is to order all of the parts you need in one kit from someone like KTA Services Inc. Their kits typically contain the parts shown in Figure 10-7; you get nearly everything you need to complete your conversion, the parts are matched to work together, and you have someone to turn to in case you need additional assistance. Buying a prepackaged kit will greatly simplify your conversion both in terms of what it takes to get the job done and the time it takes to run around and get all the parts you need from various vendors, make sure they match, and so on.

Parts to build this particular pickup truck conversion (and pickup trucks in general) are available from Jim Harris' company, Zero Emission Motorcars, Inc. Jim's company provides the built-up control-box and motor-to-transmission adapters that really save you time and guarantee results, or the entire kit for a mid-range or low-end pickup truck of your choice, which includes all heavy-duty coil springs, battery brackets, cables, mounting hardware, etc.

The other items you want to order at this time are the detailed shop maintenance and electrical manuals for your chassis. These might also be available in the library or in a technical bookstore in a larger city. Study these before starting your conversion.

Figure 10-7 Typical KTA Services Inc. electric vehicle kit.

Prepare the Chassis

The next step is to clean the chassis and make a few measurements. The first chassis cleaning step is to give its engine compartment a good steam cleaning. Well-used chassis might require extra scrubbing to remove accumulated dirt and grime. This has a two-fold purpose: it minimizes the grease and gunk you have to deal with in parts removal, and you can see what you need to look at and measure. You might even want to repaint the engine compartment area at this time.

The measuring step involves determining the position of the transmission/ drivetrain or transaxle in your internal combustion engine vehicle and reproducing this position as closely as possible in your EV. Your internal combustion engine chassis will have one of three possible engine-drivetrain arrangements:

- Lateral-mount front engine, transmission, and driveshaft
- Transverse-mount front engine with transaxle
- Rear engine with transaxle (VW-style)

Pickup truck conversion chassis will nearly always deal with one of the first two variations. Two measurements are important: vertical height from transmission housing to floor (measured on a level surface); and a vertical and horizontal reference point on transmission housing to body or frame.

Vertical clearance from the floor is important in any conversion where your motor mounts also support your transmission. You want to reproduce this as closely as possible so your tranny is not running uphill or downhill from the way it started.

A firewall or frame horizontal and vertical alignment mark is also important. It helps in obtaining proper mechanical alignment of the mating parts later on. Scribe this mark accurately with an awl or nail and highlight it with a dab of spray paint or touch-up paint so you can later find it.

Removing Chassis Parts

The next step is to drain the chassis liquids, remove all engine-compartment parts that impede engine removal, remove the engine, and then remove all other internal combustion engine-related parts.

The parts removal process starts with draining all fluids first: oil, transmission fluid, radiator coolant, and gasoline. Remember to dispose of your fluids in an environmentally sound manner or recycle them. Draining gasoline from your tank is particularly dangerous and tricky—put out your cigarette before you attempt it. Drain as much as you can before you physically remove the tank. Discharging an air-conditioning system, if your conversion vehicle has one, is a job best left to a professional.

Next, disconnect the battery and remove all wires connected to the engine. Carefully disconnect the throttle linkage—you will need it later—and set it aside out of harm's way. Then remove everything that might interfere with the engine lift-out process. The hood is a good item to start with, followed by radiator, fan shroud, fan, coolant and heater hoses, and all fuel lines. Disconnect the manifolds and remove the exhaust system at this time, too.

If your work area isn't equipped for heavy lifting (the smallest 4-cylinder engine with accessories attached can weigh around 300 lbs.), you might be best served by letting a professional handle the heavy work. Engine overhaul shops are usually glad to do the job and you might be able to cut a deal with them for the engine and parts that results in a net gain on the transaction. Plus there are a number of other aspects—draining fluids, storing parts, not breaking cables and wires—and the physical act of actually removing the engine without damage to you, your chassis, or the engine. To use an analogy, it's like changing your own motor oil at home versus at the Valvoline. It takes you several hours and you have to clean up a mess (yourself and your spot). The Valvoline people, who do it for a living, take 10 minutes and neither they nor you get dirty.

If you do the job yourself (schedule an "inside" helper for this step), pick a sturdy joist to attach your lifting winch to—not a two-by-four—or rent an engine-lifting dolly. Cover the vehicle's fenders and sides with moving pads to protect them during lift-out. Attach the chain or cable to the engine at two different points. Remove the bolts from the engine mounting brackets and those attaching the engine to the transmission bell housing. The disconnected engine slides straight back away from the transmission, then up and out (in theory). In practice, since you have to both pull down on the winch cable and pull back on the engine block, another set of hands helps greatly. The objective of the whole process is to remove the 300 lbs. of engine without banging, bending, or damaging anything inside the bell housing (transmission spline shaft, etc.) or on the now exposed business-end of the engine (clutch-flywheel assembly, etc). Carefully set the removed engine down in an out-of-the-way yet protected part of your work area.

Finally, remove everything else associated with the internal combustion engine: gas tank, gas lines, muffler, exhaust pipes, ignition, cooling and heating systems.

Mechanical

The mechanical part involves all the steps necessary to mount the motor, and install the battery mounts and any other mechanical parts. In other words, you next do all the mechanical steps necessary for conversion. You follow this sequence because you want to have all the heavy drilling, banging, and welding—along with any associated metal shavings or scraps—well cleaned up and out of the way before tackling the more delicate electrical components and tasks. Let's take a closer look at the steps.

Mounting Your Electric Motor

Your mission here is to attach the new electrical motor to the remaining mechanical drivetrain. The clutch-to-flywheel interface is your contact point. Figure 10-8 gives you an overview of your task in generalized form. Figure 10-9 shows you a great motor compartment designed by Paul Little at EVPorsche.com. When you flip it over and place it in the undercarriage, you will see how snug the motor fits later in this chapter. Four elements are involved:

- The critical distance flywheel-to-clutch interface
- Rear support for the electric motor
- Front support-motor-to-transmission adapter plate
- Flywheel-to-motor-shaft connection via the hub or coupling

We'll cover what's involved in each of these four areas in sequence. Understand that this discussion has to be generalized because there are at least a dozen good

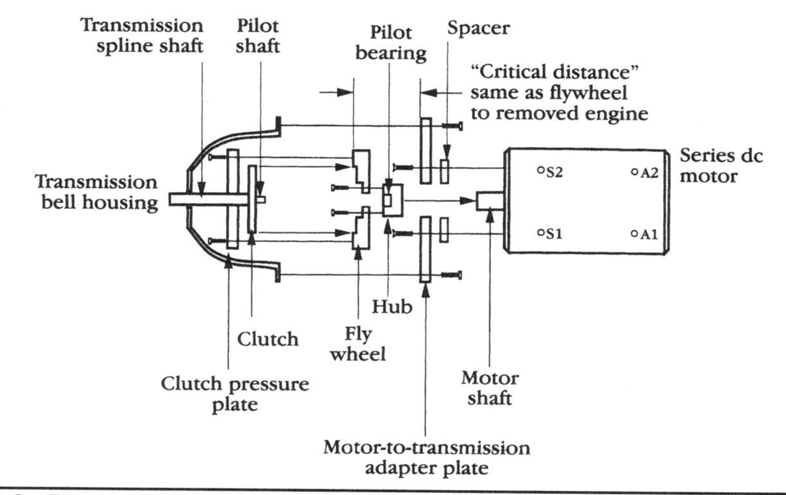

FIGURE 10-8 Elements of the electric motor to mechanical transmission connection process.

solutions for any given vehicle. So we're going to talk in general terms here. You'll have to translate them to your own unique case. And if your skills do not include precision machining of automotive metal parts, this is another good area to enlist the services of a professional such as KTA Services, Jim Harris' company, or a local machine shop.

FIGURE 10-9 Electric motor mount from EVPorsche.com.

The Critical Distance—Flywheel-to-Clutch Interface

After you remove the engine and before you take the flywheel off, carefully and accurately measure the distance from the front of the engine to the face of the flywheel (the part the clutch touches). This is the critical distance you want to reproduce in your electric motor mounting setup. In the 1987 Ford Ranger conversion, this critical distance measurement is 1.750 inches, but it will vary for different vehicles. You're going to put an adapter plate on the electric motor, put a hub or coupling on the electric motor's shaft, and attach the flywheel to this hub. When you've done this, the critical distance from the front of the flywheel to the front of the motor adapter plate (see Figure 10-8) should measure exactly the same.

Knowing your goal—the critical distance—makes it easier to navigate towards it. Whether you have a front engine plus transmission or front or rear engine plus transaxle, this goal will be the same, although the specifics will differ.

Your original flywheel might weigh 24 lbs. and have an attached ring gear (for the starter motor). Remove the ring gear that fits around the flywheel's outside edge and have the flywheel machined down to 12 lbs. or so by removing metal from its outside edge and rear (the part that faces the motor). Don't touch the flywheel's face (unless it has obvious burns or other defects). These steps don't hurt your flywheel at all, yet save weight—the most important factor in your EV conversion's performance.

This is also a good time to give the clutch (the inside layer of the clutch pressure plate/clutch/flywheel "sandwich") a thorough exam. If the clutch is old (used for more than 30,000 miles) or obviously worn, now is the time to replace it. Also take a look at transmission seals and mounts and replace them if excessively worn. Keep in mind you're going to be using your clutch in an entirely new way (*not using it* is a better description, because there's nothing to start up—you put it in gear first), so this new clutch ought to last you 200,000 miles or more.

Rear Support for the Electric Motor

This is a fairly straightforward area. You basically have two possibilities: support the motor around its middle or support it from the end opposite its drive connection to transmission or transaxle.

Figure 10-10 shows some of the middle-style motor mounts available from KTA Services accommodating different electric motor diameters. The bottoms of the mounts have bolt holes that attach to your vehicle's motor mounts. The two halves of the curved steel strap go around the motor and hold it securely in place. This is the preferred and most common method of mounting electric motors, particularly in the larger sizes.

End mounts are similar to motor faceplate adapter mounts; they bolt on to the motor's end face through mounting holes, and are then tied off to the frame through heavy rubber (old tires) shock isolators. This approach was very popular with early VW conversions that used smaller electric motors already securely attached to the transaxle housing at their front.

Front Support-Motor-to-Transmission Adapter Plate

Here's where the fun begins. Figure 10-11 shows you two motor mounting kits from KTA Services, among many possibilities. A simple VW Bug adapter kit is shown at left in Figure 10-11; a more complex VW Rabbit adapter kit is shown at right. With either kit, the large plate in the background that the other parts are resting on is the motor-to-transmission adapter plate.

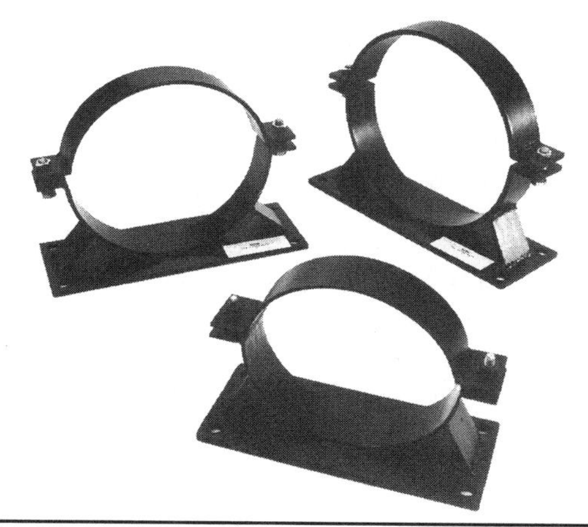

Figure 10-10 Typical KTA Services Inc. electric motor mounts for 7-inch through 9-inch motor diameters.

Notice there are two bolt hole ring patterns in the mounting plates. As shown in Figure 10-11, the inner-ring pattern with its countersunk mounting holes allows flat-head mounting bolts to attach the mounting plate to the motor and to be flush with the plate's surface on the inside when tightened. As also shown in Figure 10-11, the outer-ring pattern allows hex-head mounting bolts to be used to attach the mounting plate to the transmission bell housing and to be torqued down tight. Notice in Figure 10-11 that the center hole in each adapter is either sized to accommodate a collar around the motor, or the outside diameter of the coupling. This can vary widely from case to case.

The spacer shown in Figure 10-11 and the depth to which the hub fits on the motor shaft control the critical distance measurement; adjust these as needed.

If you are building your own, 1-inch-thick aluminum is the preferred building stock. Cut your cardboard pattern or template out to completely cover the bell housing front, exactly locate and mark the center of the transmission shaft (it might or might not be in the center area of your template), locate and mark the bolt hole locations on its front, trace the bell housing's outline on its back, and trim your pattern for 1 inch or less

Figure 10-11 Typical KTA Services Inc. electric motor mounts for VW Bug (left) and VW Rabbit (right).

overhang all around (to aesthetically suit your taste). Then transfer the electric motor mounting bolt hole patterns to this template using the transmission pilot shaft center as a reference. Take this completed template or pattern to the machine shop and ask them to reproduce it in metal.

If you are in a let's-get-it-done mode, Jim Harris' company makes a complete motor-to-transmission assembly, with adapter plate and precision-cast flywheel and hub/bushing parts, that guarantees perfect results for Ford Ranger conversion projects.

Also needed is a torque rod to support/stabilize the motor-transmission combination and prevent excessive rotation under high acceleration loads. This adjustable length rod normally attaches between one of the adapter-plate-to-transmission bolts and the frame.

Flywheel to Motor Shaft Connection

This is the least standardized area. The VW Bug hub or coupling (Figure 10-11, left) is a four-bolt affair with a keyway (square notch in central shaft opening) and set screw opening (hole in outside of coupling). The VW Rabbit hub or coupling (Figure 10-11, right) is a six-hole affair also with keyway and set screw opening. The coupling is press fit onto the electric motor's shaft, and the set screw further secures it. Figure 10-12 shows the physical positioning of the six-bolt coupling in front of the electric motor to give you a better idea. Figure 10-9 showed the motor compartment built by EVPorsche.com. They then flipped over the compartment so the motor can fit into the compartment. Then the controller and charger sit on opposite sides of the motor.

The pilot bearing shown in Figure 10-8 fits into the hub and mates with the transmission's pilot shaft (a transaxle does not require a pilot bearing). The clutch attaches to the transmission spline shaft, and the clutch pressure plate fits over it and bolts to the flywheel to complete the "sandwich."

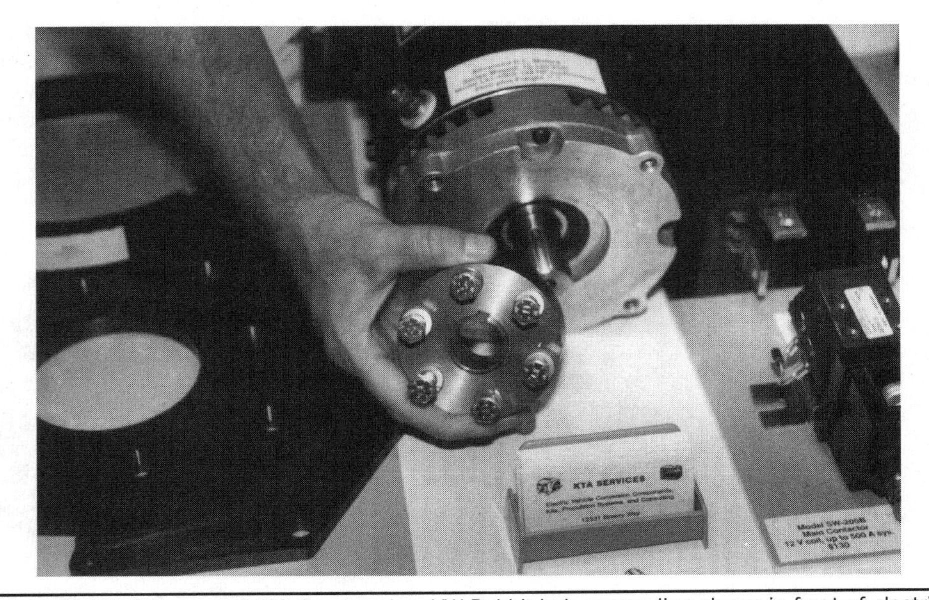

Figure 10-12 Close-up of KTA Services Inc. VW Rabbit hub or coupling shown in front of electric motor.

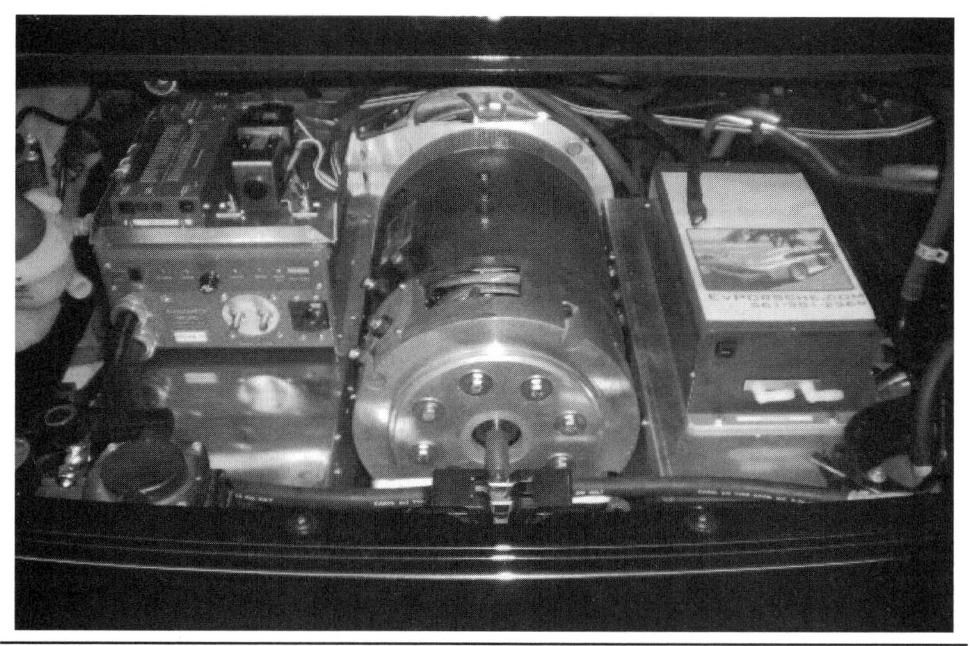

FIGURE 10-13 See how the motor is mounted into the compartment from Figure 10-9.

Hubs (with or without pilot bearing, etc.) are going to differ widely in size and shape for different vehicle, transmission, and motor types, yet their function is the same. You must determine what you need in your particular case. Call KTA Services, Jim Harris, or your local mechanic to help you if you're not sure. This is not the time and place to be shy.

In Figure 10-13, Paul Little created this great compartment that allows the motor to fit very nicely into the compartment.

Mounting and Testing Your Electric Motor

Depending on your accumulated skill and luck, your motor installation can either go smoothly the first time or involve several cut-and-try iterations. The basic approach is to mount the hub on the motor, attach the motor to its adapter plate, attach the flywheel to the hub, measure the critical distance to make sure it's exactly the same as it is for the internal combustion engine (add spacers as needed, etc.), and then attach this completed assembly to the transmission. The winch used to take out the internal combustion engine is probably the easiest way to install your motor assembly into your vehicle (the completed assembly weighs upwards of 150 lbs. with a 20-hp series DC motor), but a floor jack (elevate the front of the vehicle on jack stands first) or engine-lifting dolly will also work. The installation step is basically the reverse of the removal step: attach the assembly at two points, then slide the motor up and toward the transmission until the shafts are in alignment. When everything fits exactly, tighten the bolts on the transmission and on the motor mounts.

For a quick test, first jack up the vehicle so its front- or rear-drive wheels are off the ground. Then put a 2/0 cable in place across the motor's A2-S2 terminals. Attach another 2/0 cable from a battery's negative terminal to Sl—you can borrow the 12-volt starter battery you just removed for this purpose—and attach another 2/0 cable to the battery's positive terminal, but don't connect it yet. With the transmission in first gear, briefly touch the positive cable from the battery to Al and do two things:

- Look to see if the rear (or front) wheels move
- Listen for any strange or grinding noises, etc.

If the wheels move, good. If the wheels move and there is no strange grinding, this is doubly good and you can go on to the next step. If you hear something strange or the wheels don't turn (in this case, first ensure that the battery is charged), you need to unbutton your motor assembly from the tranny and look into the problem.

Fabricating Battery Mounts

Jim Harris' 1987 Ford Ranger pickup conversion uses 20 6-volt batteries. Jim elected to mount four of his batteries in the engine compartment area just vacated by the radiator's removal, and the 16 remaining batteries (in a four-by-four array) in the pickup bed area. In Jim's 1987 Ford Ranger pickup conversion, the rear battery bracket is attached directly to the frame for maximum rigidity and strength and lowest center of gravity. In Figure 10-14, Jim points to the chassis frame member underneath the cut-away pickup box floor. In Figure 10-15, we see a picture of Paul Little's frame rail. In Figure 10-16 Jim holds up the sturdy pieces of 2-inch by 2-inch by 1/16-inch steel angle iron that make up the battery mounting frame. The outside frame dimensions are slightly larger than the dimensions of the four-by-four battery array to allow for battery expansion. The batteries rest on a 3/16-inch-thick marine-grade plywood base, and are wedged in place inside the front and rear frames by strips of wood.

FIGURE 10-14 1987 Ford Ranger rear frame rails.

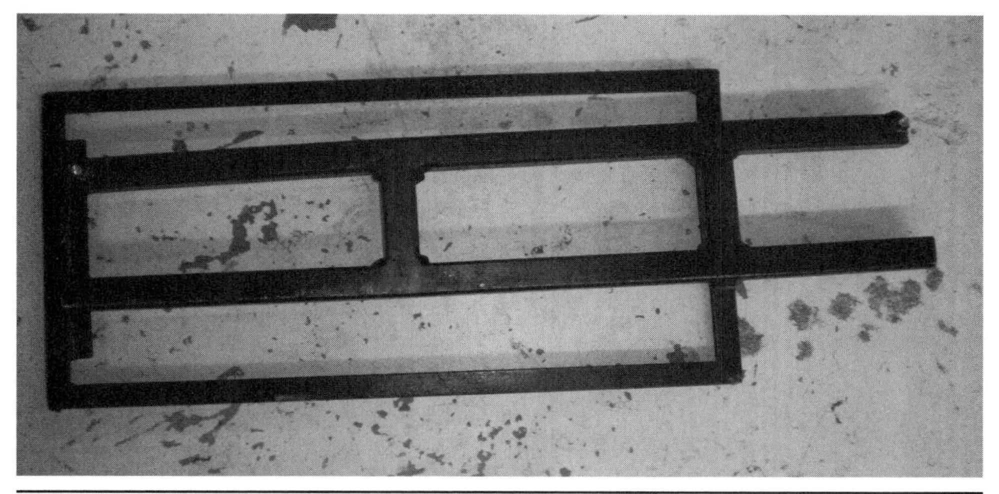

Figure 10-15 Frame rail from Paul Little, EVPorsche.com.

Additional Mechanical Components

Jim wound up adding heavier rear leaf springs for better handling to his 1987 Ford Ranger pickup conversion, and to his 1993 Ford Ranger pickup conversion he added a smooth nose in place of the stock grille for better airflow characteristics. In practice, both of these would be added at this point in the assembly process. Any metal fabricated parts inside the engine compartment (air dams, etc.) are better done at this stage because they are easier to get at before the electrical components and wiring are installed.

Figure 10-16 Battery mounting frame in place.

Shocks, coil and leaf springs, and external body parts are also best added at this time while you still have your heavy tools out.

Cleanup from Mechanical to Electrical Stage

It's a good idea to use an air hose to clean up the engine compartment and rear pickup box areas of your conversion after the mechanical phase is completed. Follow this with a broom sweep-up of the work area and/or a damp mop-up. The reason for this cleanup is to minimize the chance of metal strips or shavings finding their way into your electrical components or wiring during the electrical phase.

Electrical

The electrical part involves mounting the high-current, low-voltage, and charger components and doing the electrical wiring to interconnect them. To do the electrical wiring requires knowledge of your EV's grounding plan: the high-current system is floating, the low-voltage system is grounded to the frame, and the charging system AC neutral is grounded to the frame when in use. Doing the electrical wiring also involves knowledge of your EV's safety plan: appropriate electrical interlocks must be provided in each system to assure system shutdown in the event of a malfunction, and to protect against accidental failure modes. The electrical wiring is also greatly facilitated by using a junction box approach, so wiring is neat (no wires running every which way inside the engine compartment) and you can later trace your wiring. Let's take a closer look at the steps.

High-Current System

First you attach the high-current components, then pull the AWG 2/0 cable to connect them (another step where scheduling an inside helper is appropriate). Refer back to Figure 9-10. Notice there are seven components in the high-current line (in addition to batteries and charger—these we'll save for later):

- Series DC motor
- Motor controller
- Circuit breaker
- Main contactor
- Safety fuse
- Ammeter shunt(s)
- Safety interlock

You've already mounted the motor. In Jim's 1987 Ranger conversion, he mounted the controller on an aluminum heat sink plate directly above the motor in the engine compartment. Figure 10-17 shows the Operation Z layout of the batteries. Figure 10-18 shows the passenger-side view of Jim's series DC motor attached to the transmission (notice the earlier-mentioned torque arm between motor and frame), with its S1 power cable going up to the controller and motor, and controller power cables (with standoff clip to separate them) going down the chassis frame rail toward the rear battery pack. Figure 10-19 shows the layout of the electric motor in Paul Little's car and how the

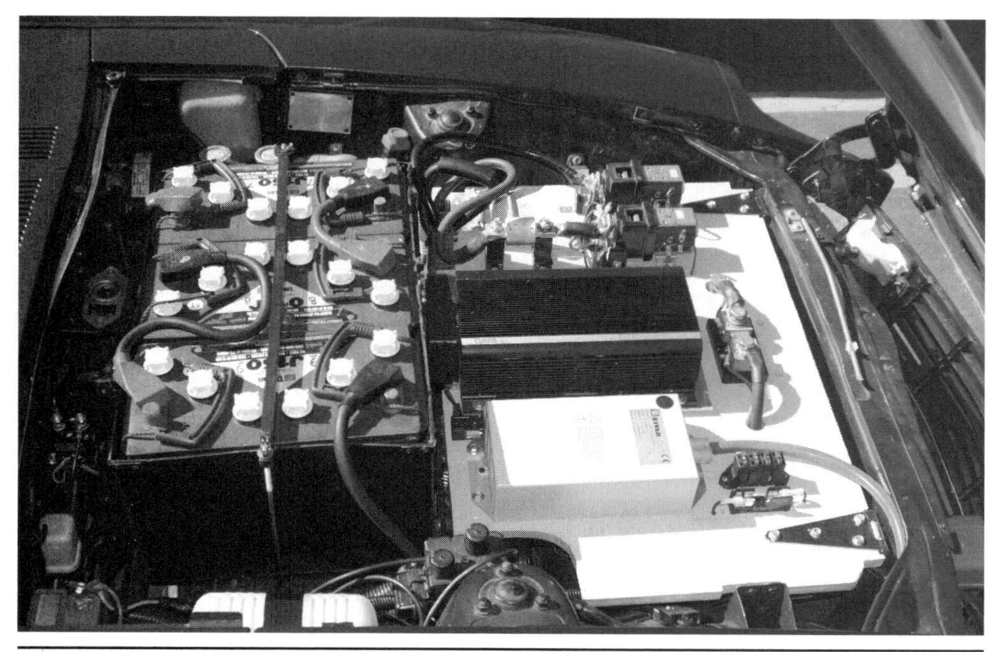

FIGURE 10-17 Operation Z battery mounted in place.

FIGURE 10-18 Ford Ranger electric motor power cabling from the passenger's side.

FIGURE 10-19 Porsche electric motor power cabling in the trunk.

cabling is laid out from the passenger side to the driver's side in the trunk (with the motor directly in the middle). Figure 10-20 shows a transverse-mounted motor from Bill Williams' 1976 Honda Civic wagon for comparison. Notice the commonality of the components—motor, adapter, controller, torque arm, and wiring but with specific differences in wiring, attachment method, and size.

Figure 10-21 (top) shows the power cabling to the mounted controller viewed from the front of the engine compartment. The M-destination of the earlier-referenced S1 power cable is clearly visible as is the B+ destination of the A1 power cable.

FIGURE 10-20 Bill Williams' 1976 Honda Civic's transverse-mounted electric motor.

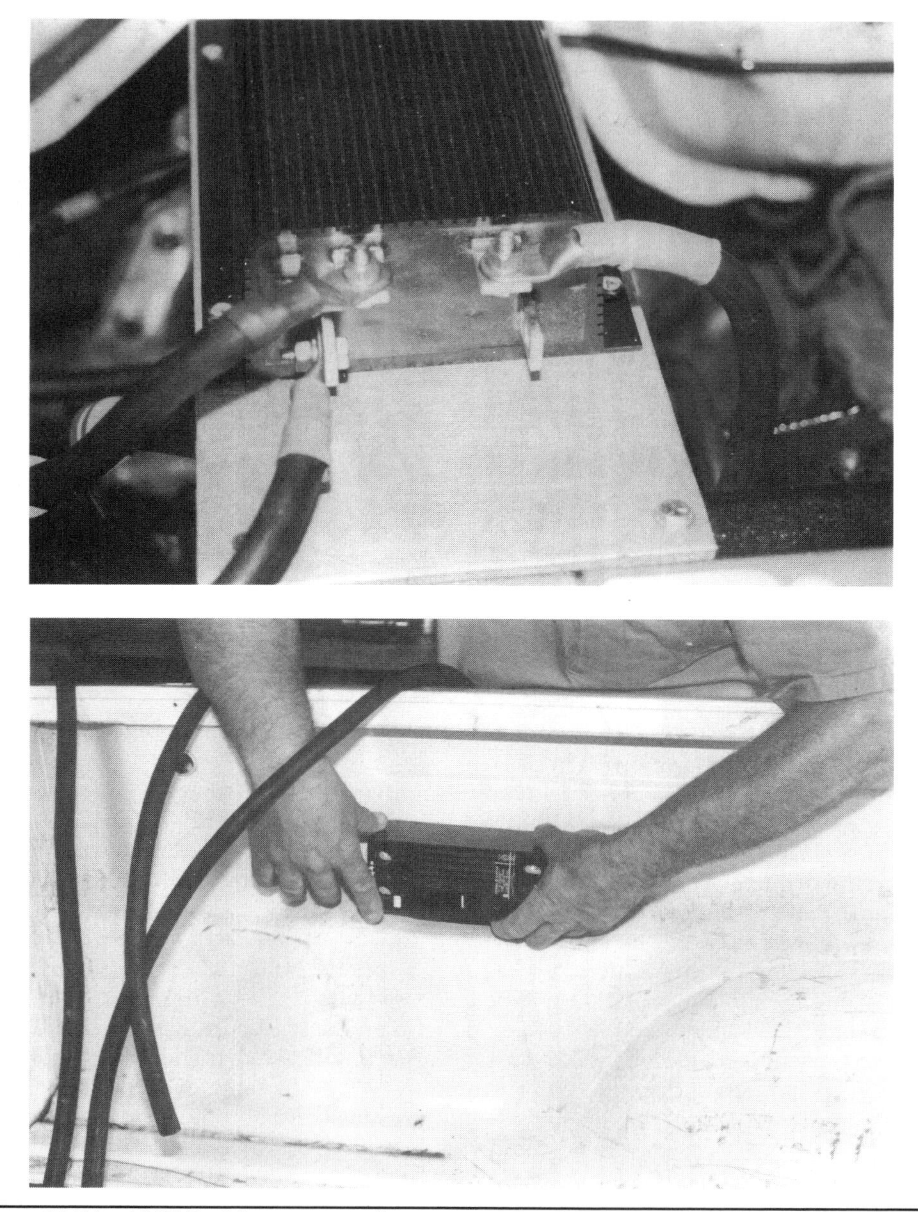

FIGURE 10-21 1987 Ford Ranger controller power cabling from box-mounted main circuit breaker (bottom), front (top), and rear pickup.

Figure 10-22 EVPorsche.com WARP motor in the vehicle prior to fastening it in place.

Figure 10-22 shows one of Paul Little's EV Porsche's having a WarP motor in the vehicle prior to fastening it in place.

This circuit breaker has two inputs and two outputs. Other EV converters prefer to mount this circuit breaker within reach of the driver. The trade-off is you have to pull heavy AWG 2/0 power cables behind the dash or some other location inside the driving compartment. The benefit is the peace of mind afforded by knowing that if everything goes haywire (contactor and relay contacts weld shut, etc.), there is one switch you can reach over and touch to shut everything down.

The high-current safety fuse, ammeter shunt(s), and main contactor are all located inside the junction box that will be discussed in a moment. Jim chose to implement a safety interlock on the low-voltage rather than the high-current side. Those who insist on a dashboard-mounted high-current safety or kill switch (usually with a big red knob on it) have to endure the pain of pulling AWG 2/0 cable over, under, around, and through the rear of the dashboard—not my idea of a good time. By the way, the AWG 2/0 being discussed here is stranded (solid wire would be nearly impossible to work with in this size), insulated copper (never aluminum) wire—and it's about ¾ inch in diameter so you have your hands full.

The high-current system has a floating ground. This means that the negative terminal of the battery pack is not connected to the frame or body at any point—it floats instead. This eliminates the possibility of an accidentally dropped tool arc welding itself to the body or chassis or, worse yet, causing your vehicle to bolt forward or backward while simultaneously causing battery pack meltdown. It also eliminates the possibility of your receiving a 120-volt electrical shock while casually leaning over the fender to measure battery voltage.

Low-Voltage System

On the low-voltage side, the idea is to blend the existing ignition, lighting, and accessory wiring with the new instrumentation and power wiring. There are six main components on the low-voltage side:

- Key switch
- Throttle potentiometer
- Ammeter, voltmeter, or other instrumentation
- Safety interlock(s)
- Accessory 12-volt battery or DC-to-DC converter
- Safety fuse(s)

Every EV conversion should use the already-existing ignition key switch as a starting point. In an EV, the key switch serves as the main on-off switch with the convenience of a key—its starting feature is no longer needed. You should have no problem in locating and wiring to this switch.

In Jim's 1987 Ranger and Paul Little's Porsche conversions, they mounted the throttle potentiometer (Figure 10-23) on the driver's side fender well (Figure 10-24, top). Figure 10-24 (bottom) shows it mounted in place and wired up.

Instrumentation wiring is simple; just be sure to observe meter polarity markings—the plus (+) marking on the meter goes to the positive terminal on battery. The ammeter is connected across the shunt(s) already wired into the high-current system. The voltmeter goes across the battery. The best solution is to wire the voltmeter so that it is

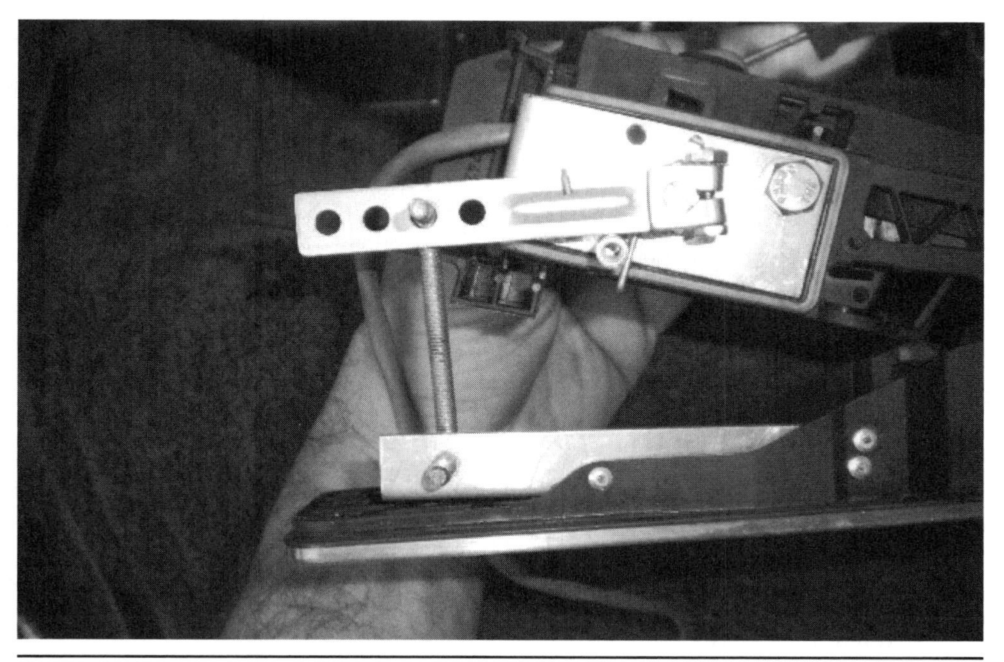

FIGURE 10-23 Electric Porsche throttle potentiometer (Courtesy of CoolGreenCars.net).

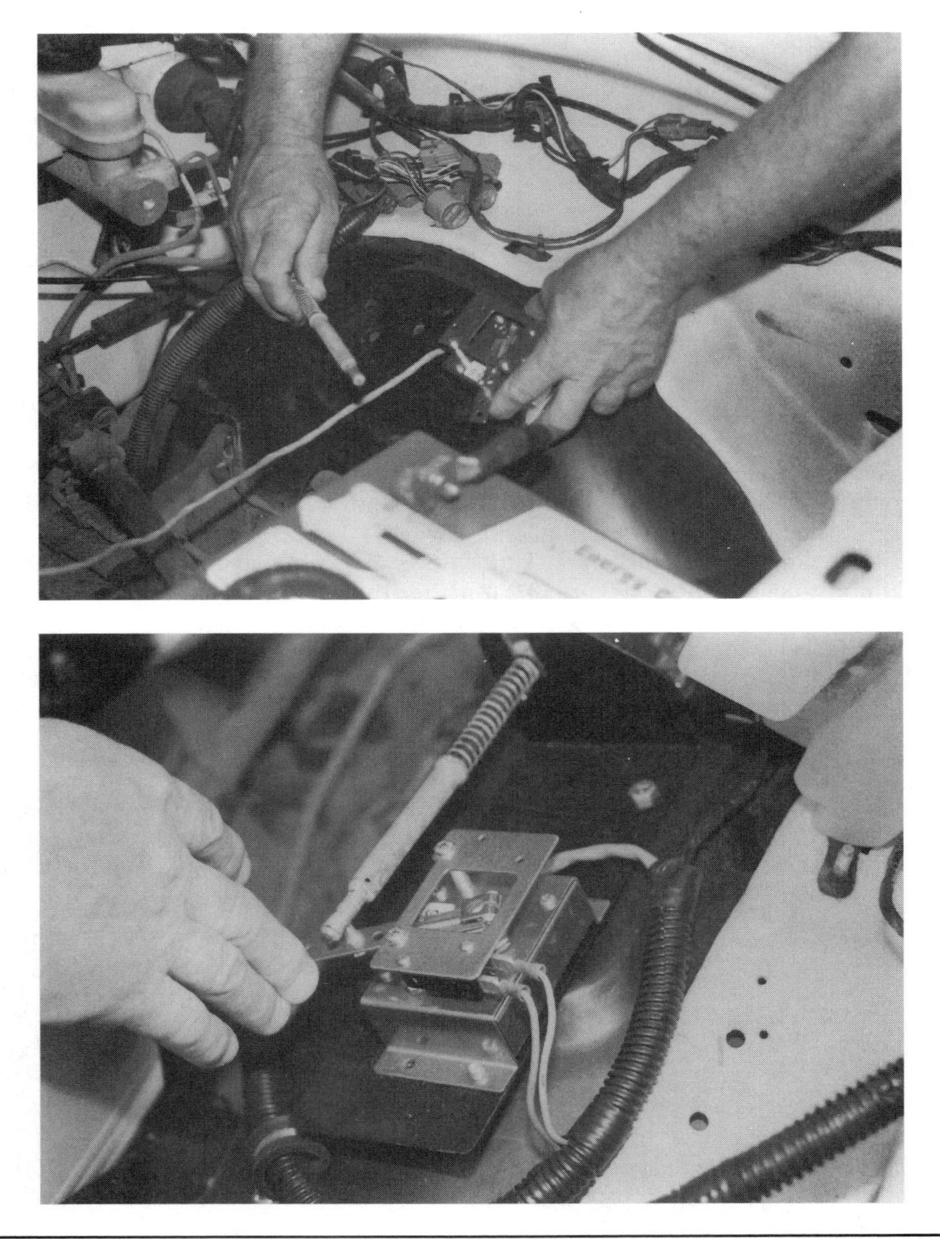

Figure 10-24 1987 Ford Ranger throttle potentiometer positioning (top) and mounted and wired throttle potentiometer (bottom).

always on, giving you a continual readout of battery status. You don't have to worry about draining the battery because a modern voltmeter's internal resistance is high enough to cause only a miniscule current drain (an order of magnitude less than the battery's own internal self-discharge rate). The battery indicator or state-of-charge

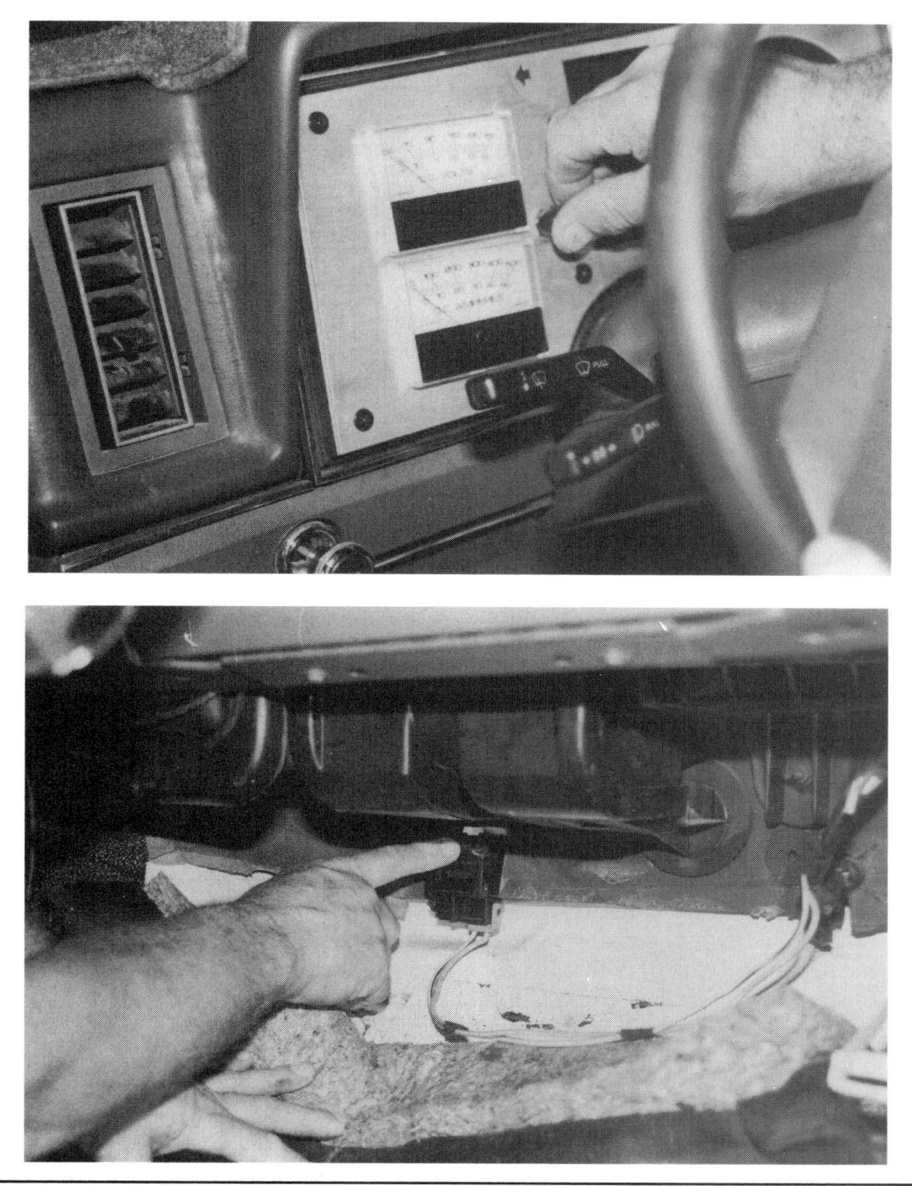

Figure 10-25 1987 Ford Ranger ammeter and voltmeter instrumentation (top) and impact cutoff switch (bottom).

meter also goes across the battery, but it should operate only when the key switch is on, so wire it to the on-off side of the main contactor (away from the battery). The temperature meter wiring is not particularly critical; just observe proper polarities and grounding, and make sure the thermistors are securely attached to whatever you are measuring. Jim utilized only ammeter and voltmeter instrumentation, with a range switch mounted on the 1987 Ranger's dashboard (see Figure 10-25, top).

If you wish to utilize a more modern digital voltmeter readout in place of the analog meters, you need to adjust the DVM's sample-and-hold circuit (it memorizes the value at any instant) either to display the average of the last few moments' sample-and-hold values, or to give a steady readout when a read button is pressed. Otherwise, the rapidly changing voltage or current will be hard to interpret.

The subject of safety interlocks is an important one. Jim's design uses three—all wired in series on the low-voltage 12-volt key-switch line: a fuel injection impact switch, a main safety cutoff switch, and a charger cutoff switch (to be covered in the "Charger System" section). The fuel injection impact switch's normal role is to shut off the fuel system in the event of a crash impact. Jim points to its location under the passenger's side of the 1987 Ranger's dashboard in Figure 10-25 (bottom). The main safety cutoff switch is a highly accessible, dashboard-mounted switch wired in series with the key switch. Punching it immediately removes energizing voltage from the main high-current contactor. A few EV converters also use a seat interlock switch that latches closed when the driver's presence in the seat is detected. You might wish to consider this as an option.

Jim opted to use a battery as the source of the 12-volt accessory system power. You can do the same or utilize the DC-to-DC converter shown in Chapter 9 that's driven from the main battery pack voltage. If you opt for the DC-to-DC converter, now is the time to install it and wire it in place; its input side goes directly across the main battery pack plus and minus terminals. Its output side provides +12 volts at its positive terminal, and its negative terminal is wired to the chassis. If you elect to use a 12-volt deep-cycle accessory battery, do the wiring for it now but wait until the battery phase to purchase, install, and connect it up.

Try to use AWG 12 (20-amp rating) or AWG 14 (15-amp rating) stranded insulated copper wire for the low-voltage system. The instrumentation gauges can be wired with AWG 16 or even AWG 18 wire.

Safety fuses of the 1-amp variety should be wired across the potentiometer and all delicate instrumentation meters. The key-switch circuit can utilize the original fuse panel but don't use the original wiring for any loads greater than 20 amps. The main fuse should be of the 10-, 15-, or 20-amp variety.

Unlike the high-current system, the low-voltage system is grounded to the frame; the negative terminal of the 12-volt battery (or DC-to-DC converter) is wired directly to the frame or body. Most internal combustion engine chassis come this way. You eliminate rewiring, extra wiring, and potential ground loops by using the existing negative-ground-to-the-frame convention.

Junction Box

A good junction box design cleans up the hodgepodge of instrumentation wiring running every which way inside the engine compartment, enables you (on anyone else) to later retrace your wiring, and provides convenient mounting and tie-off points for various components. However, not all junction boxes are created equal. Jim's and Paul's "magic boxes" are more equal than most—they combine simple design and layout with high utility (see Figures 10-26 and 10-27). The high-current safety fuse (in the center, behind the power cable), the ammeter shunts (large one on center of back wall, small one in front of main contactor), and the main contactor (on left side of box) form, along with the terminal strip, the "backbone" from which all interconnections are made. Notice all the safety fuses are located in one convenient area at the right rear of the box.

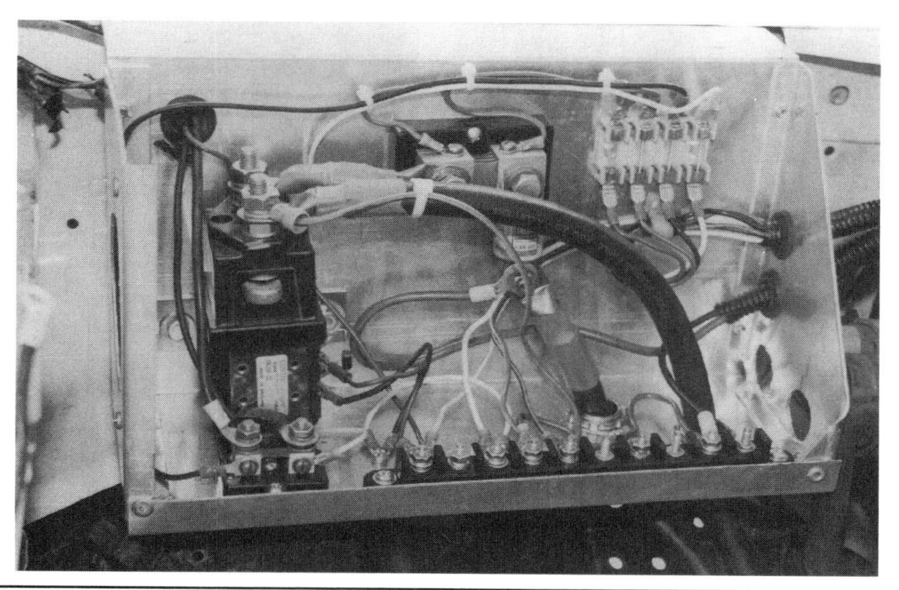

Figure 10-26 Jim Harris' prototype "magic box" for 1987 Ford Ranger.

Jim's and Paul's objectives were to simplify. You'll see the continuous design evolution of Jim's magic box in later photographs as the conversion progressed and how Paul Little made it more advanced in his design.

Charger System

The benefits of the on-board charger are convenience and the ability to take advantage of on-the-road charging opportunities as they are presented. The dual objectives in

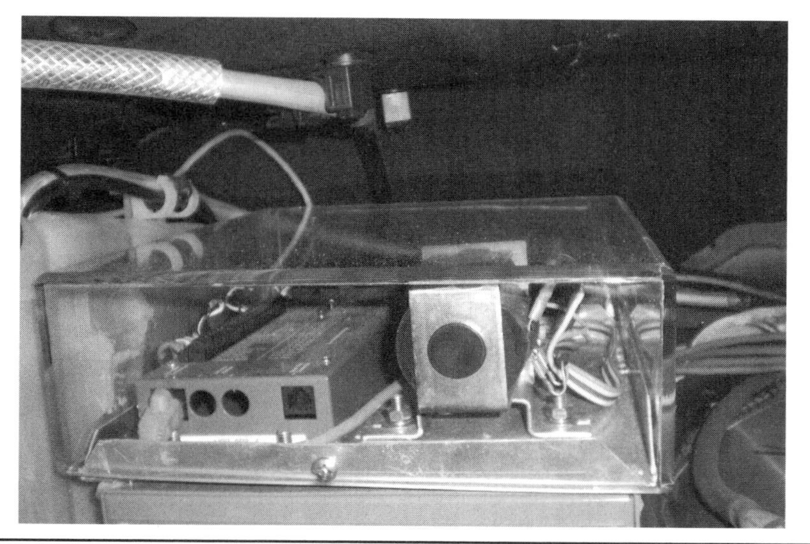

Figure 10-27 Paul Little's EV Porsche "magic box" for components.

wiring the on-board charging system are to prevent the charging routine from becoming a "shocking" experience (via proper grounding), and to prevent momentary distractions from causing you to drive away while the charging cable is still attached (via a charger safety interlock). There are four main components to the charger system:

- Compact on-board charger
- Lightweight line booster (optional)
- Safety charging interlock
- AC input system

In Jim's 1987 Ranger conversion, he mounted the on-board K & W BC-20 charger on the driver's side of the aluminum heat sink plate, directly above the motor in the engine compartment, as shown in Figure 10-28 (top). The K & W LB-20 line booster, required by the charger for 120-volt operation, is mounted next to the throttle potentiometer on the driver's side fender well (Figure 10-28, bottom).

The preferred location for most AC input charging connections in conversion vehicles is usually the location vacated by the gas tank filler neck opening. Figure 10-29 (top) shows that Jim's choice follows this pattern—it's behind the original gas cap door. Jim also chose to implement a male "twist-lock" three-prong AC charging connector, enabling him to use a standard extension cable (one end male, other end female) with the male plug on the extension cord end able to mate conveniently with the standard 120-volt AC female service outlets. (Be sure to use a three-conductor extension cable with at least AWG 12 wire in it.) Figure 10-29 (middle) shows an on-board female twist-lock three-prong AC connector. Figure 10-29 (bottom) shows that behind the gas tank filler door is not the only location; in this case, a male twist-lock three-prong AC connector has been recessed in the front bumper using a conventional outdoor AC junction wiring box. Figures 11-30 and 31 show examples of how the gas tank filler area can also be used as a great charger location. In this case, they are also using a three-prong connector and a Zivan NG3 charger.

For the charging system, you should use AWG 10 (30-amp rating) stranded insulated copper wire for both the charger-to-battery and the charger-to-AC-input receptacle connections. In order to prevent you (or anyone else) from casually driving away with the extension cord attached while charging, it's a good idea to implement a charger interlock system. Jim's approach is to use a relay whose coil is energized by the presence of 120 volts AC, and whose contacts are in series with the 12-volt keyswitch line. When the 120-volt AC line cord is plugged in, this relay latches open and keeps the main battery pack disconnected from the controller and motor—the vehicle is immobilized.

Other EV converters have also used the charger interlock feature to energize battery compartment fans (forced ventilation of the batteries) while charging. Additional interlock possibilities include: sensors that inhibit drive-away when the fault conditions of engine compartment hood open, battery compartment hood open, or AC charging connector access door open are detected; and sensors that inhibit the charging function during fault conditions such as engine or battery compartment hood open (because you don't want outsiders prying when charging currents and battery gases are present). You might consider any of these options.

In order to prevent you (or anyone else) from getting shocked when touching your EV's body while it's charging, the body and frame should be grounded to the AC neutral line (the third prong of the connector with the green wire leading to it). This neutral wire is connected between the AC input connector and the body or frame, and is utilized only when charging. The batteries should be floating—no terminals touch the frame—and might even be further isolated by locating them inside their own compartment or battery box. This is particularly appropriate for not recommended (but

Figure 10-28 1987 Ford Ranger mounted K & W battery charger (top) and mounted K & W line booster (bottom).

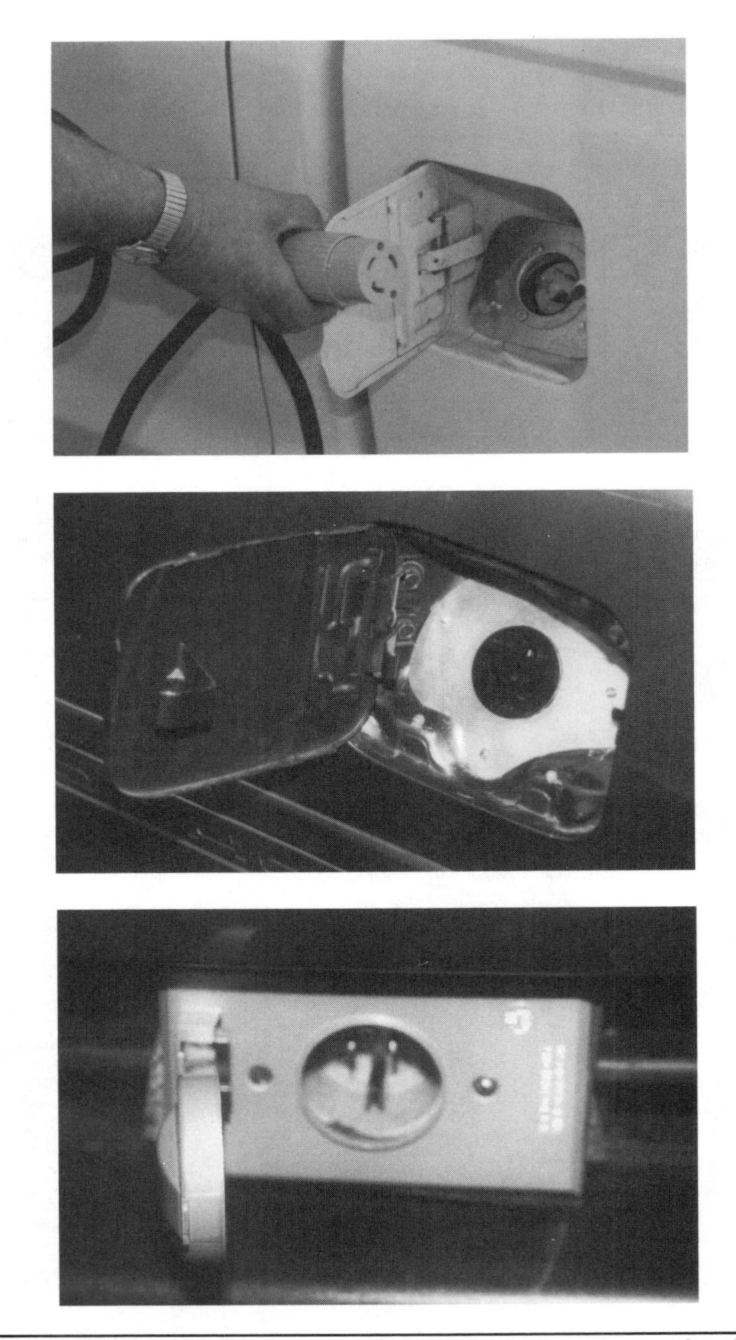

Figure 10-29 AC charging connector options—Jim Harris' male (top), optional female (middle), and optional male bumper mount (bottom).

Figure 10-30 Operation Z charging connector and charging plug installed.

Figure 10-31 Operation Z charging connection.

done anyway) inside-the-passenger-compartment battery installations. To guarantee no shocks, transformerless chargers should always have a ground fault interrupter installed, and transformer-based chargers should be of the isolation-type.

If you prefer not to use an on-board charger, this changes your wiring and interlock design plans. In this case, you are going to be providing up to a 140-volt DC input at currents up to 30 amps from a stationary charger. You will also need to charge the on-board 12-volt accessory battery (unless you utilize a DC-to-DC converter). This means your charging receptacle input needs at least two connectors: one for the high-current floating 140-volt plus and minus input leads, and the other for the low-voltage grounded-to-frame 12-volt plus and minus input leads. The charger interlock design for the off-board charger is identical to the on-board case, except that you use a relay whose coil is energized by the presence of 120 volts DC and whose contacts are in series with the 12-volt key-switch line. When the 120-volt DC or extension 120 VAC (depending on your charging preferences) extension cord from the charger is plugged in, this relay latches open, disconnects the main battery pack, and immobilizes the vehicle.

Battery

Buying and installing your main propulsion battery pack batteries is the last step in the EV conversion process. You buy them last because you don't want to be tripping over them, nor do you want to keep charging them during the weeks (or months) of assembly. When your EV conversion is nearly complete, after all the battery mounting frames are built and the wiring is done, you bring in the batteries. Let's take a closer look at the steps.

Battery Installation

The most important consideration in battery installation is to mount them in a location that will be easily accessible for servicing later. You've previously decided on your battery type and quantity, obtained their dimensions, allocated your front and rear mounting space, and designed and built your battery-mounting frames slightly oversized to accommodate battery expansion with use. All that remains is to install them.

In Jim's 1987 Ranger conversion, he mounted a bank of four 6-volt batteries in the front of the engine compartment space (Figure 10-32, top). Jim chose U.S. Battery Model 2200 batteries. These provide 220-AH capacity, slightly less than that of the Trojan T125 (235-AH) batteries recommended in Chapter 8. Size (and volume) is approximately the same. Figure 10-32 (middle) shows how a wooden spacer strip is used to wedge the batteries in place. You could also use a frame around the top of the batteries to hold them in place. Figure 10-33 also shows 12 Trojan J150, 12-volt, lead-acid, flooded batteries, but in the rear of the vehicle. Figure 10-34 shows all of Paul Little's batteries connected together and ready to be installed into his electric Porsche.

Jim's rear pickup box four-by-four of 16 6-volt batteries is shown in Figure 10-32 (bottom) along with the now wired-up main circuit breaker. Notice the plywood has been painted black to match the steel frame and chassis rails. The batteries are wedged in place with wood strips along the base as in mounting the front batteries. The top battery pack tiedown frame is again optional unless you plan on doing some rugged driving.

Figure 10-32 1987 Ford Ranger battery mounting—front batteries (top), wood strip wedges (middle), and rear batteries (bottom).

Figure 10-33 Operation Z rear batteries.

Figure 10-34 Paul Little's battery connectors ready to go!

Battery Wiring

The most important consideration in battery wiring is to make the connections clean and tight. Figure 10-32 also shows both front and rear battery packs in their wired-up condition. Check that you haven't accidentally reversed the wiring to any battery in the string as you go. Double-check your work when you finish, and use a voltmeter to measure across the completed battery pack to see that it produces the nominal 120 volts you expect. If not, measure across each battery separately to determine the problem. If reversed wiring was the culprit, the correctly installed and wired battery should fix it. If a badly discharged or defective battery is the culprit, check to see that it comes up on charging and/or replace it with a good battery from your dealer. A recharged "dead battery" will shorten the life for the entire battery pack. Please be careful to check all of the batteries. *Important: Make sure that the main circuit breaker is off before you connect the last power cable in the battery circuit. Better still, switch the main circuit breaker off and wait until the system checkout phase before final battery connection.*

Accessory Battery

This is also the time to mount your 12-volt accessory battery. Jim initially used the 12-volt starter battery that came with his conversion chassis (see Figure 10-35) for two reasons: it allowed him to do component testing during the wiring phase, and it was already mounted and wired in place (saving him a few steps). Figure 10-35 (top) shows that the inside of Jim's "magic box" is now down to two power cables (the leftmost cable's battery destination is now clearly visible) and two instrumentation cables. Figure 10-35 (bottom) shows the outside of Jim's magic box with the cover in place installed in its initial location.

After Conversion

This is the system checkout, trial run, and finishing touches stage. First, make sure everything works, then find out how well it works, then try to make it work even better. When you're satisfied, you paint, polish, and sign your work. Let's look at the individual areas.

System Checkout on Blocks

Jack up the drive wheels of your conversion vehicle (or raise them up on work stands) for this phase. The objective is to see that everything works right before you drive it out on the street. With your vehicle's drive wheels off the ground and the transmission in first gear, do the following:

- Before connecting the last battery cable, verify that the proper battery polarity connections have been made to the controller's B+ and B– terminals.
- Obtain a 100- to 200-ohm, 5- or 10-watt resistor, and wire it in place across the main contactor's terminals. With the key switch off but the last battery cable connected and the main circuit breaker on, measure the voltage across the controller's B+ and B– terminals. It should measure approximately 90 percent of the main battery pack voltage (in the neighborhood of 108 volts) with the correct polarity to match the terminals. If this does not happen, troubleshoot the wiring connections. If it does, you're ready to turn the key switch.

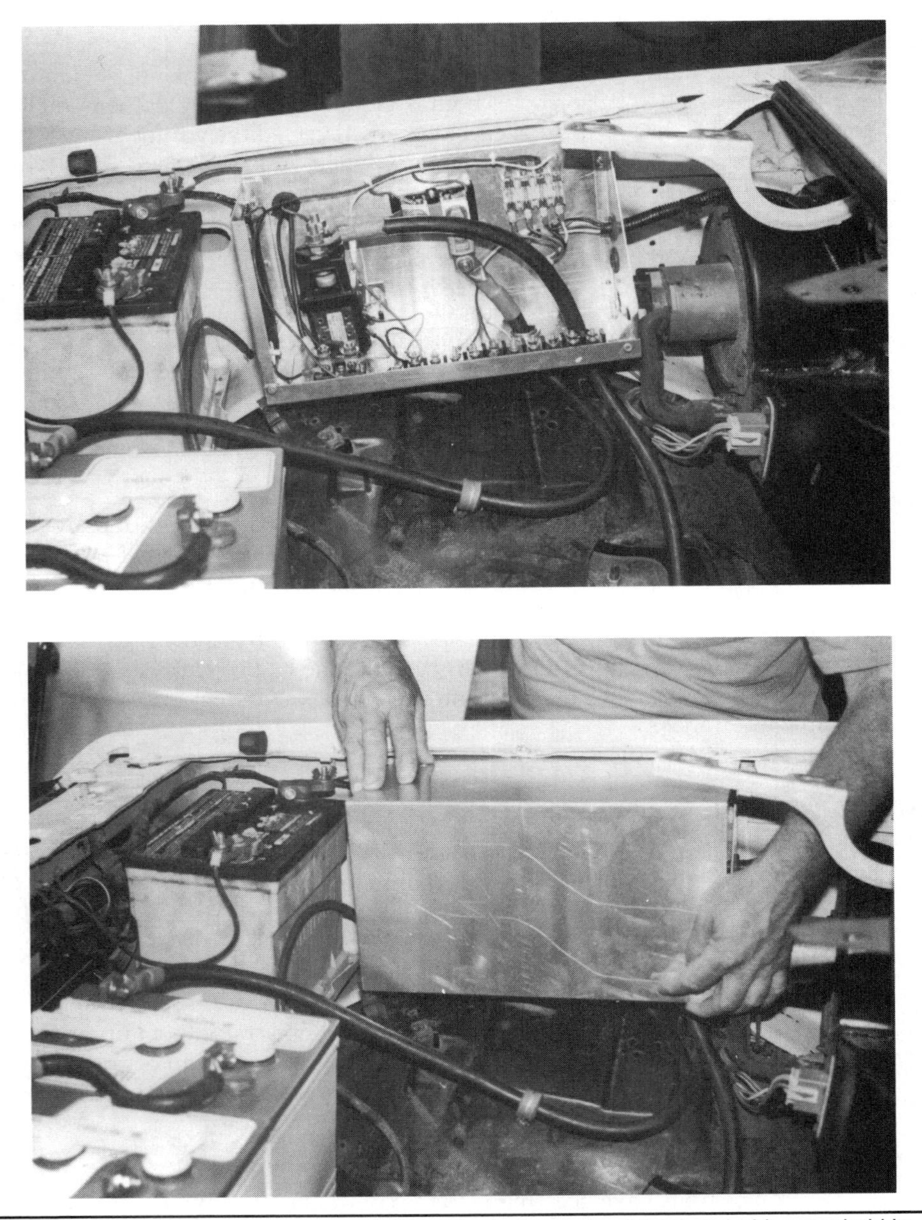

FIGURE 10-35 1987 Ford Ranger mounted Jim Harris "magic box" (top) and with cover held in place (bottom).

- Turn on the key switch with your foot off the accelerator pedal. If the motor runs without the accelerator pedal depressed, turn off the key switch and troubleshoot your wiring connections. If nothing happens when you turn on the key switch, go to the next step.
- With the transmission in first gear, slowly press the accelerator pedal and see if the wheels turn. If the wheels turn, good. Now look to see which way the wheels are turning. If the wheels are turning in the right direction, this is doubly good. If not, turn off the key switch and main breaker and interchange the DC series motor's field connections. If you are moving in the right direction, go to the next step.
- If you have the high pedal disable option on the Curtis controller, turn off the key switch, depress the accelerator pedal and turn on the key switch. The motor should not run. Now completely take your foot off the accelerator and slowly reapply it. The motor should run as before. If this does not work correctly, troubleshoot your wiring connections. If it works, you are ready for a road test. Turn off the key switch.

Neighborhood Trial Run

- Check the state-of-charge of your main battery pack. If it's fully charged or nearly full, you can proceed. If it's not charged, recharge it before taking the next step. It's just too embarrassing to run out of juice on your first neighborhood cruise—and once you drive it a little, you'll want to drive it a lot more.
- After the batteries are fully charged, remove the jack and/or wheel stands from under your EV, open the garage door (believe me, it's a necessary step), check to see that all tools, parts, and electrical cords are out of the vehicle's path, and turn on the key switch. Put it in gear, disengage the parking brake, step on the accelerator and cruise off into the neighborhood. Neat, eh?
- The vehicle should have smooth acceleration and a good top speed, and should brake and handle normally. The overwhelming silence should enable you to hear anything out of the ordinary with the drivetrain, motor, or brake linings, etc. Now, you're ready to get fancy.

First Visitor Does a Second Take

Before Jim and Paul get far in their respective electric cars, they immediately draw attention. Jim had just parked his newly completed EV conversion against the scenic backdrop of the Chula Vista marina and gotten ready to take some photographs when a city employee drove by and said, "Hey, you're not allowed to park there."

Jim replied, "I'm just taking some photographs of my electric vehicle. I'll only be a minute."

"Your what?"

"Electric vehicle."

"Where's the engine?"

One quick spin around the park loop and the city employee, now grinning ear to ear, emerged and said, "Wow, that was neat. Take as long as you like." You too can have fun and make new friends.

Everytime Paul takes out his electric Porsche's for a test, people cannot believe there is no engine or sound in the acceleration. In the Palm Beach area of Florida, his car gets noticed not just because it is a Porsche but because it is electric.

Improved Cooling

Jim noticed after a few trial runs that he was getting an unusual thermal cutout indication on the controller. Before finally isolating the culprit—it was a marginal controller that

FIGURE 10-36 Jim Harris' redesigned "magic box" and controller mounting for 1987 Ford Ranger.

Curtis quickly replaced under warranty—Jim did a thermal and mechanical redesign on the entire control electronics area.

Figure 10-36 shows Jim's "magic box" redesign. In this incarnation, everything is accessible from the outside, the two power cables bolt on, the two instrumentation cables plug in, and the two fuses are accessible from outside the case. Figure 10-36 shows a front view of the thermally redesigned controller mounting. In this design the controller is flipped over and a heat sink is mounted to its bottom (silicone thermal lubricant was spread over the area for the highest heat dissipation efficiency). The combined package was then centrally mounted in the location of maximum engine compartment airflow. Figure 10-37 shows a driver-side view: charger (in the original location), newly inverted controller with heat sink on top, and revision two of Jim's magic box.

When the replacement controller from Curtis was plugged into the new layout, everything worked fine with no more thermal cutout.

Figure 11-38 shows the result.

Further Improved Cooling

The eternal mechanic in Jim was still not satisfied—he wanted to have controller cooling independent of the vagaries of engine compartment airflow. So he added the fan and directional cooling shroud to the top of the heat sink (Figure 10-38). He also widened the mounting space between his magic box and the controller (everything is now mounted on the common heat sink aluminum plate), and relocated the battery charger to the passenger side fender well—all to maximize cooling and airflow.

Figure 10-37 "Magic box" and controller, driver-side view.

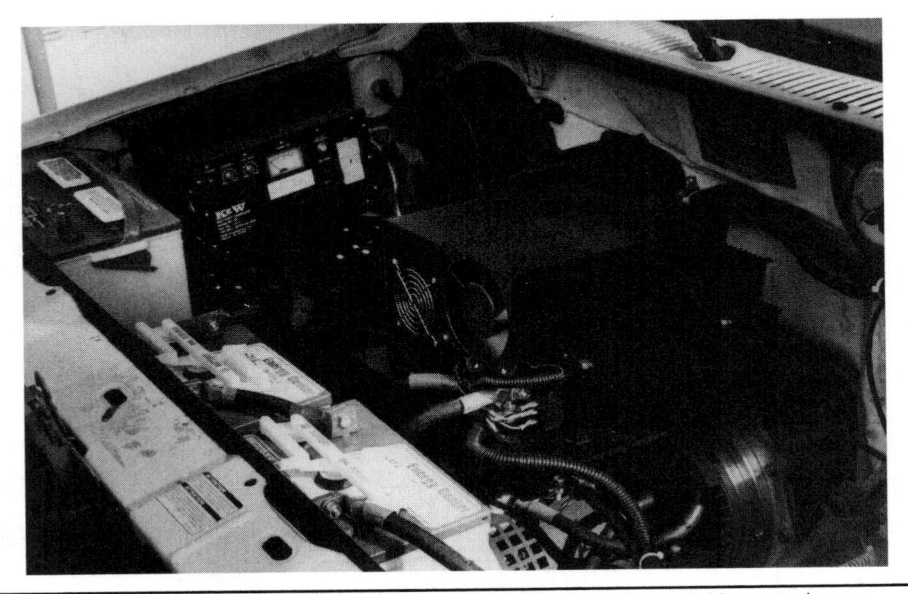

FIGURE 10-38 Jim Harris' improved "magic box," fan-cooled controller, and battery charger mounting for 1987 Ford Ranger.

Paint, Polish, and Sign

After everything is running the way you like, it's time to "pimp your ride." Why? A car is an extension of oneself. The way the exterior of a vehicle looks reflects you; you are proud of your work and want to show it off in its best light. But without an outside "ELECTRIC" sign, the only way you can tell an electric car is electric-powered is to look closely through its front grille, where the DC motor is visible just above the license plate (although the silence would definitely make you suspicious).

Because few people (if any) are likely to examine your EV conversion closely, you have to advertise instead. A couple of well-placed large letters is all it takes—or an electric vehicle license tag. Jim opted to place his letters on both front fenders and on the

FIGURE 10-39 1987 Ford Ranger with "Electric Vehicle" lettering.

FIGURE 10-40 1993 Ford Ranger "magic box."

tailgate as shown in Figure 10-39. It makes a world of difference. Be sure to carry plenty of literature because people will talk to you whether moving (neighborhood or freeway) or stopped (with the hood up or down). While you can draw a crowd whenever you visit the local convenience store—and certainly will at a gas station—you can really impress your date while waving to the folks on the freeway as you pass them at 75 mph and watch their mouths drop open.

Onward and Upward

Notice the evolution of Jim's "magic box" for the 1993 Ford Ranger in Figure 10-40. Everything—controller, junction box, heat sinks, and fans—is packaged in a protective metal enclosure that is now centrally mounted above the motor in the engine compartment. With everything in one box, you just have to connect a few wires—the conversion is even easier to do and is much more reliable an operation.

Put Yourself in the Picture

Now that you've seen how it's done, you can do it yourself. It's a very simple project that virtually anyone can accomplish today, just by asking for help in a few appropriate places and taking advantage of today's kit components and prebuilt conversion packages. Imagine yourself in Figures 10-41 and 10-42, and take the steps to make it happen.

FIGURE 10-41 Put yourself in this picture—do an EV conversion of your own.

FIGURE 10-42 Joe Porcelli and Dave Kimmins of Operation Z.

CHAPTER 11

Maximize Your Electric Vehicle Enjoyment

Now that you are driving for only pennies a day, it takes little more to make your pleasure complete.

Once you've driven your EV conversion around the block for the first time, it's time to start planning for the future. You need to license and insure it so you can drive it farther than just around the block. You also need to learn how to drive and care for it to maximize your driving pleasure and its economy and longevity.

Licensing and Insurance Overview

No one anywhere would seriously question licensing and insuring Jim Harris' 1993 Ford Ranger pickup EV conversion you met in Chapter 10. It goes 75 mph and Ford has already guaranteed its chassis compliance with FMVSS and NHTSA safety standards.

On the other hand, many jurisdictions used to balk at licensing and insuring the home-built and/or commercially produced golf-cart style EVs of the 1970s that could barely reach 45 mph and did raise issues of safety. That is why there was the creation of the FMVSS 500 ruling that allows for low-speed vehicles to be driven. They are golf cart–style EVs; however, they have some basic safety features of a regular vehicle (turn signals, brake lights, taillights, bumpers, reverse lights and sound). Those vehicles are regularly called either low-speed vehicles or neighborhood electric vehicles and cannot go more than 25 miles per hour and can only be driven on public roads that do not exceed 35 miles per hour. In addition, each state (in the United States) must also approve the use of this type of vehicle on the road. Most of the states in the United States do allow for a NEV to be driven on public roads.

Which end of the spectrum your EV conversion of today resembles directly determines its ability to clear any licensing and insuring hurdles. Let's take a closer look at each area.

Getting Licensed

Any vehicle's license falls under the jurisdiction of the state in which it resides. All state motor vehicle codes, although based on common federal standards, are just a little bit different. While your internal combustion engine vehicle conversion chassis might be fully compliant with federal FMVSS and NHTSA safety standards, you need to check out what your state's motor vehicle code says. If you're doing a from-the-ground-up EV, you'd be well-advised to check out these rules and regulations in advance.

In general, few states have specific EV regulations. You'll find that states with larger vehicle populations, such as California, New York, and Florida, are on the leading edge in terms of establishing guidelines for EVs. Check with your own state's motor vehicle department to be sure.

As for the licensing process, most of the people that work for the Department of Motor Vehicles and/or the Department of Environmental Protection for each state are far more involved with the smog-certification or DEQ (Department of Environmental Quality). It is very important that you check the vehicle compliance rules and regulations in your respective state to see what the process is for allowing a converted electric vehicle to receive license plates. However, in most states, owning an EV conversion short-circuits this process. In fact, you can offer to buy the whole local DEQ inspection team lunch if their meters find any emissions coming out of your EV at all (hybrid EV owners, please don't make this offer!).

In most states you can receive a tax credit for an electric vehicle and there is also a federal tax credit for electric vehicles. In some areas, you can be entitled to a reduction in your electric power rate. Check with your local utility and city and state governments to see if you and your EV are entitled to something similar in your area.

Getting Insured

Insurance is roughly the same as licensing. You're not likely to have any trouble with your EV if it's been converted from an internal combustion engine chassis. While it is a process to explain that the vehicle has been converted to electric with your carrier, having a vehicle identification number (VIN) and approval from the State Department of Motor Vehicles will give most large national insurance companies ease in underwriting an EV for insurance coverage. Verify that your design meets insurance requirements in advance.

Safety Footnote

My basic assumption in this section is that you would put safety high on your list of desirable characteristics for your converted EV. This line of reasoning assumes you leave intact the safety systems of the original internal combustion engine chassis: lights, horn, steering, brakes, parking brake, seat belts, windshield wipers, etc. It also assumes you are thinking "safety first" when installing your new EV components.

In particular, your electric motor control system should be failsafe and have safety interlocks, and your batteries should be mounted for minimal danger to people, other objects, or themselves—both in normal operation and in the event of an accident. Don't even think about going to a vehicle inspection appointment if your from-the-ground-up EV design or your standard EV conversion has left any of these out, or your control and/or battery systems don't emphasize safety first.

Driving and Maintenance Overview

An EV is easier to drive and requires less maintenance than its internal combustion counterpart. But because its driving and maintenance requirements are different, you'll need to adjust your acquired internal combustion engine vehicle habits. The driving part is very similar to the experience of a lifelong stick-shift driver who drives an automatic transmission vehicle for the first time. As for the maintenance, there's a whole lot less to do, but it has to be done conscientiously. Let's take a closer look at each area.

Driving Your Electric Vehicle

Your EV conversion may still look like its internal combustion engine ancestor, but it drives very differently. Here's a short list of reminders:

Starting

Starting an electric vehicle conversion is a little bit of a science. There's no need to use the clutch on startup because the motor's not turning when your foot is off the pedal. On the other hand, there is a very definite need to have it in gear because you always want to start your series DC motor with a load on it so it doesn't run away to high RPM and hurt itself. If you forget and accidentally leave the clutch in (or the transmission in neutral), back off immediately when you hear the electric motor winding up. The best analysis I have heard is that if your electric vehicle conversion is a lighter package (not a performance-based controller and heavy-duty motor) you can start the car in second gear without a problem since the torque of an electric motor will push the RPM (such as the Ford Ranger discussed in Chapter 10). Whereas the higher performance vehicles (such as the Porsche discussed in Chapter 10) can easily be started in fourth gear since the vehicles motor, controller, and batteries will have the necessary voltage requirements to accelerate from 0 to 60 in 4 to 6 seconds.

Shifting

If you do city driving, you'll wind up mostly using the first two gears. The lower the gear, the better your range, so use the lowest gear possible at any given speed. However, if you do highway driving, expect to shift gears higher.

Economical Driving

If you keep an eye on your ammeter while driving, you'll soon learn the most economical way to drive, shift gears, and brake. For maximum range, the objective is to use the least current at all times. You'll immediately notice the difference in drag racing and going up hills—either alter your driving habits or plan on recharging more frequently.

Coasting

If you don't have regeneration, coasting in an EV is unlike anything you've ever encountered in your internal combustion engine vehicle—there's no engine compression to slow you down. You need to learn how to explain how to correctly "pulse" and when. I have ridden with drivers who floor it for three seconds, then coast, floor it, and coast. Heavy pulsing is not good for the vehicle and wastes energy. In most driving, a steady foot is better. Light pulsing is only an advantage when little power is needed.

Regeneration

Regenerative braking is a mechanism that reduces vehicle speed by converting some of its kinetic energy into another useful form of energy. This captured energy is then stored for future use or fed back into a power system for use by other vehicles.

For example, electrical regenerative brakes in an electric railway vehicles feed the generated electricity back into the supply system. In battery electric and hybrid electric vehicles, the energy is stored in a battery or bank of capacitors for later use. Other forms of energy storage which may be used include compressed air and flywheels.

Many now have experienced coasting in a hybrid electric car. As with a hybrid and by definition, an EV is designed to be as frictionless as possible, so take advantage of this great characteristic. Learn to pulse your accelerator and coast to the next light or to the vehicle ahead of you in traffic.

Hybrid electric car owners (especially NYC taxi owners) understand this concept. When you accelerate, you do not need to floor it and then coast. You can slowly step on the accelerator and then take your foot off of the accelerator to coast. When you coast, you use the regenerative braking. It is much smoother and a more efficient use of the vehicle since it is moving forward and charging at the same time. While there are plenty of people who own converted EVs who like to floor it and coast, regeneration can be a very big help in most stop-and-go driving.

Determining Range

Use the battery voltmeter as a fuel gauge in conjunction with the odometer to tell you how far your batteries can take you. Tape a note to your dashboard or use a notebook to keep track of the elapsed mileage between full charges and voltmeter reading (for example, the voltmeter reading was x after you drove y miles) and you'll quickly get an idea of the pattern. Keep in mind that your battery pack will not reach its peak range until you've deep-cycled it a few times.

Running Out of Power

Before you totally run out of power, *get off the road!* If you totally run out of power and cannot find an electrical outlet, turn off your key switch and allow your batteries to rest for 20 to 30 minutes (shut down everything else electrical at this time also). Amazingly, you'll find extra energy in the batteries that may just be enough to take you to the power outlet you need. The convenience of an onboard charger is welcomed most of all in this particular circumstance. It is not good to be stuck in the middle of the busy freeway without power. You should understand that there is a significant power loss just before the no power condition. You should also know that a full discharge greatly shortens battery life.

Regular Driving

(Please note the equations are based on $4.50 per gallon. See Preface for more details.)

Drive your EV regularly—several times a week. Remember, the chemical clock inside your lead-acid batteries is ticking whether you use them or not, so use them. Better yet, think of how much money you are saving by using your EV. The 20 6-volt 220-AH capacity batteries gave Jim Harris' 1987 Ford Ranger conversion of Chapter 10 an on-board capacity of 26.6 kWh (220 AH × 120 volts).

If Jim got a 60-mile range out of one charge, his average energy use would be 0.44 kWh/mi (26.6 kWh/60 mi). At an average of $0.15 per kWh, that works out to be 6.6 cents per mile (0.44 kWh/mi × $0.15 per kWh).

If you compare that with gasoline at $4.50 per gallon and a typical 25 mpg for the internal combustion engine pickup before conversion, that works out to 18 cents per mile (1/25 gal/mi × $4.50 per gal). That's almost a three-to-one savings—take advantage of it. (Note: When we updated this to 2008 prices, the price ratio was *better* for electric vehicles versus the 1993 ratio that held steady for the past 50 years!)

Caring for Your Electric Vehicle

Now that you are driving for only pennies a day, it takes little more to make your pleasure complete. Actually, a properly designed and built EV conversion requires surprisingly little attention compared to an internal combustion engine vehicle. It comes down to the care and feeding of your batteries, minimizing friction, and preventative maintenance.

Battery Care

Of course, you are going to be charging your batteries on a regular basis, using the guidelines in Chapter 8, so battery maintenance really comes down to periodically checking to see that your batteries are properly watered. Excessive water loss is a sign of a problem, usually overcharging.

Jim Harris prefers to use the U.S. Batteries "quick disconnect" caps. All three caps on the 6-volt battery are ganged together, and a quick flip releases them all, making watering a breeze. Use distilled water only, not the water flowing from your faucet tap, which could be heavily mineralized. Observe the condition of the battery tops when watering; any acid overspray will cling to the battery tops and attract dirt. Clean this off immediately with a solution of water and baking soda after replacing all the battery caps as described in Chapter 8. Also check the battery terminals for corrosion, and correct any deficiencies using the baking soda solution. (Note: All distilled water is condensed water vapor. These days it is usually done at reduced pressure (and temperature) to save energy. Reverse osmosis is also effective in removing minerals.)

Tire Care

Tire rolling resistance is a major energy loss, so switch to a low–rolling-resistance tire (if possible) and frequently check that your tires are properly inflated using a more accurate meter-style tire gauge. Proper inflation means 32 psi and up; EV tires should be inflated hard. Special low–rolling resistance tires that can handle 50 psi (or more) are available for most cars. Talk to your local tire specialist about inflation limits versus loading. Also, listen to be sure that no brake shoes or pads are dragging, etc. (Note: 80 psi tires are available for many vehicles. Their designation starts with "LT.")

Lubricants

The weight of viscosity of your drivetrain fluids (transmission and rear axle lubricant) also contributes to losses on an ongoing basis, so experiment with lightweight lubricant in both these areas. The EV conversion puts a much smaller design load on the mechanical drive train so you ought to be able to drop down to a 50-weight lubricant for the rear axle and a lower-loss transmission fluid grade. Consider low-loss synthetic lubricants.

Checking Connections

Preventative maintenance mostly involves periodically checking the high-current wiring connections for tightness. Use your hands here. Warmth is bad—it means a loose connection—and anything that moves when you pull it is also bad. A few open-end and box wrenches ought to make quick work of your retightening preventive maintenance routine.

Emergency Kit

While carrying extra on-board weight is a no-no, carrying a small highway kit ought to be just enough to give you on-the-road peace of mind, knowing you have planned for most contingencies. At a minimum, your kit should have: a small fire extinguisher, a small bottle of baking soda solution, a small toolkit (wrenches, flat and wire-cutting pliers, screwdrivers, wire, and tape), and a heavy-duty charger extension cord with multiple adapter plugs (male and female).

CHAPTER 12

Sources

"Knowledge is but the beginning of wisdom."
—Darwin Gross, Universal Key

One of the most valuable benefits this book provides is information about where to go and who to see. While there are many sources, you'll find your needs best served by confining your initial search to just a handful until you get acquainted with the EV field. Please note that you must do your own reading, searching, listening, and talking before shopping to buy, convert, or build in order to get the best products and price.

The best guides for individuals interested in EVs are usually found in, around, and through the Electric Auto Association.

Less Is More

I have chosen to give you a few sources in each category to get you started rather than attempting to list everything and get you confused. The rest of this chapter is divided into four sections:

- Clubs, associations, and organizations
- Manufacturers, converters, and consultants
- Suppliers
- Books, articles, and literature

Clubs, Associations, and Organizations

The original Electric Auto Association, whose logo appears in Figure 12-1, has numerous local chapters. There are offshoots from the original, and new local entities having no connection with the original. There are also associations and organizations designed to serve corporate and commercial interests rather than individuals. Each one of these has its own meetings, events, and newsletter.

Electric Auto Association

www.eaaev.org

Founded in 1967, this is the oldest, largest organization, and has consistently been the best source of EV information for the individual. The newsletter subscription is well

Figure 12-1 The Electric Auto Association's logo says it all. Registered trademark of the Electric Auto Association.

worth the price of the membership dues. Recent newsletters have averaged 16 to 20 pages and provide information on current EV news and happenings. I can't emphasize enough the invaluable knowledge on tap in the members of this organization. The experience level of the various drivers who have built their own conversions can shorten one's learning curve substantially. Information on and photos of the many samples of conversions done over the years by this organization can be found at http:evalbum.com.

Here are some EAA local chapters—maybe there's one near you.

Canada

Durham Electric Vehicle Association
Web site: www.durhamelectricvehicles.com
Contact: JP Fernback <mail@durhamelectricvehicles.com>
(905) 706-6647
Mailing: P.O. Box 212, Whitby, ON L1N 5S1
Meetings: 1st Thursday of the month from September till June

Electric Vehicle Council of Ottawa
Web site: evco.ca
Contact: Alan Poulsen <info@evco.ca>
(613) 271-0940
Mailing: P.O. Box 4044, Ottawa, ON K1S5B1
Meetings: 7:30 PM–10:00 PM, last Monday of the month

Vancouver Electric Vehicle Association
Web site: www.veva.bc.ca
Contact: Haakon MacCallum <info@veva.bc.ca>
(604) 527-4288
Mailing: 4053 West 32nd Avenue, Vancouver BC V65 1Z5
Meetings: 7:30 PM, 3rd Wednesday of the month (please check web site for details)

Electric Vehicle Society of Canada—Toronto
Web site: www.evasociety.ca
Contact: Neil Gover <neil@ontarioev.ca>
(416) 255-9723
Mailing: 88 Lake Promenade, Etobicoke, ON M8W-1A3
Meetings: 7:30 PM, 3rd Thursday of the month (except July and August)

Alaska

Alaska EAA

Web site: www.alaskaEVA.org
Contact: Mike Willmon <electrabishi@ak.net>
(907) 868-5710
Mailing: Attn: Mike Willmon, 2550 Denali Suite 1, Anchorage, AK 99503
Meetings: 8:00 PM–9:00 PM, 3rd Friday of the month

Arizona

Flagstaff EAA

Contact: Barkley Coggin <cbcoggin@yahoo.com>
(928) 637-4444
Mailing: 6215 Rinker Circle, Flagstaff, AZ 86004
Meetings: 7:00 PM–9:00 PM, 1st Wednesday of the month

Phoenix EAA

Web site: www.phoenixeaa.com
Contact: Jim Stack <jstackeaa@yahoo.com>
(480) 659-5513
Mailing: Attn: Sam DiMarco, 1070 E. Jupiter Place, Chandler, AZ 85225
Meetings: 9:00 AM, 4th Saturday of the month

Tucson EVA II

Web site: www.teva2.com
Contact: John Barnes <johnjab@cox.net>
(520) 293-3500
Mailing: Attn: John Barnes, 4207 N. Limberlost Place, Tucson, AZ 85705
Meetings: 9:00 AM, 2nd Saturday of the month

California

Central Coast EAA

Web site: www.eaacc.org
Contact: Will Beckett <will@becketts.ws>
(831) 688-8669
Mailing: 323 Los Altos Drive, Aptos, CA 95003
Meetings: Call or see web site for meeting information

Chico EAA

Web site: geocities.com/chicoeaa
Contact: Chuck Alldrin <chicoeaa@sunset.net>
(530) 899-1835
Mailing: 39 Lakewood Way, Chico, CA 95926
Meetings: 11:00 AM–1:00 PM, 2nd Saturday of the month

East (SF) Bay EAA
Web site: www.ebeaa.org
Contact: Ed Thorpe <EAA-contact@excite.com>
(510) 864-0662
Mailing: 2 Smith Court, Alameda, CA 94502-7786
Meetings: 10:00 AM–12:00 noon, 4th Saturday of the month

Education Chapter: San Diego State University, College of Engineering
Contact: James S. Burns, PhD <jburns@mail.sdsu.edu>
(619) 933-6058
Mailing: 6161 El Cajon Boulevard, San Diego, CA 92115
Meetings: 4th Tuesday of each month during the academic year, except for December

EVA of Southern California
Contact: Leo Galcher <leo4marg@mac.com>
(949) 492-8115
Mailing: 35 Maracay, San Clemente, CA 92672
Meetings: 10:00 AM, 3rd Saturday of the month

Greater Sacramento EAA
Contact: Tim Hastrup <tim.hastrup@surewest.net>
(916) 791-1902
Mailing: 8392 West Granite Drive, Granite Bay, CA 95746
Meetings: 12:00 noon, 3rd Tuesday of February, May, August, and November

Konocti EAA
Web site: www.konoctieaa.org
Contact: Dr. Randy Sun <rsun@mchsi.com>
(707) 263-3030
Mailing: 800 S. Main Street, Lakeport, CA 95453
Meetings: 11:00 AM, last Friday of the month

North (SF) Bay EAA
Web site: www.nbeaa.org
Contact: Chris Jones <chris_b_jones@prodigy.net>
(707) 577-2391 (weekdays)
Mailing: c/o Agilent Technologies, 1400 Fountaingrove Parkway, Santa Rosa, CA 95403
Meetings: 10:00 AM–12:00 noon, 2nd Saturday of the month, check web site for details

EVA of San Diego
Web site: www.evaosd.com
Contact: Bill Hammons <ncsdca@att.net>
(858) 268-1759
Mailing: 1638 Minden Drive, San Diego CA 92111
Meetings: 7:00 PM, 4th Tuesday of the month

San Francisco Electric Vehicle Association
Web site: www.sfeva.org
Contact: Sherry Boschert <info2007@sfeva.com>
(415) 681-7716
Mailing: 1484 16th Avenue, San Francisco, CA 94122-3510
Meetings: 11:00 AM–1:00 PM, 1st Saturday of the month

San Francisco Peninsula EAA
Contact: Bill Carroll <billceaa@yahoo.com>
(650) 589-2491
Mailing: 160 Ramona Avenue, South San Francisco, CA 94080-5936
Meetings: 10:00 AM, 1st Saturday of the month

San Jose EAA
Web site: geocities.com/sjeaa
Contact: Terry Wilson <historian@eaaev.org>
(408) 446-9357
Mailing: SJEAA, 20157 Las Ondas, San Jose, CA 95014
Meetings: 10:00 AM, 2nd Saturday of the month

Silicon Valley EAA
Web site: www.eaasv.org
Contact: Jerry Pohorsky <JerryP819@aol.com>
(408) 464-0711
Mailing: 1691 Berna Street, Santa Clara, CA 95050
Meetings: 3rd Saturday (Jan–Nov)

Ventura County EAA
Web site: geocities.com/vceaa
Contact: Bruce Tucker <tuckerb2@adelphia.net>
(805) 495-1026
Mailing: 283 Bethany Court, Thousand Oaks, CA 91360-2013
Meetings: Please contact Bruce for time and location

Colorado

Denver Electric Vehicle Council
Contact: Graham Hill <ghill@21wheels.com>
(303) 544-0025
Mailing: 6378 S. Broadway, Boulder, CO 80127
Meetings: 3rd Saturday monthly, contact George for time and location

Florida

Florida EAA
Web site: www.floridaeaa.org
Contact: Shawn Waggoner <shawn@suncoast.net>
(561) 543-9223
Mailing: 8343 Blue Cypress, Lake Worth, FL 33467
Meetings: 9:30 AM, 2nd Saturday of the month

Georgia

EV Club of the South
Web site: www.evclubsouth.org
Contact: Stephen Taylor <sparrow262@yahoo.com>
(678) 797-5574
Mailing: 750 West Sandtown Road, Marietta, GA 30064
Meetings: 6:00 PM, 1st Wednesday every even-numbered month

Illinois

Fox Valley EAA
Web site: www.fveaa.org
Contact: Ted Lowe <ted.lowe@fveaa.org>
(630) 260-0424
Mailing: P.O. Box 214, Wheaton, IL 60189-0214
Meetings: 7:30 PM, 3rd Friday of the month

Kansas/Missouri

Mid America EAA
Web site: maeaa.org
Contact: Mike Chancey <eaa@maeaa.org>
(816) 822-8079
Mailing: 1700 East 80th Street, Kansas City, MO 64131-2361
Meetings: 1:30 PM, 2nd Saturday of the month

Massachusetts

New England EAA
Web site: www.neeaa.org/
Contact: Bob Rice <bobrice@snet.net>
(203) 530-4942
Mailing: 29 Lovers Lane, Killingworth, CT 06419
Meetings: 2:00 PM–5:00 PM, 2nd Saturday of the month

Pioneer Valley EAA
Web site: www.pveaa.org
Contact: Karen Jones <PVEAA@comcast.com>
Mailing: P.O. Box 153, Amherst, MA 01004-0153
Meetings: 2:00 PM, 3rd Saturday of the month (Jan–June; Sept–Nov)

Minnesota

Minnesota EAA
Web site: mn.eaaev.org
Contact: Craig Mueller <craig.mueller@nwa.com>
(612) 414-1736
Mailing: 4000 Overlook Drive, Bloomington, MN 55437
Meetings: 7:00 PM–8:30 PM CDT

Nevada

Alternative Transportation Club, EAA

Web site: www.electricnevada.org
Contact: Bob Tregilus <lakeport104@yahoo.com>
(775) 826-4514
Mailing: 2805 W. Pinenut Court, Reno, NV 89509
Meetings: 6:00 PM, monthly, see web site or call for details

Las Vegas Electric Vehicle Association

Web site: www.lveva.org
Contact: William Kuehl <bill2k2000@yahoo.com>
(702) 636-0304
Mailing: 2816 El Campo Grande Avenue, North Las Vegas, NV 89031-1176
Meetings: 10:00 AM–12:00 noon, 3rd Saturday of the month

North Carolina

Coastal Carolinas Wilmington

Contact: Page Paterson <pagepaterson@mac.com>
(910) 686-9129
Mailing: 1317 Middle Sound, Wilmington, NC 28411
Meetings: Please contact for time and date

Piedmont Carolina Electric Vehicle Association

Web site: www.opecthis.info
Contact: Todd W. Garner <tgarnercgarner@yahoo.com>
(704) 849-9648
Mailing: 1021 Timber Wood Court, Matthews, NC 28105
Meetings: Please contact for time and date

Electric Cars of Roanoke Valley

Contact: Harold Miller <EV@schoollink.net>
(252) 534-1258
Mailing: 567 Miller Trail, Jackson, NC 27845
Meetings: Please contact for time and date

Triad Electric Vehicle Association

Web site: www.localaction.biz/TEVA
Contact: Jack Martin <jmartin@hotmail.com>
(336) 213-5225
Mailing: 2053 Willow Spring Lane, Burlington, NC 27215
Meetings: 9:00 AM, 1st Saturday of the month

Triangle EAA

Web site: www.rtpnet.org/teaa
Contact: Peter Eckhoff <teaa@rtpnet.org>
(919) 477-9697
Mailing: 9 Sedley Place, Durham, NC 27705-2191
Meetings: 3rd Saturday of the month

Oregon

Oregon Electric Vehicle Association

Web site: www.oeva.org
Contact: Rick Barnes <barnes.rick@verizon.net>
Mailing: 19100 SW Vista Street, Aloha, OR 97006
Meetings: 7:30 PM, 2nd Thursday of the month

Pennsylvania

Eastern Electric Vehicle Club

Web site: www.eevc.info
Contact: Peter G. Cleaveland <easternev@aol.com>
(610) 828-7630
Mailing: P.O. Box 134, Valley Forge, PA 19482-0134
Meetings: 7:00 PM, 2nd Wednesday of the month

Texas

Alamo City EAA

Web site: www.aceaa.org
Contact: Alfonzo Ranjel <acranjel@sbcglobal.net>
(210) 389-2339
Mailing: 9211 Autumn Bran, San Antonio, TX 78254
Meetings: 3:00 PM CST, 3rd Sunday of the month

AustinEV: the Austin Area EAA

Web site: www.austinev.org
Contact: Aaron Choate <austinev-info@austinev.org>
(512) 453-2890
Mailing: P.O. Box 49153, Austin, TX 78765
Meetings: Please see web site

Houston EAA

Web site: www.heaa.org
Contact: Dale Brooks <brooksdale@usa.net>
(713) 218-6785
Mailing: 8541 Hatton Street, Houston, TX 77025-3807
Meetings: 6:30 PM, 3rd Thursday of the month

North Texas EAA

Web: http://www.nteaa.org/
Contact: John L. Brecher <jlbrecher@verizon.net>
(214) 703-5975
Mailing: 1128 Rock Creek Drive, Garland, TX 75040
Meetings: 2nd Saturday of the month

Utah

Utah EV Coalition

Web site: www.saltflats.com.
Contact: Kent Singleton <kent@saltflats.com>
(801) 644-0903
Mailing: 325 E. 2550 N #83, North Ogden, UT 84414
Meetings: 7:00 PM, 1st Wednesday of the month

You'll meet BYU Electric Team, WSU-EV Design Team, other land speed racing celebrities. Always a great turnout.

Washington

Seattle Electric Vehicle Association

Web site: www.seattleeva.org
Contact: Steven S. Lough <stevenslough@comcast.net>
(206) 524-1351
Mailing: 6021 32nd Avenue NE, Seattle, WA 98115-7230
Meetings: 7:00 PM, 2nd Tuesday of the month

Washington D.C.

EVA of Washington DC

Web site: www.evadc.org
Contact: David Goldstein <goldie.ev1@juno.com>
(301) 869-4954
Mailing: 9140 Centerway Road, Gaitherburg, MD 20879-1882
Meetings: 7:00 PM, 2nd or 3rd Tuesday of the month

Electric Vehicle Association of Greater Washington DC has an excellent overview, "Build an EV" at www.evadc.org/build_an_ev.html.

Wisconsin

Southern Wisconsin EV Proliferation

Web site: www.emissionsfreecars.com
Contact: Mike Turner <mike.turner@emissionsfreecars.co>
(920) 261-7057
Mailing: 808 Fieldcrest Court, Watertown, WI 53511
Meetings: Please contact for date and location

EAA Special Interest Chapters

California Cars Initiative

Web site: calcars.org
Contact: Felix Kramer <info@calcars.org>
(650) 520-5555
Mailing: P.O. Box 61045, Palo Alto, CA 94306

Plug In America
Web site: www.pluginamerica.com
Contact: Linda Nicholes <Linda@pluginamerica.com>
(714) 974-5647
Mailing: 6261 East Fox Glen, Anaheim, CA 92807
Meetings: Please contact for details

Electric Drive Transportation Association
1101 Vermont Avenue, NW, Suite 401
Washington, DC 20005
(202) 408-0774
www.electricdrive.org

"EDTA is the preeminent industry association dedicated to advancing electric drive as a core technology on the road to sustainable mobility. As an advocate for the adoption of electric drive technologies, EDTA serves as the unified voice for the industry and is the primary source of information and education related to electric drive. Our membership includes a diverse representation of vehicle and equipment manufacturers, energy providers, component suppliers and end users." (www.electricdrive.org/index.php?tg=articles&topics=1)

National Electric Drag Racing Association
3200 Dutton Avenue #220
Santa Rosa, CA 95407
www.nedra.com

The National Electric Drag Racing Association (NEDRA) exists to increase public awareness of electric vehicle performance and to encourage through competition, advances in electric vehicle technology. NEDRA achieves this by organizing and sanctioning safe, silent, and exciting electric vehicle drag racing events.

Northeast Sustainable Energy Association (NESEA)
23 Ames Street
Greenfield, MA 01301
(413) 774-6051

Organizes the annual "American Tour de Sol" and electric vehicle symposium.

Solar Energy Expo and Rally (SEER)
239 S. Main Street
Willits, CA 95490
(707) 459-1256

Host for annual "Tour de Mendo," when Willits temporarily becomes the solar capital of the world.

Solar and Electric Racing Association
11811 N. Tatum Boulevard, Suite 301
Phoenix, AZ 85028
(602) 953-6672

Organizes annual Solar and Electric 500 in Phoenix and promotes electric vehicles.

Electric Utilities and Power Associations

Any of the following organizations can provide you with information.

American Public Power Association
2301 M Street, N.W.
Washington, DC 20202
(202) 775-8300

Arizona Public Service Company
P.O. Box 53999
Phoenix, AZ 85072-3999
(602) 250-2200

California Energy Commission
1516 9th Street
Sacramento, CA 95814
(916) 654-4001

Electric Power Research Institute
412 Hillview Avenue
P.O. Box 10412
Palo Alto, CA 94303
(415) 855-2580

Director of Electric Transportation
Department of Water and Power City of Los Angeles
111 N. Hope Street, Room 1141
Los Angeles, CA 90012-2694
(213) 481-4725

Public Service Co. of Colorado
2701 W. 7th Avenue
Denver, CO 80204
(303) 571-7511

Sacramento Municipal Utility District
P.O. Box 15830
Sacramento, CA 95852-1830
(916) 732-6557

Southern California Edison
2244 Walnut Grove Avenue
P.O. Box 800
Rosemead, CA 91770
(818) 302-2255

Government

The following agencies are involved with EVs directly or indirectly at city, state, or federal government levels.

California Air Resources Board
1012 Q Street
P.O. Box 2815
Sacramento, CA 95812
(916) 322-2990

Environmental Protection Agency
401 M Street S.W.
Washington, DC 20460
(202) 260-2090

National Highway Traffic Safety Administration
400 7th Street S.W.
Washington, DC 20590
(202) 366-1836

New York Power Authority
123 Main Street
White Plains, NY 10601

New York State Energy Research and Development Authority (NYSERDA)
17 Columbia Circle
Albany, NY 12203-6399
www.nyserda.org

Manufacturers, Converters, and Consultants

There is a sudden abundance of people and firms doing EV work. This category is an attempt to present you with the firms and individuals from whom you can expect either a completed EV or assistance with completing one.

Manufacturers

This category includes the household names plus the major independents you already met in Chapters 3 and 4. When contacting the larger companies, it is best to go through the switchboard or a public affairs person who can direct your call after finding out your specific needs.

Ampmobile® Conversions LLC
P.O. Box 5106
Lake Wylie, SC 29710
(803) 831-1082 or toll free (866) 831-1082
Email: info@ampmobiles.com

Battery Automated Transportation
2471 S. 2570 W.
West Valley City, UT 84119
(801) 977-0119

Best known for its proprietary "Ultra Force" lead-acid batteries and Ford Ranger pickup truck conversions.

California Electric Cars
1669 Del Monte Boulevard
Seaside, CA 93955
(408) 899-2012

Best known for its "Monterey" electric vehicle.

Clean Air Transport of North America
23030 Lake Forest Drive, Suite 206
Laguna Hills, CA 92653
(714) 951-3983

Best known for its "LA301" electric vehicle.

Cloud Electric Vehicles Battery Powered Systems
102 Ellison Street, Unit A
Clarksville, GA 30523
(866) 222-4035

Bob Beaumont
Columbia Auto Sales
9720 Owen Brown Road
Columbia, MD 21045
(301) 799-3550

Best known as the former head of Sebring-Vanguard and its CitiCar offering, Bob Beaumont is beginning production again with another two-seater electric vehicle that he says will not repeat the mistakes of its predecessor. If true, it will be a well-priced, well-positioned offering. Just as a reference point, Bob produced the Renaissance Tropica seen in the late 1990s on the network television show *Nash Bridges*. The sleek two-seater was powered by 72-volt lead-acid batteries; only two or three survived. The program director dubbed in the sound of an internal combustion engine motor revving during scenes where actress Jasmine Blyth pulled away from the camera in the car. (Another example of EV education sorely being needed in Hollywood!)

Conceptor Industries
521 Newpark Boulevard
P.O. Box 149
New Market, ON
L3Y 4X7 CANADA
(416) 836-4611

A subsidiary of Vehma International, best known for its "G-Van" EV conversions of the General Motors Vandura van.

Cushman
900 North 21st Street
Lincoln, NE 68503
(402) 475-9581

Manufactures three-wheeler industrial and commercial electric carts.

Electric Fuel Propulsion Corp.
4747 N. Ocean Drive, #223
Ft. Lauderdale, FL 33308
(305) 785-2228

Robert R. Aronson, a long-timer in the EV field, now offers the luxury "Silver Volt," with tri-power lead-cobalt batteries.

Electric Mobility
591 Mantua Boulevard
Sewell, NJ 08080
(800) 257-7955

Manufactures electric carts, bicycles, etc.

ElectroAutomotive
P.O. Box 1113-W
Felton, CA 95018-1113
(831) 429-1989
Fax: (831) 429-1907
Email: electro@cruzio.com

EV Parts.com
108-B Business Park Loop
Sequim, WA 98382
(360) 582-1271
(888) 387-2787
Fax: (360) 582-1272
Email: sales@evparts.com

Green Motor Works
(also a Solar Electric dealer)
5228 Vineland
North Hollywood, CA 91601
(818) 766-3800

Metric Mind Engineering
9808 SE Derek Court
Happy Valley, OR 97806-7250
(503) 680-0026
Fax: (503) 774-4779
Email: ac@metricmind.com

Palmer Industries
P.O. Box 707
Endicott, NY 13760
(800) 847-1304

Manufactures an electric bicycle.

Solar Car Corporation
1300 Lake Washington Road
Melbourne, FL 32935
(407) 254-2997

Sebring Auto Cycle
P.O. Box 1479
Sebring, FL 33871
(813) 655-2131

The latest incarnation of the original Sebring-Vanguard operation, best known for its three-wheeled zipper electric vehicle.

Conversion Specialists

In this category, the line between those who provide parts and those who provide completed vehicles is blurred.

Jim Harris
Zero Emissions Motorcar Company
1031 Bay Boulevard, Suite T
Chula Vista, CA 91911
(619) 425-4221
Fax: (619) 425-2312

Jim's Ford Ranger pickup truck conversion is the centerpiece feature of Chapter 10. His company offers pre-packaged motor adapter and controller kits, and complete vehicles, components, literature, and expertise. His word is his bond.

Jeff Shumway
Ecotech Autoworks
1524-V Springhill Road
P.O. Box 9262
McLean, VA 22102
(703) 893-3045

Vehicles and components.

Bill Kuehl
Electric Auto Conversions
4504 W. Alexander Road
Las Vegas, NV 89030
(702) 645-2132

Grassroots Electric Vehicles
1918 South 34th Street
Fort Pierce, FL 34947
(772) 971-0533

Left Coast Electric
www.leftcoastelectric.com
info@leftcoastelectric.com

Vehicles and Components

Vicor Corporation
25 Frontage Road
Andover, MA 01810-5413
(800) 735-6200
Fax: (978) 475-6715

Ken Bancroft
Electric Motor Cars Sales and Service
4301 Kingfisher
Houston, TX 77035
(713) 729-8668
Vehicles and components.

Don Karner
Electric Transportation Applications
P.O. Box 10303
Glendale, AZ 85318
(602) 978-1373
Vehicles and components.

Larry Foster
Electric Vehicle Custom Conversion
1712 Nausika Avenue
Rowland Heights, CA 91748
(818) 913-8579
Vehicles and components.

Stan Skokan
Electric Vehicles, Inc.
1020 Parkwood Way
Redwood City, CA 94060
(415) 366-0643

Gene Hitney
Hitney Solar Products
655 N. Highway 89
Chino Valley, AZ 86323
(602) 636-2201

Frank Kelly
Interesting Transportation
2362 Southridge Drive
Palm Springs, CA 92264-4960
(619) 327-2864

W. D. Mitchell
20 Victoria Drive
Rowlett, TX 75055
(214) 475-0361

Ron Larrea
San Diego Electric Auto
9011 Los Coches Road
Lakeside, CA 92040
(619) 443-3017

Experienced EV Conversions and Consulting

Ed Ranberg
Eyeball Engineering
16738 Foothill Boulevard
Fontana, CA 92336
(714) 829-2011

Experienced EV conversion professional; components and consulting.

Lon Gillas
E-Motion
515 W. 25th Street
McMinnville, OR 97128
(503) 434-4332

Consultants

Companies and individuals who are more likely to provide advice, literature, or components—rather than completed vehicles—are listed.

AeroVironment
P.O. Box 5031
Monrovia, CA 91017-7131
(818) 359-9983

Developers of the GM Impact, Paul MacCready and AeroVironment need no further introduction. In September 2007 Paul passed away after a short illness, just after retiring from AeroVironment. His insight initially sparked the concept car the GM made into the EV-1, the subject of the 2006 movie *Who Killed the Electric Car?*

3E Vehicles
P.O. Box 19409
San Diego, CA 92119

Another experienced participant in the EV field, 3E offers an outstanding line of conversion booklets that (although somewhat dated today) are still highly useful.

Michael Hackleman
Earthmind
P.O. Box 743
Mariposa, CA 95338
(310) 396-1527

Author, editor of *Alternative Transportation News*, experienced EV participant, and a consultant.

Michael Brown and Shari Prange
Electro Automotive
P.O. Box 1113
Felton, CA 95018-1113
(831) 429-1989

This organization, an experienced participant in the EV field, offers books, videos, seminars, consulting, and components. Mike and Shari still supply kits for conversion builders, complete parts, and instruction manuals and are finding that with the high gasoline prices since Hurricane Katrina came ashore in 2005 that their business is brisk. They carry AC drive systems from Azure Dynamics (formerly Solectria, founded by MIT students).

Mike Kimball
18820 Roscoe Boulevard
Northridge, CA 91324
(818) 998-1677

EV technician and maintenance mechanic extraordinaire, Mike has probably forgotten more about EVs than most people will ever know.

Carl Taylor
3871 S.W. 31st Street
Hollywood, FL 33023
(305) 981-9462

EV maintenance, repair, and troubleshooting.

Bill Williams
Williams Enterprises
P.O. Box 1548
Cupertino, CA 95015

Experienced participant in the EV field, conversion specialist, and consultant, Williams offers an outstanding conversion guide that (although somewhat dated today) is still very useful.

Howard G. Wilson
2050 Mandeville Canyon Road
Los Angeles, CA 90049
(310) 471-7197

Former Hughes vice-president, Howard Wilson was the real "make it happen" factor behind GM's Impact and Sunraycer projects.

Bob Wing
P.O. Box 277
Inverness, CA 94937
(415) 669-7402

Suppliers

This category includes those from whom you can obtain complete conversion kits (all the parts you need to build your own EV after you have the chassis); conversion plans; and suppliers specializing in motors, controllers, batteries, chargers, and other components.

You can find more information about conversions and components at http://eaaev.org/eaalinks.html.

Battery Powered Systems
204 Ellison Street, Unit A
Clarkesville, GA 30523
www.beepscom.com/

EV Parts, Inc.
160 Harrison Road #7
Sequim, WA 98382
www.evparts.com/firstpage.php

They are a great component supplier.

Manzanita Micro EV components
www.manzanitamicro.com

Rich Rudman
360-297-7383 Office
360-620-6266 Cell
360-297-1660 Production Shop
360-297-3311 Metal Shop
5718 Gamblewood Road NE
Kingston, WA 98346

Canadian Electric Vehicles Ltd.
PO Box 616
1184 Middlegate Road
Errington, BC V0R 1V0 Canada
(250) 954-2230
Fax: (250) 954-2235
Email: randy@canev.com

EV Source LLC
19 W. Center
Suite 201
Logan, UT 84321
(877) 215-6781
sales@evsource.com

KTA Services, Inc.
20330 Rancho Villa Road
Ramona, CA 92065
Toll free: 877-465-8238
(760) 787-0896
Fax: (760) 787-9437
Email: wistar.rhoads@kta-ev.com
www.kta-ev.com
Provides EV components and kits

EV Parts, Inc.
160 Harrison Road, #7
Sequim, WA 98382
www.evparts.com
(888) 387-2787
Email: sales@evparts.com

Metric Mind Corporation
9808 SE Derek Court
Happy Valley, OR 97086
(503) 680-0026
(503) 774-4779 (fax)
Contact: Victor Tikhonov imports Siemens AC drives.
www.metricmind.com/

ThunderStruck Motors
3200 Dutton Avenue #319
Santa Rosa, CA 95407
(707) 575-0353 voice
(707) 544-5304 Fax
ThunderStruck Motors is a small research, development, and manufacturing company that also retails electric vehicles and components.

Conversion Kits

Companies and individuals listed here are those more likely to provide the parts that go into converting or building an EV once you already have the chassis, such as components, advice, and literature.

Ken Koch
KTA Services Inc.
944 W. 21st Street
Upland, CA 91786
(909) 949-7914
Fax: (909) 949-7916

Ken's services to those interested in converting or building EVs are in a class by themselves. His company offers everything from individual components to several levels of kits, and its products are featured throughout this book. You'll find KTA knowledgeable about your needs, prompt in its deliveries, and a pleasure to do business with. While Ken is still around, he is looking to sell his business and retire.

John Stockberger
Electric Auto Crafters
643 Nelson Lake Road, a2S
Batavia, IL 60510
(312) 879-0207

Provides parts, information, and testing for EV builders.

Bob Batson
Electric Vehicles of America
P.O. Box 59
Maynard, MA 01754
(508) 897-9393

Provides kits, components, and literature.

Steve van Ronk
Global Light and Power
55 New Montgomery, Suite 424
San Francisco, CA 94105
(415) 495-0494

Kits and components; promotes annual Clean Air Revival.

Steve Deckard
King Electric Vehicles
P.O. Box 514
East Syracuse, NY 13057

C. Fetzer
Performance Speedway
2810 Algonquin Avenue
Jacksonville, FL 32210
(904) 387-9858

Paul Schutt and Associates
673 Via Del Monte
Palos Verdes Estates, CA 90274
(310) 373-4063
Represents manufacturers who supply EV components.

Unique Mobility
425 Corporate Circle
Golden, CO 80419
(303) 278-2002
Well known for their prototype vehicles using proprietary technology, Unique offers numerous advanced EV capabilities and components.

Conversion Plans

Listed here are companies and individuals who are more likely to provide vehicle plans, kits, or components rather than completed vehicles.

Rick Doran
Doran Motor Company
1728 Bluehaven Drive
Sparks, NV 89431
(805) 546-9654
(702) 359-6735
Best known for the Doran three-wheeler and its plans.

Dolphin Vehicles
P.O. Box 110215
Campbell, CA 95011
(408) 734-2052
Best known for the Vortex three-wheeler and its plans for either internal combustion or electric propulsion.

Motors

A considerable number of companies manufacture electric motors. The short list here is only to get you started

Advanced D.C. Motors, Inc.
219 Lamson Street
Syracuse, NY 13206
(315) 434-9303

Aveox Inc.
2265A Ward Avenue
Simi Valley, CA 93065
(805) 915-0200
www.aveox.com/Default.aspx

Baldor Electric Co.
5711 South 7th
Ft. Smith, AR 72902
(501) 646-4711

Azure Dynamics
9 Forbes Road
Woburn, MA USA 01801-2103
(781) 932-9009
Fax: (781) 932-9219

NetGain Technologies, LLC
900 North State Street, Suite 101
Lockport, IL 60441
(630) 243-9100
Fax: (630) 685-4054
www.go-ev.com/

NetGain Technologies, LLC is the exclusive worldwide distributor of WarP™, ImPulse™, and TransWarP™ electric motors for use in electric vehicles and electric vehicle conversions. These powerful electric motors may also be used in the conversion of conventional internal combustion engine vehicles to hybrid gas/electric or electric-assist vehicles. Their motors are manufactured in Frankfort, Illinois by Warfield Electric Motor Company.

UQM Technologies
7501 Miller Drive
P.O. Box 439
Frederick, Colorado 80530
(303) 278-2002
Fax: (303) 278-7007

UQM Technologies, Inc. is a developer and manufacturer of power-dense, high-efficiency electric motors, generators, and power electronic controllers for the automotive, aerospace, medical, military, and industrial markets. For more information on the company, please visit its worldwide web site at www.uqm.com.

Controllers

A considerable number of companies manufacture controllers; again, this short list is only to get you started.

Café Electric LLC
(866) 860-6608
www.cafeelectric.com

Supplier of the Zilla and Baby Zilla DC motor speed controllers. This well-engineered design is proving virtually indestructible compared to some of its predecessors such as Auburn and DCP controllers. The IGBT-based solid-state controllers are favorites among street and drag race performance EVs. They can handle 2,000 amps at pack voltages to 348 volts DC, and 1,000 amps for the baby Zilla.

Curtis PMC
6591 Sierra Lane
Dublin, CA 94568
(510) 828-5001

Specialists in controllers for EV applications, this company's controller is featured in Chapter 7 and used in Chapter 10's conversion. These analog DC controllers are moderately priced, provide moderate performance, and are moderately good.

AC Propulsion, Inc.
462 Borrego Court, Unit B
San Dimas, CA 91773
(714) 592-5399

AC Propulsion's AC EV drive systems are simply a better idea whose time will come when economies of scale drive prices down. You've met them in Chapter 8. They made the tZero, two-seater sports car, and were able to prove 300 miles at freeway speeds on a single charge using commodity lithium batteries successfully.

Batteries

Technical Publications Department, Motorola Inc.
Semiconductor Products Sector
P.O. Box 52073
Phoenix, AZ 85072
(800) 521-6274

Trojan Battery Co.
12380 Clark Street
Santa Fe Springs, CA 90670
(800) 423-6569
(213) 946-8381
(714) 521-8215

Trojan has manufactured deep-cycle, lead-acid batteries suitable for EV use longer than most companies, and has considerable expertise; its batteries are featured in Chapter 9.

Alco Battery Co.
2980 Red Hill Avenue
Costa Mesa, CA 92626
(714) 540-6677

Offers a full line of lead-acid batteries suitable for EVs.

Concorde Battery Corp.
2009 W. San Bernadino Road
West Covina, CA 91760
(818) 962-4006

Offers lead-acid batteries for aircraft use.

Eagle-Picher Industries
P.O. Box 47
Joplin, MO 64802
(417) 623-8000

Offers a full line of lead-acid batteries suitable for EVs.

SAL
251 Industrial Boulevard
P.O. Box 7366
Greenville, NC 27835-7366
(919) 830-1600

Manufactures nickel-iron and nickel-cadmium batteries suitable for EVs.

U.S. Battery Manufacturing Co.
1675 Sampson Avenue
Corona, CA 91719
(800) 695-0945
(714) 371-8090

Manufactures deep-cycle, lead-acid batteries suitable for EVs. Thriving today with distributors all around.

Yuassa-Exide
9728 Alburtis Avenue
P.O. Box 3748
Santa Fe Springs, CA 90670
(800) 423-4667
(213) 949-4266

Manufactures deep-cycle, lead-acid batteries suitable for EVs. Still making flooded lead-acid batteries, manufactured in Japan and probably China too.

Chargers

There are many battery charger manufacturers; this short list is only to get you started.

Lester Electrical
625 West A Street
Lincoln, NE 68522
(402) 477-8988

Lester has been manufacturing battery chargers suitable for EV use longer than most companies, and has considerable expertise. Their charger is featured in Chapter 10's discussion.

K & W Engineering
3298 Country Home Road
Marion, IA 52302
(319) 378-0866

K & W's lightweight, transformerless chargers designed for onboard use are also featured in Chapter 10's discussion.

Avcon Corporation
4640 Ironwood Drive
Franklin, WI 53132
(877) 423-8725
Fax: (414) 817-6161
Email: powerpak@webcom.com

Manzanita Micro
Rich Rudman
5718 Gamblewood Road NE
Kingston, WA 98346
(360) 297-7383

Designer Joe Smalley and protégé Rich Rudman make a range of fully powered factor-corrected chargers that deliver 20–50 amps DC into traction packs from 48 to 312 volts, from any line source (120 to 240 volts). These units provide remarkable power and have done well in the conversion market. Improvements (such as increasing the switching frequency) would allow a second generation to be made and packaged in a much smaller container, thereby further increasing its popularity.

Other Parts

Cruising Equipment Co.
6315 Seaview Avenue
Seattle, WA 98107
(206) 782-8100

Offers the Amp-Hour+ meter for monitoring the state of battery charge. Sold to Xantrax. Very versatile and powerful SoC instrumentation.

Books, Articles, and Papers

There have been a number of books and thousands of articles and papers written about EVs, both technical and nontechnical. Here are some available related books and manuals, and a sampling of a few nontechnical articles that will give you instant expertise in the subject area.

Books

Michael Brian Schiffer, Tamara C. Butts, Kimberly K. Grimm, *Taking Charge: The Electric Automobile in America*, by Smithsonian Institution Press, 1994.
David Kirsh, *Electric Vehicle and the Burden of History,* Rutgers University Press, 2000.
Martin Sheen, *Who Killed the Electric Car?*, SONY Pictures, DVD. 2006.
Irving M. Gottlieb, *Electric Motors and Control Techniques,* McGraw-Hill/TAB Electronics, 1994.
James Larminie, John Lowry, *Electric Vehicle Technology Explained,* Wiley, 2003.
Sherry Boschert, *Plug-in Hybrids: The Cars that will Recharge America,* New Society Publishers, 2006.
Michael H. Westbrook, *The Electric Car: Development and Future of Battery, Hybrid and Fuel-Cell Cars (Iee Power & Energy Series, 38)*, by Institution Electrical Engineers, 2001.
Bob Brant, *Bob, Build Your Own Electric Vehicle,* McGraw-Hill/TAB Electronics, 1993.

Mehrdad Ehsani, Yimin Gao, Sebastien E. Gay, *Modern Electric, Hybrid Electric, and Fuel Cell Vehicles: Fundamentals, Theory, and Design (Power Electronics and Applications Series)*, CRC, 2004.

Mark Beyer, *Transportation of the Future (Life in the Future)*, Children's Press (CT), 2002.

Daniel Sperling, Mark A. Delucchi, Patricia M. Davis, A. F. Burke, *Future Drive: Electric Vehicles and Sustainable Transportation*, Island Press, 1994.

Edwin Black, *Internal Combustion: How Corporations and Governments Addicted the World to Oil and Derailed the Alternatives*, St. Martin's Press, 2006.

Stephen Carlson, *From Boston to the Berkshires: Pictorial Review of Electric Transportation in Massachusetts (Bulletin, No 21)*, Boston Street Railway Assn., 1990.

Jack La Russa, *Electric Lines: The Magazine of Electric Transportation*, 1988.

Articles

T.E. Lipman, M.A. Delucchi, *A retail and lifecycle cost analysis of hybrid electric vehicles*, Transportation Research Part D, Elsevier, 2001.

K.T. Ulrich, *Estimating the technology frontier for personal electric vehicles*, Transportation Research Part C, Elsevier, 2003.

Gale Reference Team, *NY Transportation Plan to Support Electric Vehicle Projects*, Electric and Hybrid Vehicles Today, 2005.

Lee Zion, *Ford pulls the plug on its think line of electric cars*, San Diego Business Journal, published by CBJ, L.P. on September 9, 2002.

The Associated Press, *After Bike-Sharing Success, Paris Considers Electric Cars*, July, 29, 2008.

Steven Erlanger, *Israel Is Set to Promote the Use of Electric Cars*, NY Times, January 21, 2008.

Richard S. Chang, *You Built an Electric Car ... Out of a DeLorean?*, NY Times, June 11, 2008.

Nick Bunkley, *Nissan Says Electric Cars Will Be Quickly Profitable*, NY Times, July 23, 2008.

Papers

Bradley J. Fikes, *Getting start on charge toward electric cars*, Special Report: Transportation, San Diego Business Journal.

Office of Transportation Technologies Business/Technology Books, *Encouraging The Purchase and Use of Electric Motor Vehicles: Final Report*, Electric Vehicle Information Series, 1995.

U.S. Department Of Energy Business/Technology Books, *Electric and Hybrid Vehicles Program—U.S. Department of Energy Annual Reports to Congress for Fiscal Years 1993-1995*, B/T Books, 1996.

Publishers

Here are a few companies that specialize in publications of interest to EV converters.

Battery Council International

401 N. Michigan Avenue
Chicago, IL 60611
(312) 644-6610

Publishes battery-related books and articles.

Institute for Electrical and Electronic Engineers (TREE)

IEEE Technical Center
Piscataway, NJ 08855

Publishes numerous articles, papers, and proceedings. Expensive, but one of the best sources for recent published technical information on EVs.

Lead Industries Association
292 Madison Avenue
New York, NY 10017
Publishes information on lead recycling.

Society of Automotive Engineers (SAE) International
400 Commonwealth Drive
Warrendale, PA 15096-0001
(412) 772-7129
Publishes numerous articles, papers, and proceedings. Also expensive, but the other best source for recent published technical information on EVs.

Newsletters

Here are a few companies that specialize in newsletter-type publications of interest to EV converters.

Electric Grand Prix Corp.
6 Gateway Circle
Rochester, NY 14624
(716) 889-1229

Electric Vehicle Consultants
327 Central Park W.
New York, NY 10025
(212) 222-0160

Solar Mind
759 S. State Street, #81
Ukiah, CA 95482
(707) 468-0878
Electric vehicle directories.

Online Industry Publications

These online publications report on industry activity, manufacturer offerings, and local electric drive–related activities.

EV World
evworld.com
Houses an online "library" of EV-related reports, articles, and news releases available to the general public. EV owners also can register and share their experiences with others. Visitors can sign up for a weekly EV newsletter. EV World has information about conversions, conversion suppliers, and a list of popular EV conversion vehicles (www.evworld.com/archives/hobbyists.html).

Electrifying Times
www.electrifyingtimes.com
One of the original founders of the National Electric Drag Racing Association, Electrifying Times has been chronicling the rise of EV Kar Kulture since the mid-90s, with an irregularly published print edition, Web site, and a Yahoo group.

Fleets & Fuels

www.fleetsandfuels.com

A biweekly newsletter (distributed online) providing business intelligence on alternative fuel and advanced vehicles technologies encompassing electric drive, natural gas, hydraulic hybrids, propane and alcohol fuels, and biofuels. The newsletter is dedicated to making the AFVs business case to fleets.

Hybrid & Electric Vehicle Progress

www.hevprogress.com

Formerly Electric Vehicle Progress. Follows new EV products, including prototype vehicles; provides status reports on R&D programs; publishes field test data from demonstration programs conducted around the world; details infrastructure development, charging sites, and new technologies; and includes fleet reports, battery development, and a host of other EV-related news. Published twice a month.

Advanced Battery Technology

www.7ms.com/abt/index.html

Advanced Fuel Cell Technology

www.7ms.com/fct/index.html

e-Drive Magazine

www.e-driveonline.com

Features new products, services, and technologies in motors, drives, controls, power, electronics, actuators, sensors, ICs, capacitors, converters, transformers, instruments, temperature control, packaging, and all related subsystems and components for electrodynamic and electromotive systems.

Industrial Utility Vehicle & Mobile Equipment Magazine

www.specialtyvehiclesonline.com

Dedicated to engineering, technical, and management professionals as well as dealers and fleet managers involved in the design, manufacture, service, sales, and management of lift trucks, material handling equipment, facility service vehicles and mobile equipment, golf carts, site vehicles, carts, personal mobility vehicles, and other types of special purpose vehicles.

Greencar Congress

www.greencarcongress.com

Earthtoys

www.earthtoys.com

A resource for alternative energy and hybrid transportation information and features. In addition to the bimonthly emagazine, there is also an up-to-date news page, link library, company directory, event calendar, product section, and more.

Grassroots Electric Drive Sites

The Electric Auto Association

www.eaaev.org

The California-based nonprofit group's site showcases EV technology, has a newsletter, and displays links to EV chapters and owners nationwide.

Northeast Sustainable Energy Association (NESEA)

www.nesea.org

NESEA is a nonprofit, membership organization dedicated to promoting responsible energy use for a healthy economy and a healthy environment. NESEA promotes electric-drive vehicles (EDs, HEDs, fuel cell EDs) and renewably produced fuels through its annual road rally, the NESEA American Tour de Sol; the U.S. electric vehicle championship; conferences for professionals; and K–12 education that uses sustainable transportation as a theme. NESEA maintains a web site with a listing of electric cars, buses, and bikes; K-12 educational resources; information on building and energy programs; and a quarterly magazine.

Cool Fuel Road Trip

www.coolfuelroadtrip.org

Federal Government Sites

IRS Forms–EV Tax Credits

www.irs.gov

Qualified EV Tax Credit Forms must accompany any tax returns that are claiming the ownership or purchase of a qualified EV.

Energy Information Administration

www.eia.doe.gov

International Partnership For The Hydrogen Economy

www.usea.org/iphe.htm

Serves as a mechanism to organize and implement effective, efficient, and focused international research, development, demonstration, and commercial utilization activities related to hydrogen and fuel cell technologies. It also provides a forum for advancing policies, and common codes and standards that can accelerate the cost-effective transition to a global hydrogen economy to enhance energy security and environmental protection.

THOMAS

www.congress.gov

Acting under the directive of the leadership of the 104th Congress to make federal legislative information freely available to the Internet public, a Library of Congress team brought the THOMAS World Wide Web system online in January 1995. The THOMAS system allows the general public to search for legislation and information regarding the current and past business of the U.S. Congress.

NREL Home Page

www.nrel.gov

The National Renewable Energy Laboratory (NREL) has created a web site detailing research efforts in renewable energies and alternative transportation technologies. Some key areas include hybrid vehicle development, renewable energy research, and battery technology research.

Alternative Fuels Data Center

www.eere.energy.gov/cleancities/

A comprehensive source of information on alternative fuels. Sections include an interactive map of AFV refueling stations in the U.S.; listings and descriptions of different alternative fuels and AFV vehicles; online periodicals; and resources and documents on AFV programs. The site is part of the National Renewable Energy Laboratory's (NREL) web site.

Advanced Vehicle Testing Program

www.avt.inel.gov

Office of Transportation Technologies, U.S. Department of Energy. This web site is run by the Idaho National Engineering and Environmental Laboratory (INEEL). It offers EV fact sheets, reports, performance summaries, historical data, and a kids' page. Visitors can also request information online.

Office of Transportation Technologies EPACT & Fleet Regulations

www.eere.energy.gov/vehiclesandfuels/epact/

Many public and private fleets are subject to AFV acquisition requirements under the Energy Policy Act (EPAct) regulations. These requirements differ for different types of fleets. Visit this site to obtain information on fleet requirements and the manners in which you can comply with the EPAct regulations.

The US Department of Defense Fuel Cell Program

www.dodfuelcell.com

US Department of Transportation Advanced Vehicle Technologies Program

http://scitech.dot.gov/partners/nextsur/avp/avp.html

The homepage includes links to the regional members of the Advanced Vehicle Program (AVP):

- **Mid-Atlantic Regional Consortium for Advanced Vehicles (MARCAV)**
  www.marcav.ctc.com

  A Pennsylvania-based organization that was established to organize industrial efforts to develop enhanced electric drives for military, industrial, and commercial vehicles. Visitors can review a list of MARCAV projects and research specific projects.

- **Hawaii Electric Vehicle Demonstration Project**
  www.htdc.org/hevdp

  A consortium dedicated to furthering electric vehicle development and sales in Hawaii. The web site provides visitors with background on the program and lists accomplishments.

- **Center for Transportation and the Environment (CTE)**
  www.cte.tv

  CTE is a Georgia-based coalition of over 65 businesses, universities, and government agencies dedicated to researching and developing advanced transportation technologies. The site includes industry news, studies and projects, a database of products, and a section on EV education.

- **CALSTART/WestStart**
  www.calstart.org

  A California-based nonprofit organization dedicated to "transforming transportation for a better world." Visitors can read daily and archived industry news updates and publications, search EV-related databases, and interact with other EV owners in an online forum.

- **Northeast Alternative Vehicle Consortium (NAVC)**
  navc.org

  A Boston-based association of private and public sector organizations that works to promote advanced vehicle technologies in the Northeast. Visitors can read about NAVC projects and link to related Internet sites.

State- and Community-Related Electric Vehicle Sites

California Air Resources Board (CARB)

www.arb.ca.gov

This site provides access to information on a variety of topics about California air quality and emissions. The site has general information on all types of alternative fueled vehicle programs and demonstrations. The CARB's mission is to promote and protect public health, welfare, and ecological resources through the effective and efficient reduction of air pollutants while recognizing and considering the effects on the economy of the state.

The California Air Resources Board's guide to zero and near zero emission vehicles is available at Driveclean.ca.gov.

California Energy Commission

www.energy.ca.gov

This site gives viewers access to information on a variety of topics about California's energy system. The site dedicates a page to electric vehicles, where it has general information on electric transportation, lists sellers of EDs in California, outlines state and federal government incentives for AFVs, and includes a database of contacts in the electric transportation industry.

Mobile Source Air Pollution Reduction Review Committee (MSRC)

www.msrc-cleanair.org

The MSRC was formed in 1990 by the California legislature. The MSRC web site offers information on a variety of topics regarding California air quality and programs underway to improve it, including a number of EV-related programs and incentives.

San Bernardino Associated Governments (SANBAG)

www.sanbag.ca.gov

SANBAG is the Council of Governments and Transportation Commission for San Bernardino County. The site has information about current transportation projects underway in the San Bernardino area, as well as information for commuters. Further, the web site contains funding alerts for individuals and companies looking to obtain project funding and/or assistance.

Ohio Fuel Cell Coalition

www.fuelcellsohio.org

OFCC represents the Ohio fuel cell community to multiple audiences, seeks to expand market access, fosters technological innovation, and advances the competitiveness of the Ohio fuel cell community. OFCC member organizations value their collaborative work in public education, information sharing, and better linking the academic and industrial communities. OFCC provides thought leadership on issues and policies that affect the worldwide fuel cell industry via advocacy and government relations.

General Electric Drive Information Sites

Many web sites disseminate information on EDs or report industry news and developments. A few of these, which "house" specific EV-related information, are provided here.

Advanced Transportation Technology Institute

www.atti-info.org

The Advanced Transportation Technology Institute (ATTI), a nonprofit organization, promotes the design, production, and use of battery-powered electric and hybrid-electric vehicles. The organization supports individuals and organizations interested in learning more about electric and hybrid-electric vehicles, particularly electric buses.

Alternative Fuel Vehicle Institute

www.afvi.org/electric.html

AFVI was formed by Leo and Annalloyd Thomason, who each have more than 20 years' experience in the alternative fuels industry. In 1989, following more than five years' natural gas vehicle market development work for Southwest Gas Corporation and Lone Star Gas Company, the Thomasons founded Thomason & Associates. The company quickly became a nationally known consulting firm that specialized in the market development and use of alternative transportation fuels, particularly natural gas. In this capacity, they incorporated the California Natural Gas Vehicle Coalition and worked extensively with the California Legislature, the California Air Resources Board, the South Coast Air Quality Management District, and other government agencies to establish policies and programs favorable towards alternative fuels. Thomason & Associates also conducted market research and analyses, developed dozens of alternative fuel vehicle (AFV) business plans, and assisted clients in creating markets for their AFV products and services.

Association for Electric and Hybrid Vehicles

www.asne.nl/

ASNE is the Dutch division of the Association Européenne des Véhicules Electriques Routiers (AVERE), an association founded under the auspices of the European Community. The goal of ASNE is to encourage the easy use of totally or partly (hybrid) electric vehicles and vehicles with other alternative propulsion systems in road traffic.

California Fuel Cell Partnership

www.cafcp.org

Introduced in April 1999 and comprised of the world's largest automakers, energy providers, fuel cell manufacturers, and government agencies, the California Fuel Cell Partnership (CaFCP) evaluates fuel cell vehicles in real-world driving conditions, explores ways to bring fuel cell vehicles to market, and educates the public on the benefits of the technology. The CaFCP primary goals aim to demonstrate vehicle technology by operating and testing the vehicles under real-world conditions in California; demonstrate the viability of alternative fuel infrastructure technology, including hydrogen and methanol stations; explore the path to commercialization, from identifying potential problems to developing solutions; and increase public awareness and enhance opinion about fuel cell electric vehicles, preparing the market for commercialization.

Fuel Cells 2003

www.fuelcells.org

The online Fuel Cell Information Center.

Hydrogen Now!

www.hydrogennow.org

The mission of Hydrogen Now! is to educate and motivate the public to seek and use hydrogen and renewable energy technologies for greater energy independence and improved air quality.

National Hydrogen Association

www.hydrogenus.org

The National Hydrogen Association is a membership organization founded by a group of ten industry, university, research, and small business members in 1989. Today the NHA's membership has grown to nearly 70 members, including representatives from the automobile industry; aerospace; federal, state, and local government; energy providers; and many other industry stakeholders. The NHA serves as a catalyst for information exchange and cooperative projects and provides the setting for mutual support among industry, government, and research/academic organizations.

ZEVInfo

www.zevinfo.com

Designed by the California ZEV Education and Outreach Group, which was established under the California Air Resources Board's (CARB) ZEV Program. The basis of the web site is to serve as a "one-stop-shop" for information on electric drive products in California. Moreover, the web site's goal is to inform the public of the benefits and availability of advanced electric drive technologies, from early deployment and on into the future.

Other Related Web Sites

Canadian Environment Industry Association

www.ceia-acie.ca

The Canadian Environment Industry Association (CEIA) is the national voice of the Canadian environment industry. CEIA is a business association that, along with its provincial affiliates, represents the interests of 1,500 companies providing environmental products, technologies, and services.

Fuel Cell Bus Club

www.fuel-cell-bus-club.com

The Fuel Cell Bus Club comprises the participants of the European fuel cell bus projects who intend to introduce fuel cell transit buses to their fleets and establish a hydrogen refueling infrastructure in their cities.

Natural Gas Vehicle Coalition

www.ngvc.org

The NGVC is a national organization dedicated to the development of a growing, sustainable, and profitable natural gas vehicle market. The NGVC represents more than 180 natural gas companies; engine, vehicle, and equipment manufacturers; and service providers, as well as environmental groups and government organizations interested in the promotion and use of natural gas as a transportation fuel.

Technology Transition Corporation

www.ttcorp.com

Since 1986, Technology Transition Corporation (TTC) has been creating and managing collaborative efforts to accelerate the commercial use of new technologies. They design and implement strategic initiatives to help emerging technologies move from the research and development environment to profitable and sustainable businesses.

National Station Car Association

www.stncar.com

Although closed at the end of 2004, the National Station Car Association worked for 10 years to guide the development and testing of the concept of using battery-powered cars for access to and egress from mass transit stations, and to make mass transit a convenient door-to-door service. The NSCA released a report (National Station Car Association History.doc) that gives an overview of the program's history.

Fair-PR

www.fair-pr.com/background/about.php

The largest international commercial exhibition on hydrogen and fuel cells at the Hannover Fair in Germany, featuring over 100 companies and research institutions from 30 countries.

Index

B

F

G

H

I

J

K

L

T

AN ONLINE FORUM FROM THE GREEN LIVING GUY

www.greenlivingguy.com